"*Nuclear Power in Britain and Germany* is a great book, both relevant and innovative. Focusing on six nuclear places in Britain and Germany, it shows that essential societal issues were anchored to nuclear energy. It strongly improves our understanding of the technology's society-splitting potential."

Prof. Dr Helmuth Trischler, *Deutsches Museum, Munich*

"Christian Götter has written a fascinating book on local debates and conflicts around nuclear power in Britain and Germany. With its impressive empirical depth, methodological innovativeness and comparative outlook, it is a must-read for anyone interested in the historical drama of European nuclear power in its diverse societal settings."

Prof. Dr Per Högselius, *KTH, Stockholm*

Nuclear Power in Britain and Germany: Culture, Emotions, and Rationality

Split into two volumes, volume 1 is a cultural history of technology that provides new insights into the international history of nuclear energy by examining the perspectives of six nuclear power plants' host communities in Britain and Germany from the 1950s to the late 1980s.

Based upon a continuous, qualitative analysis of the public debates in these communities, this book demonstrates that the societal conflicts about atomic power were never confined to the risks of one exceptional energy technology. Instead, they were shaped by larger societal issues, which they, in turn, heavily influenced. Therefore, the local discourses about nuclear energy provide an insight into transformations the British and German societies underwent during the Cold War.

The book will be of value for scholars and students interested in the history of technology, environmental, political, and contemporary history, as well as science and technology studies. It will be of interest to researchers and members of the public interested in the local history of Biblis, Hinkley Point, Lingen, Oldbury-on-Severn, Stade, and Torness; as well as the wider public interested in the ongoing debates about nuclear power as a source of energy and a potential answer to the challenges of climate change.

Christian Götter works at TU Braunschweig (Germany) and specialises in 20th-century British and German history. He is mainly interested in the intersection of societal conflicts and communication. His publications include *From a Militarily to a Politically Enforced Instrument* (2021) and *Technologies as Anchors for Societal Conflicts* (2022).

SPONSORED BY THE

Federal Ministry
of Education
and Research

Routledge Studies in the History of Science, Technology and Medicine

49 **Human Extinction**
A History of the Science and Ethics of Annihilation
Émile P. Torres

50 **Tore Godal and the Evolution of Global Health**
Conrad Keating

51 **William Blake, the Single Vision, and Newton's Sleep**
A History of Science, Poetry, and Progress
Keith G. Davies

52 **Pestilence, Insanity, and Trees**
How Stephen Smith Changed New York
John M. Harris Jr.

53 **Communication Maintenance in *Longue Durée***
Edited by Gabriele Balbi and Roberto Leggero

54 **Women, Gender, and Technosciences, 1900–2020**
A Beard to Govern
Edited by Grégory Dufaud and Isabelle Lémonon-Waxin

55 **Nuclear Power in Britain and Germany: Culture, Emotions, and Rationality**
Splitting Societies 1956–1989. Volume One
Christian Götter

56 **Nuclear Power in Britain and Germany: Trust, Democracy, and Futures**
Splitting Societies 1956–1989. Volume Two
Christian Götter

For more information about this series, please visit: www.routledge.com/Routledge-Studies-in-the-History-of-Science-Technology-and-Medicine/book-series/HISTSCI

Nuclear Power in Britain and Germany: Culture, Emotions, and Rationality

Splitting Societies 1956–1989. Volume One

Christian Götter

Routledge
Taylor & Francis Group
NEW YORK AND LONDON

First published 2026
by Routledge
605 Third Avenue, New York, NY 10158

and by Routledge
4 Park Square, Milton Park, Abingdon, Oxon, OX14 4RN

Routledge is an imprint of the Taylor & Francis Group, an informa business

ISBN: 978-1-032-95208-6 (hbk)
ISBN: 978-1-032-95209-3 (pbk)
ISBN: 978-1-003-58367-7 (ebk)

DOI: 10.4324/9781003583677

Typeset in Sabon
by Apex CoVantage, LLC

Contents

Preface and Acknowledgements

With Britain first and Germany trailing behind, shackled by its legacy of having plunged the world into a murderous war, the 1950s saw most industrialised societies embark on a journey to become "splitting societies": societies splitting the atom to generate electricity. About three decades later, the so-called "Peaceful Atom" they employed had, to varying degrees, split the societies themselves. Nuclear energy had turned out not only to dominate international politics in the Cold War but to provide the core of one of the most dominant conflicts within civil societies in the second half of the 20th century. Starting in the 1970s, this conflict quickly became a focus for social scientists and historians, aiming to understand and explain what was going on. Answers are still coming in, especially once one looks beyond the national plane, which has been the focus of most studies. This book takes up a perspective mostly ignored by researchers so far: that of communities living near nuclear power stations. It compares several such local perspectives internationally, specifically in Britain and West Germany, and identifies several conflicts that became anchored to that concerning the atom. While these conflicts stoked the central debate on the atom and complicated it further, they also turned the controversy surrounding nuclear energy into an indicator for themselves. The "Peaceful Atom" thus became a sensor for farther-reaching conflicts, splitting societies independently of the new technology. These connections between local, public debates about atomic energy on the one hand and, on the other, the British and German conditions of debate, the roles afforded to emotions and rationality, the fundamental importance placed on trust and perceived opportunities for democratic participation, and disputed visions of societal futures stand at the centre of the analysis presented here. It is this book's central hypothesis that these intertwined debates shaped the varying dynamics of the conflict surrounding this technology in the two nations: The conflict was not just about splitting the atom; it was at the same time anchored to a multitude of conflicts that had the potential to split societies.

Writing such a book does require a lot of time. I began to think, read, and write about and for it ten years ago, in 2014. In 2022, I was able to hand it in as a habilitation thesis at TU Braunschweig in Germany. Afterwards, it had to be turned into a publishable text. The whole process required the support of numerous people and institutions to whom I am deeply indebted.

First and foremost, I would like to thank the Federal Ministry of Education and Research, which funded this study, and the ever-helpful staff at DLR, who were always available as contacts and actively supported me with numerous questions and challenges. Alla Nevshupa deserves special mention here.

I am also indebted to the Deutsches Museum in Munich, where I was able to carry out the greater part of my research project. This was made possible above all by the Head of Research, Helmuth Trischler, who always supported me in word and deed, right from the application stage. This is no less true for the head of the Research Institute, Ulf Hashagen, who supported my project from the moment I arrived in Munich. Special thanks also go to Andrea Walther, without whom I would certainly not have been able to manage the administration of the project finances, among other things. She, Andrea Lucas, Sabine Bärmann, and Daria Schumann were irreplaceable. The staff of the museum's library have also given me unparalleled support. Florian Preiß and Eva Bunge deserve special mention here. I would also like to give special thanks not only to my student assistants Jonathan Ott and Katharina Winter, but also to my interns Jelena Ostertag, Nico Kleiner, and Lorenz Schmid, who supported me during my time in Munich. Finally, it is hard to measure the degree of gratitude I owe to the colleagues who repeatedly lent me their ears, their eyes, their time and, above all, their critical intelligence. In particular, I would like to mention Rudolf Seising, Vanessa Osganian, Alexander Gall, Fabienne Will, Helen Piel, Dinah Pfau, Ellen Harlizius-Klück, Frank Dittmann, Stefan Esselborn, Martin Meiske, Jürgen Teichmann, Michael Eckert, and Stephan Lindner.

Great thanks are also due to the Institut für Geschichtswissenschaft at the TU Braunschweig, where my project was able to mature from a wild idea to initial research to a successful funding application and beyond. The fact that this was possible is mainly thanks to Ute Daniel, Christian Kehrt, Klaus Latzel, and Heiko Stoff, who supported this process from the beginning and provided critical guidance throughout. The staff of the Institute's administration, especially Petra Fuchs, were indispensable in the implementation of the various work steps, not least in the planning and execution of research. The latter applies no less to the staff of the University Library and the Institute's student assistants, who actively supported me. I would like to emphasise Jascha Makulik in particular. Sebastian Spiller, Matthias Himstedt, and Katja Crusius, who completed an internship at the Institute, were also a great support during the initial research. But I would especially like to thank the numerous colleagues who supported me time and again with their analytical thinking and critical questions, or who simply listened when I had to talk about the various problems accompanying a new – or not-so-new – research project. This includes the participants of the History of Technology Colloquium on the one hand and, on the other, especially Lennart Bohnenkamp, Bernadette

Descharmes, Eike-Christian Heine, Daniel Jankowski, Sebastian Mönnich, Franziska Neumann, Ebbo Schröder, Frauke Schubert, and Michael Wrehde.

Without the support of the staff of the archives in Stade, Biblis, Darmstadt, Lingen, Meppen, Oldenburg, Munich, Gloucester, Haddington, Taunton, and London, which I was allowed to visit during my research, I would not have been able to compile the source base on which this book is based. In this context, special thanks are also due to the contemporary witnesses who were willing to talk to me and let me look at their material. Dr Helge Heyn, Ulrich Hemke, and Dr Tönnies Hagelstein deserve special mention here. I would also like to thank all those who sheltered me during my visits to the archives – Tim Leffler in particular.

Turning the manuscript into a successful habilitation thesis was possible only because numerous colleagues were ready and willing to get involved in the necessary work and, not least, the committee. I extend my greatest thanks to them and, especially, Charlotte Böhl, who was an indispensable help during the whole process.

Beyond that, I would like to thank the Georg Agricola Society, which honoured this work with its price for dissertations and habilitations in 2024.

Many thanks, of course, are due to the people who made it possible to turn a scholarly text into a two-volume book: the people at Routledge, in particular Louise Ingham and Max Novick.

Finally, I would like to thank the numerous colleagues with whom I was able to discuss my ideas in various places over the years. I would like to make special mention of Jacob Darwin Hamblin, Per Högselius, Karena Kalmbach, Astrid Mignon Kirchhof, Jan-Henrik Meyer, Eva Oberloskamp, Ulrich Smeddinck, and Frank Uekötter. Caroline Taylor is also to be mentioned here, without whom a readable text would probably never have emerged from an error-ridden manuscript.

Admittedly, there is no denying that despite all the help that has been given to me and to this book, it will not be completely free of errors or mistakes. I am responsible for all of them.

Lastly, but by no means least, I would like to thank the people I have neglected far too often during the time I have been working on this project and who have nevertheless always supported me, my friends, my family and especially Iris.

This book is dedicated to my teachers, first and foremost among them Ute Daniel.

Abbreviations and Translations

Abbreviation	English Term (where applicable)	German Term (where applicable)
AEA	Atomic Energy Authority	
AEC	Atomic Energy Commission	
	Administrative Advisory Board	Verwaltungsbeirat
	Administrative Board	Verwaltungsausschuss
AGR	Advanced Gas-cooled Reactor	
AGURS		Aktionsgemeinschaft für Umweltschutz Rhein-hessen-Süd e. V.
AGUSR		Aktionsgemeinschaft für Umweltschutz südliches Ried
AUN		Arbeitsgemeinschaft Umweltplanung Niederelbe
	Atomic Energy Act	Gesetz über die friedliche Verwendung der Kernenergie und den Schutz gegen ihre Gefahren (Atomgesetz)
	Atom State	Atomstaat
BANG	Bridgwater Anti-Nuclear Group	
BASF		Badische Anilin- & Soda-Fabrik Aktiengesellschaft
BB		*Bibliser Blatt*
BBU		Bundesverband Bürgerinitiativen Umweltschutz
BE		*Bergsträßer Echo*

Abbreviation	English Term (where applicable)	German Term (where applicable)
BEgA		Bürgerinitiativen Emsland gegen Atomanlagen
BgU		Bürger gegen Umweltvergiftung
BM	*Bridgwater Mercury*	
BUND		Bund für Umwelt und Naturschutz Deutschland e. V.
	Citizens' Dialogue on Nuclear Energy	Bürgerdialog Kernenergie
BZ		*Bürstädter Zeitung*
	Bursting Protection	Berstsicherung
CANT	Campaign Against Nuclear Transport	
CEA	Central Electricity Authority	
CEGB	Central Electricity Generating Board	
	Chief City Director	Oberstadtdirektor
	Chief District Director	Oberkreisdirektor
CDU	Christian Democratic Union of Germany	Christlich Demokratische Union Deutschlands
	Citizens' Application	Bürgerantrag
	City Director	Stadtdirektor
	Commission on Radiological Protection	Strahlenschutzkommission
COLA	Consortium of Opposing Local Authorities	
DAtF		Deutsches Atomforum e. V.
DE		*Darmstädter Echo*
	German Atomic Energy Commission	Deutsche Atomkommission
	Development Plan	Bebauungsplan
	District Administrator	Landrat
	District Director	Kreisdirektor
	District Farmers' Association	Kreisbauernverband
DGB	German Trade Union Confederation	Deutscher Gewerkschaftsbund

Abbreviation	English Term (where applicable)	German Term (where applicable)
Dow	Dow Chemical Company	
ELC	*East Lothian Courier*	
ELDC	East Lothian District Council	
ELN	*East Lothian News*	
	Environmental Committee	Umweltausschuss
EN	*Evening News*	
FAZ		*Frankfurter Allgemeine Zeitung*
FDP	Free Democratic Party	Freie Demokratische Partei
FNP		*Frankfurter Neue Presse*
FRG	Federal Republic of Germany	Bundesrepublik Deutschland
GAB		Gemeindearchiv Biblis
GDR	German Democratic Republic	Deutsche Demokratische Republik
	Society for Reactor Safety	Gesellschaft für Reaktorsicherheit
GKM		Großkraftwerk Mannheim AG
GlouA	Gloucestershire Archives	
	Graduate Physicist	Diplom-Physiker
GN		*Grafschafter Nachrichten*
HEW		Hamburgische Electricitäts-Werke AG
	Head of Section	Ministerialdirigent
HoNESt	History of Nuclear Energy and Society	
HStAD		Hessisches Staatsarchiv Darmstadt
IbS		Institut für biologische Sicherheit
ICRF	Imperial Cancer Research Fund	
IGFM	Interest Group Women and Mothers from Biblis and its Surroundings	IG Frauen und Mütter aus Biblis und Umgebung
	Nuclear Power Plant Information Circle	Informationskreis Kernkraftwerk

Abbreviation	English Term (where applicable)	German Term (where applicable)
	Institute for Reactor Safety	Institut für Reaktorsicherheit
IUS		Interessengemeinschaft Umweltschutz Südhessen
JCC	Joint Consultative Committee	
JGC	John Gray Centre	
Jusos	Working Group of Young Socialists in the SPD	Arbeitsgemeinschaft der Jungsozialisten in der SPD
KKE	Emsland Nuclear Power Plant	Kernkraftwerk Emsland
KKS	Stade Nuclear Power Plant	Kernkraftwerk Stade
KKS GmbH		Kernkraftwerk Stade GmbH
KLE		Kernkraftwerke Lippe-Ems GmbH
	Power Plant Talks	Kraftwerksgespräche
LBANG	Lothian and Borders Anti-Nuclear Group	
LLC	Local Liaison Committee	
LT		*Lingener Tagespost*
	Lord Mayor	Oberbürgermeister
MM		*Mannheimer Morgen*
MT		*Meppener Tagespost*
	Municipal Council	Gemeindevertretung
NA	National Archives	
NATO	North Atlantic Treaty Organization	
NFU	National Farmers' Union	
NG	*Northavon Gazette*	
NII	Nuclear Installations Inspectorate	
NIREX	Nuclear Industry Radioactive Waste Executive	
NLA OL		Niedersächsisches Landesarchiv, Abteilung Oldenburg
NLA STD		Niedersächsisches Landesarchiv, Abteilung Stade

Abbreviation	English Term (where applicable)	German Term (where applicable)
NST		*Neue Stader*
NSW		*Neue Stader Wochenblatt*
NWK		Nordwestdeutsche Kraftwerke AG
ÖTV	Public Services, Transport and Traffic Union	Gewerkschaft Öffentliche Dienste, Transport und Verkehr
PAT	Parents Against Torness	
PANDIS	People Against Nuclear Dumping in Sedgemoor	
	Permanent Secretary	Ministerialrat
PE		PreußenElektra AG
PTB		Physikalisch-Technische Bundesanstalt
	President of the Regional Council	Regierungspräsident
PWR	Pressurised Water Reactor	
RSK	Advisory Commission on Reactor Safety	Reaktorsicherheitskommission
	Regional Planning Procedure	Raumordnungsverfahren
	Regional Planning Programme	Landesraumordnungsprogramm
	Residual Risk	Restrisiko
RWE		Rheinisch-Westfälisches Elektrizitätswerk AG
	River Authorities	Wasser- und Schifffahrtsämter
SANA	Severnside Anti-Nuclear Alliance	
SANA	Somerset Anti-Nuclear Alliance	
SANE	Students Against Nuclear Energy	
SCAR	Severnside Campaign Against Radiation	
SCG	*Somerset County Gazette*	
SCOSLA	Standing Conference of Severnside Local Authorities	

Abbreviation	English Term (where applicable)	German Term (where applicable)
SCRAM	Scottish Campaign to Resist the Atomic Menace	
SDW		Schutzgemeinschaft Deutscher Wald e. V.
SGHWR	Steam-Generating Heavy Water Reactor	
SHC	Somerset Heritage Centre	
SHE	Stop Hinkley Expansion	
SNP	Scottish National Party	
SP		*Südhessische Post*
SPD	Social Democratic Party of Germany	Sozialdemokratische Partei Deutschlands
SSEB	South of Scotland Electricity Board	
StadtA LIN		Stadtarchiv Lingen
StadtA STD		Stadtarchiv Stade
STAG	Stop Torness Action Group	
ST		*Stader Tageblatt*
STPG	Severn Tidal Power Group	
SZ		*Süddeutsche Zeitung*
TC	*The Citizen*	
TCC	Torness Community Concern	
TG	*Thornbury Gazette*	
THORP	Thermal Oxide Reprocessing Plant	
TMI	Three Mile Island	
TS	*The Scotsman*	
TÜV	Technical Inspection Association	Technischer Überwachungsverein
UK	United Kingdom of Great Britain and Northern Ireland	
USA	United States of America	
VEW		Vereinigte Elektrizitätswerke Westfalen AG
	Water Management Office	Wasserwirtschaftsamt
	Water Supply Board	Wasserversorgungsverband
WZ		*Wormser Zeitung*

Introduction – A Close-Up View on a Complex Conflict

The 20th century was an age of conflict. In its second half, humankind's new-found ability to split the atom provided it with the capacity to destroy itself, and most of the planet, in a single exchange of nuclear fire. While an all-out military utilisation of humanity's recently acquired capability could be avoided, this very skill stood at the heart of a clash that split most of the industrialised, democratic countries that attempted to use it to generate electricity. Protests against a planned atomic[1] power plant in Brokdorf in northern Germany escalated into what was widely called "The Civil War in Brokdorf".[2] In France, a confrontation between the police and tens of thousands of transnational[3] protesters opposing a fast breeder reactor at Creys-Malville resulted in serious injuries and one dead activist.[4] In the United States of America (USA), about 2,000 activists protesting against the Diablo Canyon nuclear power plant were arrested.[5] In Britain, a public inquiry into the construction of the Sizewell B nuclear power plant lasted from 1983 to 1985.[6] The power to split the atom was splitting societies. The resulting rifts, however, took different shapes in the various countries that were part of the technology's transnational development. In Europe, the epicentre of the protests is commonly situated around the Swiss, French, and German border triangle of the southern Rhine Valley. In particular, demonstrations against a nuclear power plant in Wyhl are often taken to have inspired protests and various forms of direct action elsewhere, resulting, among others, in the violent clashes at Brokdorf. But while the conflict escalated in West Germany especially, it remained confined in Britain. The search for the reason behind the different degrees to which nuclear power split societies has occupied researchers for about five decades now. It also inspired the research for this book.

The study focuses on the United Kingdom of Great Britain and Northern Ireland (UK) and the Federal Republic of Germany (FRG), in short, Britain and Germany, in what can be called the first atomic age, from the beginning of the commercial use of atomic energy in the middle of the 1950s to the preliminary end of its expansion at the end of the 1980s. It employs them as prominent examples for two nations that share many commonalities as atom-splitting societies but that went almost opposite ways in their debates about nuclear power. In fact, Britain is often considered a nation where there was

DOI: 10.4324/9781003583677-1

next to no active resistance to the technology, at least none worth mentioning in comparison to other Western European nations, especially Germany. This narrative was established in the 1970s and has been kept up since, with a prominent iteration stemming from the European research project "History of Nuclear Energy and Society" (HoNESt).[7]

The book's *first goal* is to contribute to answering the question as to the reasons behind these perceived differences in Britain's and Germany's nuclear histories and to re-evaluate them by way of its *main hypothesis*: The conflict about nuclear power was not dominated by some outstanding characteristics inherent to the technology. It was not confined to differing costs and benefits, risks and opportunities of one energy technology among many others.[8] The conflict was more complex because other, fundamental societal conflicts, issues of societal organisation and self-perception, which were hardly publicly debateable on their own, became anchored to it and were substantiated in the process so that they could be publicly fought over. Based upon the empirical analysis carried out here, the six most prominent among them concerned the conditions of debate prevailing in both nations; the role emotions played in political arguments; the question as to what role rationality had to play in them; the relationships of trust between citizens, politicians, and experts; the constitution of democratic decisions between participation and representation; and the conception of the future for which both societies were heading. Because of their connection to the debates about nuclear power, these issues interacted with it and with one another. They entered a phase of co-evolution, shaping the course of the debates *and* being themselves shaped by it.[9] In sum, to varying degrees, they represent and constitute wider differences between Britain and Germany. Specifically, these differences concern the way political, administrative, and economic decision-makers from the local to the national level interacted with people affected by their decisions or, more generally, less formally powerful people. These could themselves be in decisive positions, albeit on a subordinated level, but most prominently were citizens publicly engaging in political questions. These differences could be described as ones of democratic culture, with Britain tending to be open for divergent opinions, while a trend towards top-down control was observable in Germany. Yet, directly focusing on such wider differences, be they called democratic cultures or otherwise, tends to produce analytical results too abstract to understand and explain them and their interaction with the nuclear debates satisfactorily.[10] However, they do become tangible via the analysis of the anchored issues focused on here, with each pointing to, but also going beyond them. In sum, the anchored issues all but determined the course of nuclear history in the two countries and the different degrees to which both societies were split by the debates about atomic power. They even put this difference between Britain and Germany into a new perspective. They emphasise acceptance of public critique in Britain, which led to compromises and calmed the conflict, and resistance to citizen-"interference" by pro-nuclear actors in Germany, resulting in the conflict's escalation, as critical citizens had to fight not only against the technology but also against an aversion to grassroots politics.

The entanglement between nuclear power and such fundamental, societal issues is not specific to Britain or Germany. At the latest, nuclear technologies were discursively supercharged and thereby transformed into an "exceptional issue"[11] when the USA's President Dwight D. Eisenhower initiated the "Atoms for Peace" campaign. Aiming to divert public attention from the technology's military core, he created what is often called, including by this study,[12] the "Peaceful Atom": nuclear power, ostensibly applied for civilian purposes as a very potent technological fix.[13] During the following decades, it was further enriched when basic societal questions like the ones mentioned earlier became anchored to it internationally.[14] Consequentially, the conflict about nuclear energy can be used as a tool to analyse the societal issues anchored to it. It can be employed as a sensor for the sociopolitical conditions of the countries engaged in it, for their self-diagnoses and visions for their future. In fact, the potential offered by analysing the issues anchored to the atom has often been pointed out. It has rarely been exploited, however, which is, in all likelihood, due to the fact that all of the anchored issues became entangled with the question of nuclear energy and with one another, thereby becoming almost unresolvable – and hard to analyse, especially on an aggregated, national level of debate.

It is this book's *second goal* and its *methodological hypothesis* to demonstrate that such an analysis of Britain's and Germany's nuclear history is rendered possible by a hermeneutically guided, diachronic as well as synchronous, international comparison of regionally embedded (trans-)nationally[15] entangled, comprehensive, *local* debates[16] about nuclear power. By focusing on public debates around the sites where atomic reactors were planned, constructed, and operated, where the "Peaceful Atom" became a part of everyday life.[17] By way of a close-up analysis of the public discourses at these sites, it becomes possible to discern the individual strands of the complex mesh of topics that were, to varying degrees, part of and conditional for the debates. By focusing on the perspectives of the people living their everyday lives with nuclear power plants, local politicians, administrators, journalists, and citizens from all walks of life – those opposing the power plants as well as those supporting them or working for them, those who participated in the debates without a clear-cut position, and even on those contributing to the debates from further away, by invitation or because of their own initiative – it becomes possible to move the complex issues out of their black box, to highlight their dynamic interactions throughout the conflict, and thereby decipher their contribution to it – and its effects on them. So far, while this potential of local analyses has been pointed out often, in practice, research has mostly been confined to wider areas instead of the power plants' host communities themselves – and to places that were dominated by spectacular developments that gained the attention of contemporary mass media and politicians at that, instead of "ordinary" ones.[18]

It is this book's *third goal* to correct this imbalance. To do so, it focuses on the public debates about nuclear energy in six communities that were home to atomic power plants but have neither seen sensational developments nor been prominent in historical studies so far. The six communities chosen are those of Hinkley Point, Oldbury-on-Severn, and Torness in the UK, and Stade, Biblis, and Lingen (Ems) in the FRG. This book answers the question as to what their nuclear histories looked like. It takes them seriously as places where nuclear power plants were an ongoing concern for the historical actors living with them, for the "Peaceful Atom's" local stakeholders detailed earlier; as places that formed the ground zero of the debates about nuclear energy, showing and shaping the conflict about and the history of nuclear energy as a "public technology";[19] and as places locked in often conflictual relationships with the better-known regional or even (trans-)national areas.

This cultural history of technology,[20] by aiming for these goals and arguing for these hypotheses, provides an example for the potential of local historical perspectives to better understand large-scale societal conflicts temporarily anchored to and co-evolving with a technology, by focusing on their discursive public construction, on the perceptions, experiences, expectations, deliberations, and actions of people living their everyday lives with the technology close by. This is particularly the case for conflicts that were, still are, and will continue to be as complex as that concerning nuclear power. This technology was central to the broad changes the West underwent in the 1970s and harboured implications far beyond them, spatially as well as temporally, not least because of its inextricable connections to global, environmental, and sociopolitical issues. Consequently, even though, in keeping with the dominant local interests, some aspects often brought up in the debates are left out or at least not focused on by this analysis, including national political programmes or legal texts, nuclear energy's military side, the technological details and working practices of nuclear power plants, nuclear physics, health effects, risk assessment, dual use, the full fuel cycle, and attributions of gender characteristics to atomic energy, this study touches upon issues relevant to the history of technology, political history, social history, environmental history, emotional history, media history, the history of science and science communication and, especially, the history of the 1970s as a hinge decade within the field of contemporary history. Mayhap it can even contribute to avoiding future Brokdorfian escalations on kindred fields of strife. To call something so far-reaching this book's *fourth goal* would be presumptuous. It is what a publicly funded historian should aim for, nonetheless.[21]

In the remainder of this introduction, I first cover the state of research into the history of the "Peaceful Atom" in Britain and Germany, without which this study would not have been possible. Based on this, I also argue that the local approach adopted here has the potential to make a meaningful contribution to an issue that is still dominated by national perspectives, especially regarding the still relevant search for the reasons behind the internationally differing dynamics of the public debates about nuclear power. Afterwards, I introduce the methodological concept behind this study, its spatial and temporal layout, the concepts applied, and the source documents used. Based upon these steps, I finally discuss in more detail the aforementioned main hypothesis and, derived from it, the book's structure and the further chapter-specific arguments supporting, broadening, and differentiating it.

The National, the Spectacular, and the Complex Atom: State of Research

Since the 1970s, the history of nuclear energy has been intensively researched by historians, often building on earlier work by social scientists. The following historiographical overview cannot, of course, be all-encompassing.[22] Rather, it aims to show the existence of two important desiderata that motivate this book's analysis and establish its goals. One regards the perspectives on the history of nuclear energy that were considered, the other the explanations for the way the conflicts about the technology developed. A particular focus lies on the different dynamics that occurred in different countries, specifically in Britain and Germany. Further literature of specific interest for individual chapters that helps to locate these in their individual historical contexts is discussed in the respective chapters' introductions.

This analysis of the state of research is divided into three main sections. *First*, I outline the most comprehensive studies into the British and German history of nuclear energy to demonstrate the ongoing dominance of the national perspectives. While these national histories are essential to grasp the general framework in which the local histories are situated, this analysis aims to improve upon them by focusing on the local debates, which, in turn, did shape them. In a *second* step, I present the arguments made for local perspectives in the history of nuclear energy and demonstrate that the researchers who have taken up this call so far have mostly concentrated on places of special interest. Therefore, locales where nuclear power came to be part of everyday life without extraordinary developments were all but ignored – and, with them, the insights into the dynamics of the nuclear debates to be gained here.[23] The *third* section demonstrates that from the late 1970s onwards, social scientists and historians took note when participants in the debates about nuclear energy (which these researchers, in turn, also became)[24] claimed that the conflict around the technology was also about other issues. A varying number of such issues tended (and still tend) to be identified, but actual research was mostly concentrated on single ones out of the many, deemed to be of overarching importance, if they were examined at all. Consequently, only partially satisfactory explanations of the atomic debates' dynamics resulted, prompting the research undertaken here.

The Prevalent National View of the Atom

So far, the national perspective has dominated the history of nuclear energy,[25] with most studies focusing on leading political, scientific, industry or military actors, or protest movements.[26] For Germany's nuclear history, Joachim Radkau's seminal work "Aufstieg und Krise der deutschen Atomwirtschaft" remains an essential text for anybody interested in the topic, in particular since he teamed up with Lothar Hahn to publish an updated and revised version in the wake of the Fukushima disaster, aimed at a wider public.[27] A similar bird's eye perspective was chosen by Wolfgang D. Müller, long-time editor of the journal "Die Atomwirtschaft". His three-volume history includes one on the German Democratic Republic (GDR).[28] Paul Laufs' study "Reaktorsicherheit für Leistungskernkraftwerke" emphasises technological and regulatory developments in the FRG.[29] Earlier overviews of the country's nuclear history were presented by participants in the conflict about the technology,[30] and of course by social scientists, including, prominently, Dieter Rucht.[31] With a view to the British case, Margaret Gowing, assisted by Lorna Arnold, covered the country's early nuclear history, which was dominated by secrecy and military as well as power-political considerations. Still, these beginnings laid the foundation for the later civilian nuclear power programme – initially aiming at the co-production of electricity and plutonium.[32] Christoph Laucht studied the early years and, in particular, scientists behind the atom and Britain's "nuclear culture", with a certain dominance of nuclear energy's military side.[33] For later years, comprehensive overviews were presented by critics of nuclear power, prominently including Friends of the Earth's Walter Cram Patterson[34] and Michael Flood,[35] as well as Ian Welsh, who had participated in the conflict as a member of SCRAM, the Scottish Campaign to Resist the Atomic Menace.[36] There were also more balanced approaches[37] and overviews from the pro-nuclear side.[38] More recently, Charles N. Hill produced a detailed study of British nuclear history from a technological angle,[39] while Karena Kalmbach studied the debates about Chernobyl.[40] Following a path sketched out by Jeff Hughes,[41] Jonathan Hogg investigated the UK's "official" and "unofficial" national narratives regarding the atom, primarily oriented around the military side of Britain's nuclear history. His investigation highlighted these narratives' interconnections with "*localized* nuclear *cultures*"[42] as a prime example for the value of taking local perspectives seriously. In fact, corresponding calls to do so and, based upon them, to differentiate and improve our understanding of the national nuclear histories have been made throughout the last half century of research.

The Potential of Local Perspectives

Beginning in 1975, Shirley van Buiren, who researched the debates about nuclear power in Germany for the Battelle Institute at the federal government's behest, pointed out two things. First, the topic of nuclear energy was more often broached in local media than in national media, based in the country's big cities. Second, local media followed the debates more continuously than the national outlets, which focused on singular events.[43] This observation was repeated multiple times by other researchers.[44] It was on the local level that the debates originated and were shaped.[45]

Consequently, several spatially focused studies into the history of nuclear energy have been undertaken. However, most of these valuable analyses have concentrated on places that were special because they were the sites of exceptional incidents or accidents, spectacular protests, or public disputes. They followed the attentional logic of the national media at the time in question. As a result, their focus often remained guided by the national level of the dispute, which tended to omit some of the locally visible complexities. As Ian Welsh observed, the inadequate consideration of local nuclear histories even generated the wrong impression that nuclear energy had been widely accepted in Britain before the 1970s.[46]

When it comes to places singled out for in-depth analyses, the German example of Wyhl in Baden-Württemberg has been examined most often. The protests arising against the nuclear power plant planned near this small community on the Franco-German border have often been described as the catalyst, turning localised opposition to nuclear power in Germany into a national movement. It has also been seen as a central hub in a transnational network of opponents,[47] spanning, at least, Germany, France, and Switzerland.[48] In recent years, Stephen Milder pointed to these aspects of the debates around Wyhl, identifying a transnational, imagined community, ostensibly focused on environmental questions, which served as a role model for Germany's anti-nuclear protest movements and sparked further initiatives.[49] He also pointed to the limits of the transnational element, which he deemed relevant primarily at a grassroots level. Beyond that, to him, the transnational dimension of Wyhl was mostly confined to the role of a kind of creative pool for the various national movements.[50] Jan-Henrik Meyer and Astrid Mignon Kirchhof emphasised the importance of Wyhl's transnational dimension for experts and some exceptional actors.[51] To Andrew S. Tompkins, Wyhl was one local hub within a wider network of protest-movements,[52] enabling transnational transfers of knowledge, practical support, or even personal relocation in some cases.[53] Wyhl also played a central role in Natalie Pohl's (now: Frickel-Pohl) analysis of anti-nuclear protest in the upper Rhine area.[54]

Further local focal points within the German history of nuclear energy were Gorleben and Wackersdorf, the two most contested sites for a planned reprocessing plant.[55] Highlighting Gorleben, Anselm Tiggemann presented a history of the German opposition to nuclear energy and especially the problem of nuclear waste,[56] while Janine Gaumer concentrated upon the last attempt made in Germany to close the nuclear fuel cycle by way of a reprocessing plant.[57] Other places have been far less prominent, although there have been studies, for example, on Karlsruhe[58] and Greifswald.[59]

In the British case, the nuclear complex of Windscale (or Sellafield) and protests against it have been extensively researched. Lorna Arnold focused on the 1957 Windscale Fire, one of the three large nuclear accidents occurring within this book's scope.[60] The other two were the accident of Three Mile Island, near Harrisburg, in the USA, where a partial core meltdown occurred in March 1979,[61] and the catastrophe of the Soviet reactor at Chernobyl in 1986.[62] Windscale, at any rate, was not only the site of a spectacular incident. Already in 1978, journalist Ian Breach described the public inquiry into the construction of a Thermal Oxide Reprocessing Plant as a rallying point for the British anti-nuclear movement.[63] Veronica McDermott later investigated how the complex became a bone of contention between Britain and Ireland.[64] An oral history project documenting the voices of local people, organised by the Whitehaven Archive and Local Studies Centre, was presented by Hunter Davies.[65] Besides the Windscale/Sellafield complex, the Sizewell B power plant and the public inquiry into its construction have also received scholarly attention.[66]

There is, in other words, no lack of local studies in the history of nuclear energy. However, these were mostly concerned with places of special interest for the (trans-)national debates about nuclear energy. This demonstrates the local perspective's value for analytical purposes and its importance for understanding the wider conflict. That notwithstanding, it was not applied to "normal" atomic power plants' host communities, where nuclear history was continuously written, albeit without exclamation marks. By largely skipping the local pro-nuclear grassroots perspectives in this way, the conflict about the technology acquired the semblance of a struggle between local citizens and national actors, while its character as a multi-level conflict between different groups of citizens was sidelined.

The Topics Anchored to the Atom

A similar situation applies to other issues that became anchored to the debates about nuclear power. Diagnoses of this phenomenon were often observed, as they were very widespread. This held especially true for the concept of a totalitarian and authoritarian Atomstaat (Atom State), as Robert Jungk famously put it in 1977.[67] The concept was based on the conviction that a near total control of the population, and thereby the end of liberal democracy, would be necessary to safeguard the increasingly large amount of plutonium wanted as fuel for breeder reactors by nuclear power's proponents.[68] These reactors, in turn, were deemed essential to unleash the full potential of atomic energy in the form of an oft-imagined closed fuel cycle, including the reprocessing of used fuel elements.[69] To nuclear power's critics, such a "Plutonium Economy" implied nothing less than a global system of repression.[70] In fact, Jungk was far from the only one with this vision.[71] The argument was made in Britain by Friends of the Earth and other anti-nuclear groups.[72] It had even come up in the Royal Commission on Environmental Pollution's famous sixth report, the "Flowers Report" (cf. "Rational Atoms"), which warned against an unrestrained expansion of nuclear fission in 1976, not least because of this conundrum.[73] The concept also conformed to the French "électrofascisme".[74] Obviously, connections between the issue of nuclear energy and wider, societal questions could not be missed. Especially in the case of the Atom State, they caught the eye of historians of technology early on, most prominently Langdon Winner.[75] Overall, however, the role of these connections in the nuclear debates and the effect the latter had on them was rarely systematically explored. As a result, they continue to be highlighted as an important, but often rather vague, one could even say black-boxed, part of the explanation for the dynamics of debates around nuclear energy after half a century of research.

Among the earliest researchers to analyse the public debates about nuclear energy were social scientists, who studied them, in part, at the behest of governments interested in ways to convince the public of a more positive view of the technology.[76] More generally, researchers were looking for models to explain and predict the internationally varying development of protests. To do so, they early on considered the fact that public debates about nuclear power were not confined to the pros and cons of one energy technology among many, nor to special characteristics inherent to the "Peaceful Atom".[77] While the latter claim was made at times, it could hardly explain the diverse courses the nuclear debates took in different societies.[78] As Helga Nowotny put it in 1979: "[T]he atomic theatre is the stage for a socio-political conflict".[79] Similarly, Germany's leading nuclear historian Radkau hinted at the wider implications of the nuclear debates from the very first – while warning that this argument was often used by the technology's proponents to disqualify the valid technical critique of its opponents.[80]

Most prominently, in 1980, political scientist Herbert P. Kitschelt pointed out the importance of a wide range of questions attached to the nuclear debates. These included, for instance, the role played by emotions, but his main focus was on connections to societal conflicts about the distribution of wealth, the application of democracy and the spread of rational knowledge.[81] On this basis, he suggested a model to explain the dynamics of the debates about nuclear power, which, however, would have had to be adapted for most individual case studies.[82] Aiming to avoid a similar calamity, political scientist Joseph A. Camilleri, in his 1984 study of the debates about nuclear energy, concentrated on aspects of the political systems of the states he analysed – which included the FRG and the UK.[83] He also saw the strife surrounding atomic power as part of a dichotomous "larger conflict of values, an ideological confrontation" about capitalism,[84] between a *nucleocracy*[85] and its opponents, who rejected an "increasing incursion of the state . . . into formerly private realms" necessitated by nuclear technologies and "the socio-cultural order which had given birth to it".[86] His explanation, however, focused on the degree of a given state's centralisation, its investment in the technology's application, and its interconnection with images of national grandeur.[87] Kitschelt presented a similarly focused and still influential analytical tool in 1986 with his "Political Opportunity Structures", a model concentrated mainly on the openness of a political system to outside inputs on the one hand and its ability to implement its own goals on the other.[88]

While this model was taken up time and again during the following years, it was also criticised, for example, by Rucht, the author of an early comprehensive study of German anti-nuclear movements, especially in their transnational dimension around Wyhl.[89] He argued that to understand the conflict and its connections to wider societal ones, not least concerning "the capitalist state or technocracy in general", which "at least partly explains the intensity of the conflict",[90] it could be necessary "to shift to a more concrete analytical level".[91] To focus, in other words, on a local level, where the complexities of the topics anchored to the "Peaceful Atom" would be more readily discernible. Rucht was not alone in demanding that one consider other factors. Ortwin Renn emphasised the technology's connection to questions of societal development.[92] Spencer R. Weart, author of the seminal 1988 cultural history of the atom "Nuclear Fear",[93] which was updated and revised in the wake of the Fukushima disaster, reasoned that "[the] history of attitudes toward nuclear energy is a sensitive indicator of something larger: attitudes toward all rational knowledge, technological progress, and organized decision making."[94] Thomas P. Hughes saw, in nuclear energy, one example for a large technological system's societal entanglement beyond the specific artefacts.[95] The position that the debates about nuclear power were anchored to much wider questions has been pointed out more or less constantly by researchers, from the 1970s up to today.[96] In 1994, Peter Wagner even went so far as to claim that nuclear energy itself was only of secondary relevance to the anti-nuclear movements, which primarily aimed for larger societal changes, regarding the role of the individual, social networks and participatory elements within the democracy.[97] This, of course, was still often implied by proponents of the technology, attempting to discredit their critics' arguments.[98] Consequently, most analysts of the conflict did not go that far.

Nonetheless, the relevance of larger societal conflicts for the debates about nuclear energy was continuously pointed out. Most prominently, Gabriele Hecht developed the concept of "technopolitics" in her 1998 formative study "The Radiance of France". She uses it to describe how, in France, the country's image as a great power became connected to its mastery of the atom,[99] and how technological choices regarding the nuclear were dominated by political considerations, in turn narrowing down alternative policies.[100] Two years later, Welsh interpreted the British nuclear power programme as part of an attempt to consolidate the country's national grandeur and to turn the society into a more scientifically minded community by a "social scientific movement".[101] Building, in a way, upon what Flood saw as an "Old Boys Network",[102] he emphasised that the various societal dimensions attached to the conflict about nuclear energy had to be included in its analysis if one wanted to be able to understand it.[103] Three years later, in his study on

Kaiseraugst, Patrick Kupper found that "democracy, rule of law, federalism, economic model, technological progress or environmental protection" were coupled with the debate about nuclear energy in Switzerland, showing that the anchoring phenomenon was widespread.[104] Gabriele Metzler pointed to the anchoring of political concepts,[105] and Holger Nehring emphasised that the democratic dimension of the anti-nuclear movement was especially important in Germany,[106] as did Albrecht Weisker[107] and Thomas Dannenbaum.[108] Jens Ivo Engels saw an anchoring of societal questions to the debates about nuclear energy with a similarly national focus.[109] Highlighting the transnational network around Wyhl, Milder emphasised the nuclear conflict's connection to a longer drawn-out struggle for more participatory forms of democracy.[110] Michael Schüring, analysing the Protestant church's role in the nuclear debates, emphasised the importance of the conflict's "cultural and religious" dimension.[111] Silke Mende and Birgit Metzger saw the nuclear debate connected, not least, to questions about democratic practices and the societal role played by science and scientists.[112] Tompkins highlighted various topics intertwined with anti-nuclear protest, including industrialisation, growth-orientation, health, use of force by the state, and the functioning of democracy.[113] Wolfgang C. Müller and Paul W. Thurner looked into differences between political systems once more in 2017,[114] while Gaumer, in her 2018 study of the Wackersdorf debates, emphasised the role played by questions of applied democracy.[115] Analysing the success of the German anti-nuclear movement in a study including the GDR's nuclear history, Dolores L. Augustine pointed to a number of related conflicts, including the role of emotions, namely utopian hopes[116] and dystopian fears, as well as the importance of experts within the discourse.[117] A broad list of "numerous factors influencing and shaping perceptions and values" was also presented by Kirchhof in a comparative study based on the HoNESt project.[118] Summing up, she concluded that to understand the conflicts about nuclear energy, one has to consider "trust in institutions and socio-cultural contextual factors",[119] of which the "Political Opportunity Structures" were explicitly highlighted.[120] A similar position was taken up in a publication by Arne Kaijser and others, also resulting from the HoNESt project,[121] as well as by Eva Oberloskamp, who spoke of differences in the two countries' "political culture", among others.[122] Most recently, in an anthology focusing on Wyhl, Frickel-Pohl highlighted the "political contexts",[123] while Carol Hager pointed to connections to questions of energy, economy, and democracy.[124] In a way, then, research into the debates about nuclear energy has come full circle, with some of the most intriguing questions still unanswered.

This overview clearly shows how, on the one hand, the importance of the wider societal conflicts anchored to the debates about nuclear energy has constantly been observed. "It therefore seems", to use the wording of Joop van der Pligt, J. Richard Eiser, and Russell Spears in 1984, "impossible to detach the issue of nuclear energy from questions about the kind of society in which we want to live."[125] The fact, however, that these observations of wider societal conflicts anchored to the debates about nuclear energy continue to be made, and that questions continue to be asked regarding the varying dynamics of the debates, clearly indicates that no satisfactory answers have so far been found. One reason surely is that, up to now, the diverse issues anchored to the debates about nuclear energy and their interplay with it, as well as with one another, have not been turned into the focus of analysis, and that focusing on single issues, even the important political opportunity structures, is not sufficient to understand the complex conflicts. As Ioan Charnley-Parry, John Whitton, Pieter Cools, and Anne Bergmans concluded in a comparative report of the HoNESt-Project: "Finally, we find that the UK and Germany provide the clearest indication that political system, as a single criterion, does not determine effective engagement or dictate particular nuclear-related decisions or pathways."[126] On the other hand, the existence of the manifold studies into various individual aspects provides an indispensable starting point for this cultural history of nuclear technology. It aims to provide a more satisfactory answer to the conundrum of the conflicts anchored to the debates about atomic energy with an international comparison of the local nuclear histories of six non-exceptional communities with nuclear power plants in Britain and Germany, focusing on the aspects most prominently occupying them. In the process, I hope to shed further light on the phenomenon of technologies serving as anchors for societal conflicts in general. How these goals are to be reached is discussed in the following section of this introduction.

Coming to Terms with Conflicts Anchored to the Atom: Methods

Based upon the state of research and the highlighted desiderata, it is clear that local actors' perspectives, and the local debates about nuclear power in communities around the sites of nuclear power plants that were not the scene of outstanding events, are key to understanding the complex connections between the conflict about nuclear energy and the societal issues that became anchored to it, that shaped the debates about the technology, just as they themselves were shaped by them. This section of the introduction explains how exactly this is to be done. To this end, it first explicates the temporal and spatial cut-out of this synchronous as well as diachronic *comparison*.[127] That the study must be a comparative one to work out the different dynamics of the debates about nuclear energy is rather obvious. This necessity brings with it the benefit of any comparison in sharpening the analysis of each individual case by contrasting it with the others, allowing for the identification of similarities and differences between the cases in question over time as well as from a synchronous perspective. This is elementary for averting false generalisations or the overhasty construction of exceptionalisms while offering contextualisation that is of fundamental importance for broader insights to be gained from the individual examples.[128] With regard to the study's spatial dimension, the demarcation between the local and the regional applied here is also briefly discussed. It is a question that has spelled trouble for the analysis of numerous studies on the history of nuclear energy, as demonstrated in the state of research. In this study, it is highlighted regarding one of the issues anchored to the debates about the atom (cf. "Nuclear Democracy"). Subsequently, the selection of the source material and its evaluation are explained, summing up the methodological approach of this cultural history of nuclear energy.

From the 1950s to the 1980s: The Temporal Dimension

In terms of its *time frame*, this book covers the *diachronic* development of the public debates and their dynamics during what can be called the first nuclear age, from about 1956 to 1989. It thus starts with the early debates about commercial nuclear power at the sites analysed and concludes with the end of atomic energy's great expansion in Western societies, when, with few exceptions, new constructions failed to appear and many countries were debating or enacting nuclear phase-outs, while the level of protests against nuclear power plants declined.[129] In fact, after a slowdown of new orders since 1974, no new reactors had been ordered in the USA after 1978. In Britain, a moratorium on new constructions was officially declared in November 1989, because the technology had failed to attract willing investors when the government pursued its goal of privatising the electricity industry. In Germany, while there was in effect a phase-out after 1982, the conflict about its official status dragged on beyond 1989. Nonetheless, the last nuclear power plant built during the period of nuclear expansion was connected to the grid in that year,[130] and the nuclear share in global electricity generation levelled out to start falling from the middle of the 1990s.[131] There is an additional reason for ending the analysis in 1989: Following unification, the nuclear installations in the former GDR and their neighbouring communities, which had a completely different history of debate, also became part of a unified Germany's story. A study going beyond 1989 consequently would have necessitated including case studies from the GDR, where public debates were ruled by conditions completely divergent from the ones found in the democracies allowing free speech chosen for this analysis.[132] In addition, in Germany, most documents become available in public archives only after a retention period of 30 years.

The period between about 1956 and 1989, therefore, delineates an outer frame for the individual sites analysed here. While each site fits this frame, they do not occupy identical spaces within it. In two out of the three British cases, the debates about nuclear power started long before a commercial reactor could be built in Germany. They also occurred against the background of a much livelier presence of positive images of nuclear power transnationally and long before critical positions rose to prominence on the transnational level of debate during the 1970s. A similar asynchrony applies to the construction times of the individual power plants that started in different decades. These asynchronies are consequential for this study's *synchronous* dimension in the comparison between the cases under consideration. Conspicuous parallels in the debates at different sites can be observed, largely independent of the calendrical date and dependent instead on the relative project phase of a particular local power plant – from planning to construction to operation, or "permanent development", as the power plants were constantly revised during their operating years.[133] This seeming "non-simultaneity of the simultaneous"[134] does not hinder the comparison. For one, it is a phenomenon not unknown for comparisons.[135] More importantly, it stresses the importance of the interlocked diachronic and synchronous dimension of the comparison applied here, not least by raising the question of the relevance of local developments in relation to the ongoing (trans-)national debates, and by highlighting the fact that it is necessary to explain the early problems faced by the proponents of nuclear power in Britain, and indeed the lack of similar obstacles met at many early sites in Germany. This can be done once one questions the importance of the seeming neutrality of calendrical dates often employed by national comparisons, and considers the relative time frame of the development of each local power plant and the experiences of their respective communities rendered possible by focusing on local perspectives – thereby introducing a temporally "inclined plane" into the analysis.

Six Communities in Two Countries: The Spatial Dimension

The next decision to be discussed is that for the diachronic and synchronous comparison of local communities in the UK and the FRG, two countries that were entangled via a transnational plane of nuclear debates. The specific *nations* chosen had a number of similarities, rendering a comparison of their public debates about nuclear power meaningful by providing a number of basic aspects as tertium comparationis: Both countries were industrialised, market-oriented, European societies, allowing free speech in a representative, democratic system, albeit with differing voting systems.[136] Both enjoyed a comprehensive offering of newspapers, enabling diverse, open and widely accessible public debates, and the nuclear power industry came to be an influential voice in both countries, especially in the political arena. In addition, both countries interacted institutionally and on global, European, and local levels with direct contact between individual communities. They were also connected in wider, international, as well as transnational discourse on nuclear energy, with their pro- and anti-nuclear actors often referring to the same sources of information or events or interacting directly.[137] Advocates and opponents of atomic power plants reacted to one another, visited one another, and learned from one another. This was especially true early on, when Britain was seen as a role model aspired to by many advocates of nuclear energy in Germany.[138]

On the other hand, the two countries were different in many key aspects of their nuclear history. One could even go so far as to locate them at opposite ends of numerous narrative strands common in histories of nuclear energy:

Britain was one of the first nations to develop, produce and test nuclear weapons, which provided an important reason for its nuclear engagement, while Germany abstained from them.[139] Britain was also the first nation to utilise nuclear power for commercial electricity generation.[140] Germany had a late start, as it was only able to set up a nuclear power programme after it had regained most of its sovereignty, around the time the first British nuclear power plant became operational. Consequently, Britain's first phase of reactor construction was already a thing of the past when environmental questions and doubts regarding nuclear reactors gained prominence transnationally during the 1970s, just as Germany was about to expand its stock of reactors. This difference in timing regarding the most intensive building periods is often presented as a major factor behind the different courses taken by the nuclear debates in the two countries.[141] Britain is widely regarded as a nation where no large-scale anti-nuclear movement developed,[142] and when it did, it is regarded as rather co-operative with state actors.[143] In Germany, on the other hand, the movement is held to have been highly prominent and successful, contributing widely to the establishment of the Green party.[144] Another difference besides the nuclear power programmes' timing was population density, which was lower in Britain, enabling for reactors to be located in rather remote areas, far from population centres, which was quite impossible in Germany,[145] where a number of reactors even had to be constructed close to borders.[146] In addition, Britain tended to construct its power plants at the coast, where no controversial cooling towers were necessary,[147] and, for most of the time under consideration here, stuck to its own reactor technology. Germany, meanwhile, was an early adopter of American light water reactor types. These were the target of much critique transnationally; hence, this adoption is often taken as another reason for the different levels of escalation – although British anti-nuclear activists were quite willing to be sceptical about British technology, too.[148] It should also be noted that, in the late 1980s, the German nuclear industry became entangled in scandals involving corruption, nuclear waste, and transparency regarding incidents at German power plants.[149] Finally, in view of the climate crisis, Britain is once again developing nuclear power as part of the answer,[150] while Germany is set to do without the technology.

Limiting the number of countries compared to two allows for a detailed study of the spectrum of the public debates about nuclear power within them. With a view to researching local debates about nuclear power, it allows for the detailed analysis of a total of six communities. As this means that three out of each country's just about a dozen places where commercial nuclear power plants were actually in operation are analysed, it stands to reason that even conclusions regarding the national levels of debate can be drawn.[151] Nonetheless, the *local perspective*, as argued earlier, is the one that is elemental for this study, not only because of the desideratum regarding this perspective but also because only local approaches allow for the necessary attention to detail over the whole course of the debates, to identify the interwoven topics and the specific reasons behind the differing shifts in their dynamics.[152]

As pointed out earlier, there was a certain lack of continuity and detail on the national level of debate, as it focused mostly on places and times with newsworthy events or developments. This does not mean that, on a local level, debates went on uninterrupted for decades. Here, too, topics sprang up and faded out of the public view again, as did interest in the local reactors and nuclear power overall. Nonetheless, the "Peaceful Atom" was constantly present. Some groups or people were always willing to make their points, to challenge or reply to those on the other side of the argument. There was a kind of nuclear background noise. And while this is not reconstructed in toto in the following chapters, it is referenced regularly, as it provides the vivid basis on which the more concise phases of active debates grew. At the same time, on the national level, a kind of blended version of the more specific debates and arguments in individual places and situations was generated. It was, however, obscuring details and differences in favour of a nationalised average, as Jeff Hughes and others aptly observed.[153] In this regard, too, the local debates provide a more insightful perspective, as they allow for a detailed analysis of individual arguments and conflicts. At the same time, the local episodes must not be misunderstood as isolated quarrels without impact on the higher planes of debate. On the contrary, the local debates formed the basis for the latter.[154] They also provide the most precise insight into the wider (trans-)national arguments over the technology, allowing for a level of understanding that cannot be gained in other ways. As Engels put it, "the conflict about atomic energy was substantially based on local oppositional centres".[155] From here, as Nehring demonstrated, networks developed, connecting the local and global levels of debate and, thereby, constructing a transnational one,[156] which functioned as a living memory for the lower strata of debates. And while, admittedly, some important national developments, from the large-scale upheavals of the 1970s to more constrained developments such as changes to laws or governmental attitudes, as well as technological specifics, only came into view when they gained relevance on the spot, the insights to be gained from studies taking local perspectives seriously within the transnational history of nuclear energy are enormous. This has been most famously shown by Kate Brown's study of two nuclear communities in the USA and the Soviet Union, "Plutopia",[157] and by Hecht's work on the uranium mining in Africa, "Being Nuclear".[158]

Given these advantages regarding consistency and detail, the focus of this analysis clearly had to be local debates, to carve out the complex dynamics of the various debates anchored to that surrounding atomic energy. The *specific sites chosen* for this analysis, then, had to be locations where nuclear power was a relevant presence to become a topic for discussion in the first place. They were not to be those, however, that had become fixtures of the national debates due to spectacular incidents, scandals, or events, as, in this case, these special features would have overlaid the interplay of the debates that are to be analysed here. They had to stem from the number of "ordinary", unspectacular nuclear power plant sites that have been largely ignored by historians of nuclear energy.

The sites selected were Oldbury, Hinkley Point, and Torness in Britain, and Stade, Biblis, and Lingen in Germany. These six sites represent different starting points for the nuclear power plant construction, different sizes and types of reactors, and different surroundings in terms of the local and regional situation. In Britain, *Oldbury* belonged to the first generation of nuclear power plants, using Magnox technology. Within this group, it was one of the last to be constructed and deemed to set standards for the next generation of reactors, not least regarding safety. In addition, it was situated on the banks of the Severn, not far from Bristol, and in the vicinity of Berkeley, where not only one of Britain's oldest nuclear power plants, but also a nuclear research laboratory was located, clearly influencing the public debates about the Oldbury power plant. *Hinkley Point*, further down the river in the Severn estuary, was the site of one of Britain's first nuclear reactors, its construction starting only a few months after Berkeley. However, most importantly, it also saw the construction of a power plant from Britain's second generation of reactors and was chosen to be the second site for a third generation using US-American technology. After an intensely fought over public inquiry, this new reactor did not materialise at first. Thereby, it became the first nuclear power plant not to be built because of the British government's 1989 decision to pause its nuclear power programme. Today, however, the construction of the newest nuclear power reactor in Britain is underway in this precise location. Finally, the Scottish *Torness* was one of the last British nuclear power plants to be constructed, signalling the end of its second generation of reactors. Situated on the coast of the North Sea, it was also one of the British reactors held to be most intensely contested by protesters.[159]

In Germany, *Stade* was one of the first two commercial-scale nuclear power plants (with a reactor greater in size than the experimental or demonstration power plants, that is),[160] situated in the vicinity of Hamburg, one of the country's largest cities, on the Lower Elbe. Because of its age, Stade's nuclear power plant was hotly contested throughout the 1980s and, in the early 2000s, the first reactor to be taken off the grid in the wake of Germany's decision to phase out the technology. The Hessian *Biblis* was the site of two reactors delivering more than one gigawatt of electricity. With them, its owner, the Rheinisch-Westfälisches Elektrizitätswerk AG, established a standard for Germany for reactor size and type. Geographically, Biblis is situated in the middle of the FRG, on the Rhine. *Lingen*, finally, had first been the site of an older demonstration power plant before becoming home to the Emsland nuclear power plant, one of the last to be constructed in Germany. It was part of the so-called convoy concept, which was conceived to shorten the application processes with standardisations. More importantly, it was situated close to the Dutch border, introducing a transnational component to the local and regional debates.

The individual nuclear histories of these six diverse places are not merely presented for their own sake. When taken together, they generate insights into the larger developments of the (trans-)national nuclear debates, because they are, in two ways, exceptional. On the one hand, by representing the "ordinary" when it comes to nuclear power plants' host communities, they are exceptional among the places most often analysed. On the other hand, they were exceptional within the oft-examined (trans-)national debates about nuclear power, because of their special situation relative to the technology: They were close by, and the "Peaceful Atom" became a fixture of their everyday lives. With these two dimensions of exceptionality, they help to highlight the contours of the nuclear debates overall, far beyond their own borders.[161] As a consequence, they perfectly demonstrate the analytical potential of an international comparison of local histories.

The Local and the Regional: An Explorative Separation

This leads to the question as to what "local" means in this study, especially in relation to "regional". As indicated earlier, to a certain degree these categories merge in many of the available studies.[162] That notwithstanding, most of the local studies discussed so far did observe an "inner area", where the technology tended to be welcomed, even if critique was dominant in the area surveyed overall.[163] Most prominently, this kind of tension has been pointed out in the case of Wyhl. Here, a citizens' decision about the sale of municipal land in January 1975 found a (rather slim) majority of 55% backing the controversial plans for the construction of a nuclear power plant in the Baden-Württembergian community.[164] Similar observations have been made for other nuclear power projects, too. One well-known example is that of Brokdorf, where initial resistance, not least by the mayor, soon turned into support.[165] Other sites of nuclear power plants with similar experiences include Esenshamm, Grafenrheinfeld, Krümmel, Wackersdorf, and Windscale.[166] A split between an "inner area" and its surroundings has also been pointed out for a number of the locales examined in this study, most prominently so in the German cases of Biblis and Stade, and in the British cases of Hinkley Point and Oldbury.[167]

The explanations given for this difference are usually either of an economic or psychological nature. The economic explanation tends to point out that the places chosen for nuclear projects were structurally weak and that they offered only a few jobs, which generated a rather small income and resulted in little leeway for local councils to provide or uphold infrastructure and services. Consequently, large-scale projects such as nuclear power plants were deemed especially attractive to them, as they would profit economically, while their neighbours would not.[168] This argument was made for Biblis, Brokdorf, Grafenrheinfeld, Hinkley Point, Karlsruhe, Wackersdorf, and Windscale/Sellafield, among others.[169] The psychological explanations mostly come down to effects of the neighbouring communities' experience and familiarisation with the power plants in question, or to the impact of their repression of fears or worries about the reactors' risks. Studies made in this regard tend to go along with observations of specific physical distances to the power plants where these effects were dominant.[170] Corresponding observations have been made in the cases of Biblis, Hinkley Point, and Oldbury.[171]

Both explanatory approaches lack conviction. When it comes to the economy, the main argument is jobs. These projects offered thousands of jobs during the many years of construction and hundreds once finished, while also regularly drawing in thousands of specialists for inspections. Yet these direct benefits simply cannot be absorbed by small places like Wyhl, Brokdorf, or Seascale (near Sellafield).[172] The people working at the sites regularly came from or took residence in, shopped in, and frequented the recreational facilities and services offered in a wider area.[173] Hence, during the long-drawn-out construction of Torness, a conflict developed about what was "local" when it came to "local recruiting" promised by the industry (cf. "Atomic Trust(s)"). Also, increases in local tax income would have to be calculated against higher spending to cater for the needs of additional families living in the area and against potential reductions in subsidies,[174] if the taxes benefited the communities at all. In the case of Biblis, for example, while tax income generated during the construction of the power plants rose considerably, putting the municipality in a strong financial position, subsidies from the government shrank.[175] Once the plants were finished, said tax incomes receded once more,[176] not least because the operating company had its headquarters in Essen in North Rhine-Westphalia, not in the small Hessian community.[177] Consequently, when specific effects were analysed, the concrete economic impact was often rather small.[178] In the case of Stade, even the city director was of the opinion that to lose the additional tax income from its nuclear power plant would "not knock down" the town's finances.[179] Of course, such assessments of limited profits tended to be highlighted by critics of nuclear power.[180]

Regarding the psychological explanations, the frequent observations made relative to specific distances hardly seem helpful when it comes to understanding the dynamics of the debates about nuclear power. For one, as Welsh put it, defining an area up to a specific distance from a power plant as "local" seems rather simplistic and tends to cut off connections and interactions that were quite important for the course of the conflicts.[181] And in the end, most of the specific models discussed, while seeming quite clear-cut and therefore potent analytical tools, would have had to be adopted for each individual case. Even the rather complex ones that pointed to three different zones – an inner zone with high acceptance rates, a middle one with high rejection rates, and an outer one where the negative stance receded[182] – would have been unable to explain cases like Biblis. Here, the people living closest to the power plant – but outside the municipality – were among the project's staunchest critics (cf. "Nuclear Debates").[183] What is more, observed radii changed over time, with people changing their outlook on nuclear power according to their experiences with the technology.[184] While studies on British nuclear power plant sites, including Hinkley Point, found the highest level of acceptance within a five mile or eight-kilometre radius in the early 1980s,[185] some of the most outspoken critics of the initial development in the 1950s had been those living closest to the site – hand in hand with others residing dozens of kilometres away (cf. "Nuclear Debates"). In view of such diachronic changes, teams around Nick Pidgeon and Karen Anne Parkhill have pointed to shifting relationships of trust for an explanation – in other words, to one of the broader issues anchored to the nuclear debates (cf. "Atomic Trust(s)").[186]

This study therefore takes another approach to explain the different perceptions of nuclear power observable in the "inner areas" and their wider surroundings, based upon one of the topics anchored to the "Peaceful Atom". To this end, it actively highlights the spatial differences by understanding the "inner areas", where nuclear reactors were mostly regarded in a positive light, as "local", and the surrounding ones, where critique tended to be dominant, as "regional". Thus, the tensions between the local and the regional are emphasised and can be used productively for historical analysis – in a way going one step beyond Radkau's call to make use of the tensions between different regions for the history of technology.[187]

The empirical insights gained through the analysis of the six public debates about nuclear power under consideration clearly show that the local, tending to a positive view of "its" nuclear power plants, largely corresponded to the administrative borders of the local authorities responsible for the sites. In the FRG, this mostly meant the municipalities or city areas,[188] but also, to some degree, the districts they belonged to. In the UK, on the other hand, the situation was somewhat more complicated due to the distribution of planning and administrative powers on site, differences between England and Scotland, and a local government reform in the 1970s that reorganised responsibilities and denominations.[189] Somewhat loosely, however, the equivalent to the German municipalities regarding the perception of nuclear power plants could be found on the district level (in Scotland, up to said reforms, the county level). The counties they belonged to (in Scotland, after the reforms, the regions) would then correspond to the German districts (cf. "Nuclear Democracy" for more details).

Importantly, this conception of the local and the regional has crucial consequences for historical analysis. It introduces a spatially "inclined plane" into the comparison by orienting the conception of what was "local" or "regional" on communities defined by sociopolitical criteria, instead of seemingly neutral physical distances. Put simply, the former was the local authority areas where nuclear power plants were situated, the latter the neighbouring ones. Therefore, the size of the areas considered as local in this analysis differs not only between the two countries studied, but also between the individual cases of the sample. This might seem less precise than a concept based on measurable units of distance. However, this approach fits better with the empirical observations. It is also more fruitful analytically. On the one hand, it allows for a better understanding of the different developments of the nuclear debates in Britain and Germany, based on the different sizes of the local authority areas in both countries and the position of the power plants therein. On the other hand, by highlighting differences in the perceived participatory opportunities of people living in communities with differing levels of responsibility for the construction and oversight, it also provides a partial explanation for the occurrence of the different perceptions of the nuclear power plants in the first place. Both aspects and the corresponding arguments are discussed in more detail in "Nuclear Democracy". Here, it is important to note that the spatially inclined plane furthers this comparison's analytical potential and improves the understanding of the dynamics of the conflict about nuclear power overall – and that, to understand the dynamics of the local debates, these must be read in contrast with the surrounding, regional ones, while both stood in a constant exchange with one another and the (trans-)national levels.

Letters, Reports, Protocols: The Source Material Chosen

Finally, it is important to consider the *source material* this analysis is based upon: archives of local and regional parliamentary bodies, and local and regional media – meaning, simply put, newspapers – with an emphasis on the latter.[190]

Local and regional parliamentary bodies mean, in the German case, the municipal, town or city councils of the communities where nuclear power plants were to be erected. They represent the local level of representative democracy; the district councils encompassing or surrounding the former provide the intersection to the surrounding region. In the British case, parish councils were originally thought to ideally represent the local level of debates. It turned out, however, that the district councils were more of an equivalent to the municipal councils in Germany. Therefore, these two sources were mainly consulted for the local dimension and county councils for transition to the regional one. The level of detail regarding public council meetings in the official files, however, proved to be rather low in all cases, up to the identification of individual councillors contributing to a discussion. The number of questions asked via these councils by members of the public regarding nuclear power was even lower. In fact, a much livelier debate took place in the columns of local newspapers – where, in addition, quite detailed reports about council meetings, often allowing the identification of contributors, were the rule rather than the exception.[191] In the case of Hinkley Point, in 1970, the Williton Rural District Council even congratulated a local reporter who had made it to the post of editor and pointed out in its official protocol that he "had reported proceedings of this Council since 1936 and the Chairman spoke in appreciation of the accuracy of his reports."[192]

In other words, the local and regional press reports turned out to be a highly suitable source for covering the public debates in the communities around nuclear power plants. These newspapers continuously accompanied the nuclear power plants, from the first announcement to the very end. They covered the main points of discussion as well as small events, even including protests staged by a handful of school pupils, for example. To put it in the somewhat romanticising, though apt words of one journalist working for a British local newspaper: "If there is any important matter which the public would think it was entitled to know, and which the Press ignored, then clearly the Press would be failing in its duty."[193] The local and regional papers do provide a very reliable frame of reference for analysing public debates, notwithstanding the fact that not every single event was covered.[194] It should also be noted that local and regional media tended to be accused of bias by both sides, while they themselves often complained about a lack of information.[195] Nevertheless, when it comes to consistency of reporting and openness to voices and seemingly marginal aspects that were otherwise barely given an outlet, they differ by far from most national media, which tended to cover specific nuclear installations only in the light of special events taking place there.[196] The local and regional press, on the other hand, covered the local debates extensively, frequently looking into topics linked to nuclear power which tended to be ignored in the national press.[197] They also took up the national and transnational debates about nuclear power, especially when these debates happened to point to "their" local power plants, or came to be seen as relevant for the communities in any other way.[198] Even when the local or regional newspapers did not refer to the higher strata of the debates directly, they generally covered gateways from the (trans-)national to the local debates, for example, talks given locally by (trans-)national experts speaking for or against nuclear power, thereby allowing for an analysis of the connections between the different planes of debate.

Of course, the analytical potential of the content of local and regional newspapers as sources for the local and regional public debates varied, depending not least on the category of the articles in question. Standard news and reports, an important reference point for information about nuclear power and the debates about it that could be assumed to be part of the public debates, in most cases, were kept in as "neutral" a perspective as possible. Nonetheless, the newspapers' general stances are expressed here and can be derived from the articles, especially when considering a larger number of them.[199]

Regarding leading articles and comment pieces, the often marked continuity among the editorial personnel allows for the analysis of the development of their position towards nuclear power as one strand of the local and regional public debates. This is especially valuable as their opinions, in most cases, were voiced because of their position within the newspaper, which required them to write about nuclear power because it was deemed newsworthy, not necessarily because they had a strong, personal opinion about it and wanted to make themselves heard. Therefore, contrary to most members of the industry on the one hand and critics of nuclear power on the other, it is possible to observe changes of position regarding the technology, which were at times even openly expressed. Furthermore, statements by journalists were often explicitly constructed with the aim of taking up a "neutral" perspective, considering arguments made within the debate by critics as well as by proponents.

Even more important for this analysis was the newspapers' function as a platform for letters to the editors.[200] They often covered a wide range of opinions, with exchanges of correspondence between various readers sometimes going on for weeks or being taken up again after months or even years had passed. Certainly, one must consider potential editorial interventions, from abbreviations to the outright omission of whole letters. The empirical data suggests, however, that this was rather rare: There is evidence of readers' complaints about the way the newspapers handled certain comments, suggesting that no position was systematically and completely suppressed. What is more, the number of letters published varied widely, sometimes even covering multiple pages, with the staff occasionally pointing out the necessity to distribute some letters over time. In general, then, it can be assumed that most people who wanted to be published had a fair chance and that, if they were not published or their position was tampered with in a substantial way, this would have become known.[201] The letters to the editor in local newspapers are therefore a treasure trove of source documents allowing insights into a broad sample of individuals' opinions that could hardly have been accessed in any other way.[202]

Finally, interviews often allowed for an insight into the opinions of people belonging to some of the local or regional "elites", be it political representatives or important members of the administration or the power plants. At the same time, they shed another light on the newspapers' positions regarding the topic, which often became quite transparent in the questions asked or the timing of and partners chosen for interviews. The often very detailed coverage of local and regional parliamentary debates can loosely be sorted in this category, as they mostly consisted of a contextualised list of distinctive quotes of the participants.[203]

Taken together, local and regional mass media may not offer an insight into the historical actors' minds and mostly remain on the public side of the possible strategies behind publicly made statements and openly performed actions, but they do constitute one of the most important sources of information for the communities studied here and their public construction of nuclear power and its meanings.[204] They provide, as van Buiren argued for mass media in general, in a way, "a mirror image of the public debates", while also functioning as a "filter, able to reinforce or attenuate tendencies".[205] This should not imply any kind of direct, controlling influence over opinions.[206] Such an influence could only be claimed with a view to people who believed in it and (re-)acted accordingly.[207] Nor should it imply that a complete picture of all the heterogenous voices speaking up on the topic of nuclear energy in the places studied here could be drawn. Nonetheless, the media provided an important arena for the local and regional public debates to play out, generating a space for different strata of the communities to express their opinion, albeit in different ways.[208] As Tibor Kliment pointed out, proponents of nuclear energy in general had more opportunities to speak up because they often occupied official positions, while the technology's critics tended to weigh in more voluminously.[209] Beyond that, the newspapers themselves, via their editors and journalists, were prominent participants in the debates, at times doing hardly more than chronicling what was deemed newsworthy for their area, at others becoming some of the most recognisable voices in disputes about nuclear energy and the wide range of topics anchored to it.[210] In sum, especially in combination with the local parliamentary bodies, the local media thus allow access to and the analysis of the local communities, at least the varying politically active part of them.[211] One can even claim that they constituted a form of "participatory public",[212] as close as it gets to the "village square" from where Albert Einstein, in 1946, wanted the decision about "atomic energy" to arise.[213]

All these sources, together with information gained from the available literature, bring to life the voices and positions of the people whose everyday life was dominated by a technology to which some of the 20th century's most intense societal conflicts were anchored. They are the voices of people who have, so far, most often been neglected by historical research for the benefit of those offering – sometimes intentionally, but often as victims of events outside their control – a more newsworthy story. Therefore, they form the basis of this cultural history of technology. Based primarily upon the local historical debates, it asks for the historical actors' conception of nuclear power, for the technology's discursive construction in relationship to the issues of Britain's and Germany's conditions of debate, the roles played by emotions and scientific-technical rationality, relationships of trust, practices of democratic participation, and visions of the future. These six topics, as already mentioned, were not the only ones discussed in the six locales. Other than, for example, the atom's military side, which was prominently brought up at British sites in the early and later years under consideration here, while less so at other times and places, these issues were most prominently and comprehensively debated throughout. Some, the issue of democracy most of all, are also prominent in existing research, as indicated earlier. Here, however, their selection for the analysis as well as their (re)construction from the source material is based upon the hermeneutic technique of an empirically guided, qualitative examination of public statements made in and around the chosen nuclear power plants' host communities about the perceptions, experiences, expectations, deliberations, and actions regarding this "public technology", which was of interest – and had implications – far beyond the local areas in focus here. For the analysis, precedence was given to those contributions to the public discourse that were either typical or outstanding, that shaped the course of the historical debates at the respective locations with its continuities as well as changes, that represented and upheld the former or brought about or prevented the latter, often in spite of other, competing voices; to the contributions, in other words, that represented and determined the specifics of the local positions as such, as well as relative to higher levels of the debate, from the regional to the (trans-)national level.[214]

The Archival Situation(s)

The *practical availability* of the source documents is quite similar for the six case studies, with one important difference between the British and German cases. In both countries, local archives hold documents from the relevant parliamentary bodies, covering their public debates, and several additional documents regarding the local power plants, mostly including administrative questions, communication, and emergency planning. The local and regional newspapers are also available there. The difference between the British and German situations in this regard is that all relevant German papers were daily newspapers, while the British ones were published weekly. In the latter case, it was possible to analyse all issues published during the time under consideration. This was impossible for the German newspapers, which were not only more numerous because of their frequency, but also, in general, more voluminous. In all three German cases, however, collections of press cuttings regarding the local power plants are available, covering most of the period under scrutiny here,[215] which were complemented by choice samples from the archived papers. Similar holdings were used for the British locales, too, whenever available.[216] In both cases, selected documents covering key events from different angles not publicly available at the time – for example, the view of administrators – were also taken into consideration. This holds true also for a few documents of more general relevance for the course of atomic energy's history.[217] In detail, the following archival collections were used:

Biblis

The most important archive for public debates about nuclear power in and around Biblis is the municipal archive located in Biblis' City Hall.[218] Here, the minutes of municipal council meetings from the late 1960s to the end of the 1980s were analysed, as were collections of press cuttings covering the same period. These consist mainly of articles from the *Darmstädter Echo*, *Mannheimer Morgen*, *Südhessische Post*, and *Wormser Zeitung*, the dominant newspapers in the region. Additionally, selected articles from other regional and national newspapers are included in these collections. As they hold little material for the periods 1983–1985 and 1987–1989, the municipal *Bibliser Blatt*, which is also kept in the archive, was analysed for these years, too. Additional sources regarded the Informationskreis Kernkraftwerk (Nuclear Power Plant Information Circle), consisting mostly of representatives of the operators and regulatory authorities, local politicians, and campaigns by local anti-nuclear activists. Covering the region around Biblis on the administrative level, the Bergstraße District Council meeting minutes were consulted. They are available in part in the Biblis Municipal Archive and in the Hessian State Archives in Darmstadt.[219] The rather small number of publications focusing on Biblis' nuclear history were also considered.[220]

Lingen

Lingen City Archive holds the minutes of the city council meetings, which were systematically analysed for the years from 1977 to the end of the 1980s,[221] supplemented by selected files from the city's administrative board and its environmental committee.[222] Press cuttings, mostly from the dominant *Lingener Tagespost*, are also available for the period from 1977 till the late 1980s.[223] Since these are at times less comprehensive than those available for Stade and Biblis, they were supplemented by studying the archived copies of the *Lingener Tagestpost* for 1984 and April to July 1986. Also available are several issues of the *KLE-Blatt*, a brochure published by the operator of the local power plant developed for local consumption.[224] The archive of the Emsland District, to which Lingen belongs, keeps the files of the district council. The minutes of council meetings considering the power plant were consulted in this collection, as were the minutes from the public hearing as part of the licensing procedures in the Oldenburg section of the State Archives of Lower Saxony.[225] Publications covering the local history rounded off the picture.[226]

Stade

Regarding Stade, the minutes of the city council from the beginning of the 1960s to the end of the 1980s can be found in the Stade City Archive. More detailed protocols regarding special hearings on the topic of the atomic power plant supplement the standard minutes.[227] In addition, collections of newspaper cuttings are available concerning the local nuclear power plant, and nuclear energy as such, or local industries, sometimes with references to the power plant.[228] Relevant files from the city's administrative director were also consulted,[229] as were select samples of the complete documentation of the *Stader Tageblatt*. With a view to the Stade section of the State Archives of Lower Saxony, the minutes of the district council are the most significant of the documents analysed for this study,[230] supplemented with files stemming from the licensing authorities.[231] As Stade was the first of the selected sites, other types of sources in addition to these publicly archived files were also sampled. Dr Tönnies Hagelstein kindly allowed me to use his files covering some of the activities of the Arbeitsgemeinschaft Umweltplanung Niederelbe in the 1970s and 1980s, while Dr Helge Heyn and Ulrich Hemke kindly made themselves available for an interview regarding the activities of critics of nuclear power from the Stade area. A small number of publications also provided further insights, especially those by local actors from the city or district administration and the electricity supply industry.[232]

Hinkley Point

In the case of Hinkley Point, most relevant files are available at Taunton's Somerset Heritage Centre (SHC). Its collections hold the minutes of Somerset County Council from 1956 to the end of the 1980s,[233] as well as those of Williton Rural District Council from 1956 till 1974,[234] when its competences were transferred to West Somerset District Council. The latter's files were accessed via the National Archives (NA) in London.[235] The files of Stogursey Parish Council for the years after 1947 were untraceable. Its meetings were regularly taken up by the local press, however, as were those of the other councils. Here, the *Bridgwater Mercury* turned out to contain the most intensive coverage and liveliest debates about the power plant, while samples showed that the *West Somerset Free Press* was less active in this regard. The SHC also offers many files with correspondence and debates concerning the power stations, especially regarding the conflict about the construction of a third power plant during the 1980s and the corresponding public inquiry. These include the official report, protocols, statements, promotional material and studies by the Central Electricity Generating Board, files from the Consortium of Opposing Local Authorities and other groups of critics, for example, the biweekly *Hinkley Inquirer*.[236] In sum, Somerset County Council took in 16,000 pages produced during the inquiry.[237] Also of interest were the documents produced by Hinkley Point's Local Liaison Committee (LLC).[238] In addition, files from the Bristol Archives and the NA were taken into account, with the majority concerning the Hinkley Point C public inquiry, including further publications of the critical Stop Hinkley Expansion group[239] and administrative files covering the earlier power stations.[240] With respect to scholarly analyses of the debates about Hinkley Point, studies regarding the acceptance of risks were undertaken[241] and the power plant's early history sketched out.[242]

Oldbury

Most files relevant to the Oldbury nuclear power plant are held by the Gloucestershire Archives in Gloucester. Here, the minutes of Oldbury-on-Severn Parish Council are available,[243] as are the files of Thornbury Rural District Council and its successor, Northavon District Council.[244] Regarding local and regional media, the locally dominant *Thornbury Gazette* (reformed into the *Northavon Gazette*, Thornbury Edition for a decade after a 1974 local government reform, before returning to the former nomenclature) is available.[245] Additional newspaper articles are accessible via collections of press cuttings.[246] Further relevant items are available here, as well as in the

Bristol Archives and the NA, not the least documents regarding the early planning for the power station and the public inquiry,[247] and those covering questions of the treatment of nuclear waste, a relevant topic for the local and regional debates,[248] and files of the LLC.[249] Publications on local history and studies about the perception of nuclear power and risks also mention the power plant.[250]

Torness

In the case of the Torness atomic power station, the John Gray Centre in Haddington is the most important institution holding relevant source documents. Here, the minutes of East Lothian District Council from the middle 1970s to the end of the 1980s could be analysed, while files from Innerwick Parish Council were unavailable. The *East Lothian Courier* could be studied in full, as could parts of the *East Lothian News* and scrapbook collections regarding the power plant.[251] Additionally, further documents were reviewed concerning materials from proponents and critics of the power plant, among others,[252] a Joint Committee of Local Authorities,[253] and the LLC.[254] Finally, some relevant material is available in the NA, not least regarding the public inquiry.[255] In addition, Torness was the topic of contemporary and recent analyses of Britain's anti-nuclear movement.[256]

Debates Anchored to the Atom: Structure

Based upon the source documents and literature discussed earlier, this book is divided into six main chapters, distributed over two volumes. Each of the main chapters is dedicated to one of the six main issues anchored to the public debates about nuclear energy. These were, of course, intertwined in practice. They are treated separately for the sake of analytical clarity, as this allows me to establish each topic's co-evolution with the "Peaceful Atom" more clearly. The price for this is some repetition and the separation of connections between issues that were clearly present in the debates. References in the text point out important ones, however. Loosely arranged, on trend, from the older sites and more basal issues to the younger and more complex ones, the six issues and chapters are: the conditions of debate ("Nuclear Debates"), the roles played by emotions ("Emotional Atoms"), and scientific rationality ("Rational Atoms") in the first volume, and relationships of trust ("Atomic Trust(s)"), practices of participation ("Nuclear Democracy"), and visions of the future ("Nuclear Futures") in the second. Taken together, as well as on their own, the chapters contribute to the main hypothesis: The debates about nuclear energy were not confined to one energy technology among many. Instead, several complex and far-reaching societal issues became anchored to it, which made them appear to become more concrete themselves and thus publicly debateable. However, they also became connected to one another, forming an increasingly complex network of debates about the atom, and about themselves. This made it difficult to resolve any one of these issues. Nonetheless, in this way, they turned the nuclear debates into a valuable sensor for the historical analysis of the self-perception, the condition and development of the communities in which they took place. Focusing on these six issues anchored to the nuclear debates, therefore, gives an insight into the course, dynamics, and outcome of the disputes about nuclear energy, the growing and shrinking of the groups opposing or supporting the technology, and to explain the similarities and differences in the nations under consideration. In sum, the analysis also reveals differences between the two countries' democratic cultures, and the way political decisions were discussed and implemented in both countries. In Britain, leading voices were often rather open to oppositionally inclined citizens, which contributed to an early settlement of the remarkable conflicts brought about by nuclear technologies. In Germany, where there was a strong tendency to top-down control, on the other hand, opposing positions became entrenched and nuclear power's splitting potential was realised in an escalating conflict. Consequently, the established reading of both countries' nuclear histories must be modified to incorporate the occurrence and effects of early British protests – most prominently the enduring reduction of nuclear power's splitting potential in comparison with Germany, where it was potentiated by its entanglement with a top-down style increasingly rejected by citizens politicised by this very conflict.

Each of the six chapters compares the public debates about one British and one German nuclear power plant in their local and regional surroundings, with short references to the other sites from the sample group added in each chapter's conclusion. This method leads to a certain level of repetition as each location is highlighted twice. It has the advantage, however, of allowing for the narration of more aspects of the local histories and the perspectives and contributions of the people living with nuclear power in their everyday lives, who have largely been ignored so far. As these local histories vary greatly as regards their individual time frames and geographical situations, the sizes and designs of the local power plants, and because of the varying emphases put on the different topics anchored to the nuclear debates, the fine structures of the local sub-chapters vary. Taken together, however, they always show how the issue covered in the chapter was anchored to the nuclear debates, how it influenced them and was influenced in return – and indicate how this interaction was further influenced by the other anchored topics. Thus, each of the chapters presents additional arguments that take up, broaden, complement, or differentiate the book's main hypothesis, thereby demonstrating the potential for insights to be gained via an internationally comparative local analysis. They also offer tentative findings regarding the consequences of the nuclear debates for the topics anchored to them beyond the time and places under consideration here. This shows that these topics were not brought up with the rise of the "Peaceful Atom", nor were they solved when it lost its momentum. They were anchored to it, co-evolving for a time, but were important on their own.

The insights gained in the six main chapters are then taken up for a combined discussion in the concluding one ("Conclusion – Two Democratic Cultures") at the end of volume two. There, I recapitulate the nuclear histories of the selected sites, consider the varying interactions between the anchored issues and the debates about atomic energy, and argue for the plausibility of deriving insights into the national nuclear histories from the local perspectives. I then present the modified reading of these very histories that is based upon this study's combined findings. Finally, I briefly discuss its further implications.

Before entering the analysis, however, a technical remark should be made: Throughout the book, in the case of multiple quotations from the same source document following one another, the source they stem from is annotated at the last instance in a row.

Notes

1 Nuclear power, derived from the nuclear force within an atom's nucleus, was initially called atomic power, reflected in names like the United Kingdom Atomic Energy Authority, the German Atomministerium, or the 'Atoms for Peace' campaign. The switch to the 'more precise' phrase of nuclear energy was understood as an attempt to separate the civilian use of nuclear energy from atomic weapons. During the heyday of the conflict about the technology, 'atomic' was most often used by its opponents, 'nuclear' by its advocates. Here, both terms are used indiscriminately outside of direct quotations; cf. Patterson, Walter Cram: Nuclear Power. Harmondsworth ^{2}1986 (Original 1976), p. 14.

2 Lutterbeck, Claus: Der Bürgerkrieg in Brokdorf, in: *Stern*, 18.11.1976, pp. 26–32, here p. 31; translation, as all translations here, by CG. When out of my depth, I looked to leo.org or deepl.com for inspiration; cf. Winkler, Heinrich August: Der lange Weg nach Westen. Vol. 2. Vom 'Dritten Reich' bis zur Wiedervereinigung. Munich 2014 (Original 2000), pp. 349–350; critical: Uekötter, Frank: Atomare Demokratie. Eine Geschichte der Kernenergie in Deutschland. Stuttgart 2022, p. 138.

3 The term 'transnational' is used to describe non-state actors or perspectives that go across national borders, while 'international' is used in the case of nation states and their relations.

4 Tompkins, Andrew S.: Transnationality as a Liability? The Anti-Nuclear Movement at Malville, in: Revue belge de philologie et d'histoire 89 (2011), 3–4, pp. 1365–1379; cf. Lehtonen, Markku: France Short Country Report (2018), in: History of Nuclear Energy and Society (HoNESt) Consortium Deliverable 3.6, pp. 14–15, 75–77, www.honest2020.eu/sites/default/files/deliverables_24/FT.pdf (23.1.2023).

5 Josephson, Paul: United States Short Country Report (2018), in: HoNESt Consortium Deliverable 3.6, pp. 35–37, www.honest2020.eu/sites/default/files/deliverables_24/USA.pdf (23.1.2023).

6 O'Riordan, Timothy, Ray Kemp, Michael Purdue: Sizewell B. An Anatomy of the Inquiry. Basingstoke 1988, pp. viii, 1–7, 400.

7 Cf., for a more detailed discussion of similarities and differences between the two countries, 'Coming to Terms with Conflicts Anchored to the Atom: Methods'.

8 Cf. the discussion of the state of research in 'The Topics Anchored to the Atom'.

9 For the concept of Technologies as Anchors for Societal Conflicts (TASC): Götter, Christian: Technologies as Anchors for Societal Conflicts. An Outline

of the Concept, in: Technikgeschichte 89 (2022), 1, pp. 9–34; cf. Doering-Manteuffel, Anselm, Lutz Raphael: Nach dem Boom. Perspektiven auf die Zeitgeschichte seit 1970. Göttingen [3]2012, p. 30; Kehrt, Christian, Peter Schüßler, Marc-Denis Weitze: Einleitung. Neue Technologien in der Gesellschaft, in: Kehrt, Christian, Peter Schüssler, Marc-Denis Weitze (Eds.): Neue Technologien in der Gesellschaft. Akteure, Erwartungen, Kontroversen und Konjunkturen. Bielefeld 2011, pp. 11–25, here pp. 11–13; Kaijser, Arne et al.: Future Challenges for Nuclear Energy and Society in a Historical Perspective, in: Kaijser, Arne et al. (Eds.): Engaging the Atom. The History of Nuclear Energy and Society in Europe from the 1950s to the Present. Morgantown 2021, pp. 279–302, here p. 279.

10 Cf., for similar conceptions of more abstract differences and the questions left open by directly focussing on them: 'The National, the Spectacular, and the Complex Atom: State of Research' and 'Nuclear Debates'.

11 Hecht, Gabrielle: Being Nuclear. Africans and the Global Uranium Trade. Cambridge 2012, pp. 6–15; Joachim Radkau saw a parallel between nuclear energy's singular risks and the singular character of the debates about it: Radkau, Joachim: Aufstieg und Krise der deutschen Atomwirtschaft 1945–1975. Verdrängte Alternativen in der Kerntechnik und der Ursprung der nuklearen Kontroverse. Reinbek 1983, pp. 455–458.

12 At least, that is, when, beyond the commercial technology as such, the propagandistic tales woven around it are to be implied as well.

13 Oelschlaeger, Max: The Myth of the Technological Fix, in: The Southwestern Journal of Philosophy 10 (1979), 1, pp. 43–53.

14 Cf. Hamblin, Jacob Darwin: The Wretched Atom. America's Global Gamble with Peaceful Nuclear Technology. New York 2021; Kaijser, Arne et al.: Nuclear Energy and Society in Postwar Europe, in: Kaijser, Arne et al. (Eds.): Engaging the Atom. The History of Nuclear Energy and Society in Europe from the 1950s to the Present. Morgantown 2021, pp. 1–24, here pp. 9, 15.

15 By '(trans-)national', I mean national and transnational.

16 For the relationship between the local and the regional, cf. 'Coming to Terms with Conflicts Anchored to the Atom: Methods'.

17 For the argument that the local level can help to understand wider developments: Geertz, Clifford: Dichte Beschreibung. Beiträge zum Verstehen kultureller Systeme. Frankfurt/Main [2]1991, pp. 31–32; Heßler, Martina: Die kreative Stadt. Zur Neuerfindung eines Topos. Bielefeld 2007, pp. 13, 28–29; for an emphatic call to consider 'unofficial' contributions to the nuclear debates: Hogg, Jonathan: British Nuclear Culture. Official and Unofficial Narratives in the Long 20th Century. London 2016, pp. 174–175; cf. Eiser, J. Richard, Joop van der Pligt, Russell Spears: Nuclear Neighbourhoods. Community Responses to Reactor Siting. Exeter 1995, p. xi; Siegrist, Hannes: Perspektiven der vergleichenden Geschichtswissenschaft. Gesellschaft, Kultur und Raum, in: Kaelble, Hartmut, Jürgen Schriewer (Eds.): Vergleich und Transfer. Komparatistik in den Sozial-, Geschichts- und Kulturwissenschaften. Frankfurt/Main 2003, pp. 305–340, especially p. 333; Landwehr, Achim: Historische Diskursanalyse. Frankfurt/Main 2008, pp. 165–166; Baer, Willi, Jens Renner: Vorwort, in: Borchers, Uli (Ed.): Die AKW-Protestbewegung von Wyhl bis Brokdorf. Hamburg 2011, pp. 15–18; Arapostathis, Stathis, Robert Bud, Helmuth Trischler: Nuclear Energy in Europe. A Public Technology, in: Kaijser, Arne et al. (Eds.): Engaging the Atom. The History of Nuclear Energy and Society in Europe from the 1950s to the Present. Morgantown 2021, pp. 230–253, here p. 230.

18 Cf., for an appreciation of "the ordinary man" in democracies: de Certeau, Michel: The Practice of Everyday Life. Berkeley 1988, p. v.

19 Trischler, Helmuth, Robert Bud: Public Technology. Nuclear Energy in Europe, in: History and Technology 34 (2018), 3/4, pp. 187–212; cf. Kraft, Hans: Der Einfluß des öffentlichen Bewußtseins auf die Wissenschafts- und Technikentwicklung, in: Gatzemeier, Matthias (Ed.): Verantwortung in Wissenschaft und Technik. Mannheim 1989, pp. 157–164; Wendland, Anna Veronika: Reaktorsicherheit als Zukunftskommunikation. Nuklearpolitik, Atomdebatten und kerntechnische Entwicklungen in Westdeutschland und Osteuropa 1970–2015, in: Kampmann, Christoph, Angela Marciniak, Wenke Meteling (Eds.): 'Security Turns Its Eye Exclusively to the Future'. Zum Verhältnis von Sicherheit und Zukunft in der Geschichte. Baden-Baden 2017, pp. 305–352, here p. 330.

20 For an introduction: Daniel, Ute: Kompendium Kulturgeschichte. Theorien, Praxis, Schlüsselwörter. Frankfurt/Main [5]2006; Heßler, Martina: Kulturgeschichte der Technik. Frankfurt/Main 2012.

21 The research project was funded by the German Federal Ministry for Education and Research as part of the 'Kleine Fächer – Große Potentiale' programme from 2017 to 2021 under the funding code 01UL1706X and conducted at the Technische Universität Braunschweig and the Deutsches Museum in Munich.

22 For a wider overview: Kalmbach, Karena: Revisiting the Nuclear Age. State of the Art Research in Nuclear History, in: Neue Politische Literatur 62 (2017), 1, pp. 49–69; for one focussing the 1970s and most of the 1980s, from a German perspective with international highlights: Radkau, Joachim: Die Kernkraft-Kontroverse im Spiegel der Literatur. Phasen und Dimensionen einer neuen Aufklärung, in: Hermann, Armin, Rolf Schumacher (Eds.): Das Ende des Atomzeitalters? Eine sachlich-kritische Dokumentation. Munich 1987, pp. 307–334.

23 Biblis and Stade were identified as such places by: van Buiren, Shirley: Die Kernenergie-Kontroverse im Spiegel der Tageszeitungen. Inhaltsanalytische Auswertung eines exemplarischen Teils der Informationsmedien. Empirische Untersuchung im Auftrag des Bundesministers des Inneren. Munich 1980, p. 11.

24 Therefore, literature published up to 1989 is grouped with published sources in the bibliography.

25 In a comparative approach in 2011, national perspectives were even analysed as micro-level: Presas i Puig, Albert: As an Introductory Reflection, in: Presas i Puig, Albert (Ed.): A Comparative Study of European Nuclear Energy Programs. Berlin 2011, pp. 3–6, here p. 5; cf. Kalmbach, Revisiting, pp. 50, 52.

26 In the British case, this was explicitly bemoaned by Jonathan Hogg and Christoph Laucht, who argued for a stronger focus on 'nuclear culture': Hogg, Jonathan, Christoph Laucht: Introduction. British Nuclear Culture, in: The British Journal for the History of Science 45 (2012), 4, pp. 479–493; cf. Hughes, Jeff: What Is British Nuclear Culture? Understanding Uranium 235, in: The British Journal for the History of Science 45 (2012), 4, pp. 495–518, especially pp. 503–506, 517–518.

27 Radkau, Aufstieg und Krise; Radkau, Joachim, Lothar Hahn: Aufstieg und Fall der deutschen Atomwirtschaft. Munich 2013; for a chronicle from the critics' viewpoint, cf. Paul, Reimar (Ed.): Chronologie einer Bewegung. Hamburg 2012. Radkau was not the first historian presenting a study of Germany's nuclear history; cf. Bieber, Hans-Joachim: Zur politischen Geschichte der friedlichen Kernenergienutzung in der Bundesrepublik Deutschland. Heidelberg 1977.

28 Müller, Wolfgang D.: Geschichte der Kernenergie in der Bundesrepublik Deutschland. 3 Vols. Stuttgart 1990–2001; for the GDR, cf. Reichert, Mike: Kernenergiewirtschaft in der DDR. Entwicklungsbedingungen, konzeptioneller Anspruch und Realisierungsgrad (1955–1990). St. Katharinen 1999.

29 Laufs, Paul: Reaktorsicherheit für Leistungskernkraftwerke. Die Entwicklung im politischen und technischen Umfeld der Bundesrepublik Deutschland. 2 Vols. Berlin ^{2}2018.

30 Cf. Winnacker, Karl, Karl Wirtz: Das unverstandene Wunder. Kernenergie in Deutschland. Düsseldorf 1975.

31 Rucht, Dieter: Von Wyhl nach Gorleben. Bürger gegen Atomprogramm und nukleare Entsorgung. Munich 1980.

32 Gowing, Margaret (assisted by Lorna Arnold): Independence and Deterrence. Britain and Atomic Energy 1945–52. 2 Vols. London 1974; cf. Hennessy, Peter: The Secret State. Whitehall and the Cold War. London 2002, pp. 44–119. In Germany, secrecy regarding the early nuclear industry was pointed out by Radkau, Aufstieg und Krise, especially pp. 411–418.

33 Laucht, Christoph: Elemental Germans. Klaus Fuchs, Rudolf Peierls and the Making of British Nuclear Culture, 1939–1959. Basingstoke 2012; Laucht, Christoph: Scientists, the Public, the State, and the Debate over the Environmental and Human Health Effects of Nuclear Testing in Britain, 1950–1958, in: The Historical Journal 59 (2016), 1, pp. 221–251; cf., for the general connection between 'nuclear culture' and the atom's military side: Messmer, Michael W.: Nuclear Culture, Nuclear Criticism, in: Minnesota Review 30 (1988), pp. 161–180; note 42.

34 Patterson, Walter Cram: Going Critical. An Unofficial History of British Nuclear Power. London 1985; for Patterson, cf. Oberloskamp, Eva: Ambiguities of Transnationalism. Social Opposition to the Civil Use of Nuclear Power in the United Kingdom and in West Germany During the 1970s, in: European Review of History. Revue Européenne d'Histoire 29 (2022), 3, pp. 417–451, p. 429; for more details on the group: 'Rational Atoms'; for the critical reception of Patterson's publications: Williams, Roger: The Nuclear Power Decisions. British Policies, 1953–78. London 1980, p. 267.

35 Flood, Michael: The End of the Nuclear Dream. The UKAEA and Its Role in Nuclear Research and Development. London 1988.

36 Welsh, Ian: Mobilising Modernity. The Nuclear Moment. London 2015 (Original 2000).

37 Williams, Decisions; O'Riordan, Timothy: The Politics and Economics of Nuclear Electricity, in: Catalyst 5 (1986), 12, pp. 41–54; focussing mostly on economic aspects: Burn, Duncan: Nuclear Power and the Energy Crisis. Politics and the Atomic Industry. London 1978.

38 Pocock, Rowland F.: Nuclear Power. Its Development in the United Kingdom. Old Woking 1977.

39 Hill, C. N.: An Atomic Empire. A Technical History of the Rise and Fall of the British Atomic Energy Programme. London 2013; cf. Hall, Tony: Nuclear Politics. The History of Nuclear Power in Britain. London 1986.

40 Kalmbach, Karena: The Meanings of a Disaster. Chernobyl and Its Afterlives in Britain and France. New York 2021; cf. Kalmbach, Karena: Meanings of a Disaster. The Contested 'Truth' about Chernobyl. British and French Chernobyl Debates and the Transnationality of Arguments and Actors. Diss, European University Institute, Florence 2014.

41 Hughes, British; cf. Zeman, Scott C., Michael A. Amundson: Introduction, in: Zeman, Scott C., Michael A. Amundson (Eds.): Atomic Culture. How We Learned to Stop Worrying and Love the Bomb. Boulder 2004, pp. 1–9, here p. 2.

42 Hogg, British, quote p. 4, emphasis by Hogg; cf. Hogg, Jonathan: 'The Family That Feared Tomorrow'. British Nuclear Culture and Individual Experience in the Late 1950s, in: The British Journal for the History of Science 45 (2012), 4, pp. 535–549.

43 Van Buiren, Shirley et al.: Bürgerinitiativen im Bereich von Kernkraftwerken. Bericht für das Bundesministerium für Forschung und Technologie. Bonn 1975, pp. 293–294; van Buiren, Kernenergie-Kontroverse, p. 140; cf. Hoffmann-Riem, Wolfgang, Susanne Rubbert: Atomrechtlicher Erörterungstermin und Öffentlichkeit. Zum Verhältnis von Bürgerbeteiligung und Öffentlichkeit. Heidelberg 1984, pp. 72–74, 83; Hocke, Peter: Massenmedien und lokaler Protest. Eine empirische Fallstudie zur Medienselektivität in einer westdeutschen Bewegungshochburg. Wiesbaden 2002, especially p. 281.

44 Cf. Rucht, Dieter: Wyhl. Der Aufbruch der Anti-Atomkraftbewegung, in: Linse, Ulrich et al. (Eds.): Von der Bittschrift zur Platzbesetzung. Konflikte um technische Großprojekte. Laufenburg, Walchensee, Wyhl, Wackersdorf. Berlin 1988, pp. 128–164, here pp. 154–155; Arnold, Lorna: Windscale 1957. Anatomy of a Nuclear Accident. New York 1992, p. 69; Kliment, Tibor: Kernkraftprotest und Medienreaktionen. Deutungsmuster einer Widerstandsbewegung und öffentliche Rezeption. Wiesbaden 1994, p. 351; Welsh, Mobilising, pp. 11, 211; Gleitsmann, Rolf-Jürgen, Günther Oetzel: Fortschrittsfeinde im Atomzeitalter? Protest und Innovationsmanagement am Beispiel der frühen Kernenergiepläne der Bundesrepublik Deutschland. Berlin 2012, pp. 41–42; Pohl, Natalie: Atomprotest am Oberrhein. Die Auseinandersetzung um den Bau von Atomkraftwerken in Baden und im Elsass (1970–1985). Stuttgart 2019, pp. 23–24.

45 Camilleri, Joseph A.: The State and Nuclear Power. Conflict and Control in the Western World. Brighton 1984, pp. 87–88; Welsh, Ian: Nuclear Nation. Local Reaction, in: Journal of Regional and Local Studies (1989), 9, pp. 1–28; Welsh, Ian: The NIMBY Syndrome. Its Significance in the History of the Nuclear Debate in Britain, in: The British Journal for the History of Science 26 (1993), pp. 15–32, here pp. 17–18, 25–26, 28–31; Roth, Roland: Lokale Bewegungsnetzwerke und die Institutionalisierung von neuen sozialen Bewegungen, in: Neidhardt, Friedhelm (Ed.): Öffentlichkeit, öffentliche Meinung, soziale Bewegungen. Opladen 1994, pp. 413–436, here p. 415; Irwin, Alan, Brian Wynne: Conclusions, in: Irwin, Alan, Brian Wynne (Eds.): Misunderstanding Science? The Public Reconstruction of Science and Technology. Cambridge 1998, pp. 213–221, here p. 218; Wynne, Brian: Misunderstood Misunderstandings. Social Identities and Public Uptake of Science, in: Irwin, Alan, Brian Wynne (Eds.): Misunderstanding Science? The Public Reconstruction of Science and Technology. Cambridge 1998, pp. 19–46, here p. 21; Engels, Jens Ivo: Naturpolitik in der Bundesrepublik. Ideenwelt und politische Verhaltensstile in Naturschutz und Umweltbewegung 1950–1980. Paderborn 2006, p. 348; Uekötter, Frank: Fukushima and the Lessons of History. Remarks on the Past and Future of Nuclear Power, in: Temple, Samuel (Ed.): Europe after Fukushima. German Perspectives on the Future of Nuclear Power. Munich 2012, pp. 9–31, here pp. 22–23; Hogg, British, pp. 4–6, basing his assessment on an observation made by Jeff Hughes: Hughes, British, pp. 496–497. For this study's local framing, cf. 'Coming to Terms with Conflicts Anchored to the Atom: Methods'.

46 Welsh, Mobilising, pp. 2, 16–17; cf. Herring, Horace: From Energy Dreams to Nuclear Nightmares. Lessons from the Anti-Nuclear Power-Movement in the 1970s. Charlbury 2005, especially pp. 2, 4, 9, 68–69, 80–83, 210.

47 On a side note, Dolores L. Augustine and Dick van Lente even argued that nuclear energy was essential for bringing about a "world opinion" during the 20th century: Augustine, Dolores L., Dick van Lente: Conclusion. One World, Two Worlds, Many Worlds?, in: van Lente, Dick (Ed.): The Nuclear Age in Popular Media. A Transnational History, 1945–1965. New York 2012, pp. 233–247; nuclear energy was also argued about as a tool to push a European integration: Trischler, Helmuth, Hans Weinberger: Engineering Europe. Big Technologies and Military Systems in the Making of 20th Century Europe, in: History and

Technology 21 (2005), 1, pp. 49–83; Radkau, Joachim: Technik in Deutschland. Vom 18. Jahrhundert bis heute. Frankfurt/Main [2]2008, pp. 364–365.

48 Not the least by Dieter Rucht, cf. note 89. Wyhl as a role model for anti-nuclear activists was analysed, among others, by Foley, Michael Stewart: No Nukes and Front Porch Politics. Environmental Protest Culture and Practice on the Second Cold War Home Front, in: Conze, Eckart, Martin Klimke, Jeremy Varon (Eds.): Nuclear Threats, Nuclear Fear, and the Cold War of the 1980s. Cambridge 2017, pp. 186–205, here p. 190; cf. Hillengaß, Christian: Atomkraft und Protest. Die politische Wirkung der Anti-AKW-Bewegung in Deutschland, Frankreich und Schweden. Munich 2011; Meyer, Jan-Henrik: Challenging the Atomic Community. The European Environmental Bureau and the Europeanization of Anti-Nuclear Protest, in: Kaiser, Wolfram, Jan-Henrik Meyer (Eds.): Societal Actors in European Integration. Polity-Building and Policy-Making 1958–1992. Basingstoke 2013, pp. 197–220.

49 Milder, Stephen: 'Today the Fish, Tomorrow Us': Anti-Nuclear Activism in the Rhine Valley and Beyond, 1970–1979. Diss., Chapel Hill 2012, especially pp. 1–27, 71–148, 232–278; cf. Milder, Stephen: Between Grassroots Protest and Green Politics. The Democratic Potential of 1970s Antinuclear Activism, in: German Politics and Society 33 (2015), 4, pp. 25–39; Milder, Stephen: The 'Example of Wyhl'. How Grassroots Protest in the Rhine Valley Shaped West Germany's Antinuclear Movement, in: Conze, Eckart, Martin Klimke, Jeremy Varon (Eds.): Nuclear Threats, Nuclear Fear, and the Cold War of the 1980s. Cambridge 2017, pp. 167–185; Milder, Stephen: Greening Democracy. The Anti-Nuclear Movement and Political Environmentalism in West Germany and Beyond 1968–1983. Cambridge 2019.

50 Milder, Stephen: Between Grassroots Activism and Transnational Aspirations: Anti-Nuclear Protest from the Rhine Valley to the Bundestag, 1974–1983, in: Historical Social Research 39 (2014), 1, pp. 191–211; in general, the latter function was also observed by van Lente, Dick: Introduction: A Transnational History of Popular Images and Narratives of Nuclear Technologies in the First Two Postwar Decades, in: van Lente, Dick (Ed.): The Nuclear Age in Popular Media. A Transnational History, 1945–1965. New York 2012, pp. 1–17.

51 Meyer, Jan-Henrik: 'Where Do We Go from Wyhl?' European and International Organizations as Focal Points for Transnational Anti-Nuclear Protest in the 1970s, in: Historical Social Research 39 (2014), 1, pp. 212–235; Kirchhof, Astrid Mignon, Jan-Henrik Meyer: Global Protest against Nuclear Power. Transfer and Transnational Exchange in the 1970s and 1980s, in: Historical Social Research 39 (2014), 1, pp. 165–190.

52 Tompkins, Andrew S.: Better Active Than Radioactive! Anti-Nuclear Protest in 1970s France and West Germany. Oxford 2016; cf. Tompkins, Andrew S.: Towards a 'Europe of Struggles'? Three Visions of Europe in the Early Anti-Nuclear Energy Movement 1975–79, in: Wenkel, Christian et al. (Eds.): The Environment and the European Public Sphere. Perceptions, Actors, Policies. Winwick 2020, pp. 124–146.

53 Tompkins, Andrew S.: Grassroots Transnationalism(s). Franco-German Opposition to Nuclear Energy in the 1970s, in: Contemporary European History 25 (2016), 1, pp. 117–142.

54 Pohl, Atomprotest.

55 Both were taken up, for example, in: Kliment, Kernkraftprotest.

56 Tiggemann, Anselm: Die 'Achillesferse' der Kernenergie in der Bundesrepublik Deutschland. Zur Kernenergiekontroverse und Geschichte der nuklearen Entsorgung von den Anfängen bis Gorleben. 1955 bis 1985. Lauf an der Pegnitz [2]2010.

57 Gaumer, Janine: Wackersdorf. Atomkraft und Demokratie in der Bundesrepublik 1980–1989. Munich 2018.

58 Gleitsmann et al., Fortschrittsfeinde; cf. Müller, Kernenergie. Vol. 1, pp. 216–237.

59 Högselius, Per: Die deutsch-deutsche Geschichte des Kernkraftwerks Greifswald. Atomenergie zwischen Ost und West. Berlin 2005.

60 Arnold, Windscale.

61 Cf. Walker, J. Samuel: Three Mile Island. A Nuclear Crisis in Historical Perspective. Berkeley 2004; Zaretsky, Natasha: Atomic Nightmares and Biological Citizens at Three Mile Island, in: Conze, Eckart, Martin Klimke, Jeremy Varon (Eds.): Nuclear Threats, Nuclear Fear, and the Cold War of the 1980s. Cambridge 2017, pp. 55–78.

62 Cf. Brüggemeier, Franz-Josef: Tschernobyl, 26. April 1986. Die ökologische Herausforderung. Munich 1998; Arndt, Melanie: Tschernobyl. Auswirkungen des Reaktorunfalls auf die Bundesrepublik Deutschland und die DDR. Erfurt 2011; Radkau, Joachim: Dreißig Jahre Tschernobyl. Zur Aktualität der Reaktorkatastrophe, in: Zeitgeschichte Online 2016, www.zeitgeschichte-online.de/kommentar/dreissig-jahre-tschernobyl (23.1.2023).

63 Breach, Ian: Windscale Fallout. A Primer for the Age of Nuclear Controversy. Harmondsworth 1978; cf. Wynne, Brian: Rationality and Ritual. The Windscale Inquiry and Nuclear Decisions in Britain. Preston 1982: Wynne took part in the inquiry, delivering in his study an insider-view as well as an argument to see it as part of a staging of democratic procedures.

64 McDermott, Veronica: Going Nuclear. Ireland, Britain and the Campaign to Close Sellafield. Dublin 2008.

65 Davies, Hunter (Ed.): Sellafield Stories. Life in Britain's First Nuclear Plant. London 2012.

66 Davies, Richard: The Sizewell B Nuclear Inquiry. An Analysis of Public Participation in Decisionmaking about Nuclear Power, in: Science, Technology, & Human Values 9 (1984), 3, pp. 21–32; O'Riordan et al., Sizewell.

67 Jungk, Robert: Der Atomstaat. Vom Fortschritt in die Unmenschlichkeit. Munich [2]1977, especially pp. 190–197; cf., for the observation of an unavoidable connection between states commanding nuclear weapons and totalitarianism in 1958: Anders, Günther: Der Sprung, in: Anders, Günther: Endzeit und Zeitenende. Gedanken über die atomare Situation. Munich 1972, pp. 11–23, here pp. 17–19; for Jungk, cf. Seefried, Elke: Zukünfte. Aufstieg und Krise der Zukunftsforschung 1945–1980. Berlin 2015, pp. 135–143, 167–170; Radkau, Joachim: Geschichte der Zukunft. Prognosen, Visionen, Irrungen in Deutschland von 1945 bis heute. Munich 2017, pp. 165–168.

68 Cf. Dannenbaum, Thomas: 'Atom-Staat' oder 'Unregierbarkeit'? Wahrnehmungsmuster im westdeutschen Atomkonflikt der siebziger Jahre, in: Brüggemeier, Franz-Josef, Jens Ivo Engels (Eds.): Natur- und Umweltschutz nach 1945. Konzepte, Konflikte, Kompetenzen. Frankfurt/Main 2005, pp. 268–286, here pp. 274–275.

69 Cf. Matthöfer, Hans: Interviews und Gespräche zur Kernenergie. 'Den unsterblichen Tiger am Schwanz gepackt'. Karlsruhe 1976, p. 94.

70 Jungk, Robert: Weltweiter Widerstand. Umrisse einer informellen Internationale, in: Mez, Lutz (Ed.): Der Atomkonflikt. Berichte zur internationalen Atomindustrie, Atompolitik und Anti-Atom-Bewegung. Mit einer Einleitung von Robert Jungk. Reinbek 1981, pp. 8–11.

71 Klaus Traube remarked rather critically that the idea had been taken up belatedly in Germany, which, to him, was typical for any of the more subtle, societal aspects anchored to nuclear power: Traube, Klaus: Müssen wir umschalten? Von den politischen Grenzen der Technik. Reinbek 1978, p. 248; cf., arguing that Jungk had been inspired by debates in the United States of America, Oberloskamp, Ambiguities, p. 429–430; mentioning similar warnings in the 1950s: Radkau, Zukunft, p. 141

72 Flood, Michael, Robin Grove-White: Nuclear Prospects. A Comment on the Individual, the State and Nuclear Power. London 1976, pp. 1–29; Patterson, Nuclear, pp. 11–12; Breach, Windscale; Williams, Decisions, pp. 268–269; cf., for a critical perspective: Wendland, Reaktorsicherheit, pp. 317–319.
73 Royal Commission on Environmental Pollution. Sixth Report. Nuclear Power and the Environment. London 1976, pp. 200–205, especially 202, 204.
74 Koopmans, Ruud, Jan Willem Duyvendak: The Political Construction of the Nuclear Energy Issue and its Impact on the Mobilization of the Anti-Nuclear Movements in Western Europe, in: Social Problems 42 (1995), 2, pp. 235–251, here p. 244.
75 Winner, Langdon: Do Artifacts Have Politics?, in: Daedalus 109 (1980), 1, pp. 121–136, here pp. 133–134; Winner, Langdon: The Whale and the Reactor. A Search for Limits in an Age of High Technology. Chicago 1986, p. 175.
76 Flam, Helena: Political Responses to the Anti-Nuclear Challenge. (2) Democratic Experiments and the Use of Force, in: Flam, Helena (Ed.): States and Anti-Nuclear Movements. Edinburgh 1994, pp. 329–354, here pp. 335–337.
77 Cf. Espluga et al., Risky, p. 148.
78 Cf. Schumacher, Ernst Friedrich: Small Is Beautiful. Economics as If People Mattered. 25 Years later . . . with Commentaries. Vancouver 1999 (Original 1973), pp. 110–111; Radkau, Aufstieg und Krise, pp. 455–458; Roßnagel, Alexander: Spaltung des Volkes durch Kernenergie, in: Hermann, Armin, Rolf Schumacher (Eds.): Das Ende des Atomzeitalters? Eine sachlich-kritische Dokumentation. Munich 1987, pp. 175–184, here p. 176. With a view to nuclear weapons, for example: Anders, Sprung, pp. 19–21.
79 Nowotny, Helga: Kernenergie. Gefahr oder Notwendigkeit. Anatomie eines Konflikts. Frankfurt/Main 1979, p. 28.
80 Radkau, Joachim: Die Geschichte der Kerntechnik, in: Varchmin, Jochim, Joachim Radkau (Eds.): Kraft, Energie und Arbeit. Energietechnik und Gesellschaft im Wechsel der Zeiten. Munich 1979, pp. 169–234, here pp. 232–233; Radkau, Aufstieg und Krise, pp. 13–14, 398, 455–458, 471–474; cf. Traube, umschalten, p. 15; Meyer-Abich, Klaus Michael, Bertram Schefold: Wie möchten wir in Zukunft leben. Der 'harte' und der 'sanfte' Weg. Munich 1981, pp. 15–19; Meyer, Jan-Henrik: Kleine Geschichte der Atomkraft-Kontroverse in Deutschland, in: Aus Politik und Zeitgeschichte 71 (2021), 21–23, pp. 10–16, here p. 10.
81 Kitschelt, Herbert P.: Kernenergiepolitik. Arena eines gesellschaftlichen Konflikts. Frankfurt/Main 1980; questions of applied democracy were studied by: Beker, Gabriele: Zumutbarkeit von Kernenergierisiken. Sozialwissenschaftliche Analyse der Kernenergie-Debatte. Diss., Heidelberg 1978, pp. i, 210–211, 260–262.
82 Kitschelt, Kernenergiepolitik, pp. 252–256.
83 Camilleri, State.
84 Camilleri, State, pp. 280–281, quote p. 281.
85 Camilleri, State, pp. 13–26, quote p. 26 (emphasis in the original).
86 Camilleri, State, p. 282. As is done here, in the case of multiple quotations from the same source, the citation follows the last instance for reasons of clarity.
87 Camilleri, State, pp. 115–116; the argument that more centralised states were better equipped to implement and hold on to nuclear technologies is mirrored in the concept of nuclear power as an authoritarian technology: Winner, Artifacts; Winner, Whale, p. 175; Campbell, John L.: Collapse of an Industry. Nuclear Power and the Contradictions of U.S. Policy. Ithaca 1988, pp. 178–180; Sovacool, Benjamin K., Scott Victor Valentine: The National Politics of Nuclear Power. Economics, Security, and Governance. London 2012, p. 247; Uekötter, Lessons of History, p. 26; Uekötter, Frank: Fukushima, Europe, and the Authoritarian

Nature of Nuclear Technology, in: Environmental History 17 (2012), 2, pp. 277–284; Mez, Lutz: Atomkraft und Demokratie, in: Mohr, Roland, Hannes Utikal (Eds.): Zukunft. Chemie. Perspektiven auf die Welt von morgen. Frankfurt/Main 2013, pp. 184–189, here p. 189; Müller, Wolfgang C., Paul W. Thurner: Nuclear Energy in Western Europe. Revival or Rejection? An Introduction, in: Müller, Wolfgang C., Paul W. Thurner (Eds.): The Politics of Nuclear Energy in Western Europe. Oxford 2017, pp. 1–20, here p. 10; cf. Mumford, Lewis: Authoritarian and Democratic Technics, in: Technology and Culture 5 (1964), 1, pp. 1–8. My argument, on the other hand, centres on the observation that conflicts escalated in places where participatory elements were perceived as wanting; cf. 'Nuclear Democracy' – and Camilleri, State, p. 293.

88 Kitschelt, Herbert P.: Political Opportunity Structures and Political Protest. The Anti-Nuclear Movements in Four Democracies, in: British Journal of Political Science 16 (1986), 1, pp. 57–85.

89 Rucht, Von Wyhl nach Gorleben; cf. Rucht, Wyhl. Rucht further studied the anti-nuclear movement as a 'New Social Movement': Rucht, Dieter: Neue soziale Bewegungen oder: Die Grenzen bürokratischer Modernisierung, in: Hesse, Joachim Jens: Politikwissenschaft und Verwaltungswissenschaft. PVS Sonderheft 13 (1982), pp. 272–292.

90 Rucht, Dieter: Campaigns, Skirmishes and Battles: Anti-Nuclear Movements in the USA, France and West Germany, in: Industrial Crisis Quarterly 4 (1990), pp. 193–222, here p. 216; cf., as an example for corresponding claims: Bürgerinitiative gegen Atomenergie Osnabrück (Ed.): Fortschritt oder Zerstörung? Das Emsland. Informationen zur Erschließung und Industrialisierung einer Region. Osnabrück 1980, pp. 15–23.

91 Rucht, Campaigns, p. 214; the model's weakness was also pointed out by: Diani, Mario, Hein-Anton van der Heijden: Anti-Nuclear Movements Across States. Explaining Patterns of Development, in: Flam, Helena (Ed.): States and Anti-Nuclear Movements. Edinburgh 1994, pp. 355–382, here 378–380; Helena Flam presented a model that included further factors: Flam, Helena: A Theoretical Framework for the Study of Encounters between States and Anti-Nuclear Movements, in: Flam, Helena (Ed.): States and Anti-Nuclear Movements. Edinburgh 1994, pp. 9–26. Another modification was suggested by: Koopmans et al., Political; Koopmans, Ruud: Democracy from Below. New Social Movements and the Political System in West Germany. Boulder 1995. Political scientist Wolfgang Rüdig was sceptical about putting too much emphasis on empirical work: Rüdig, Wolfgang: Anti-Nuclear Movements. A World Survey of Opposition to Nuclear Energy. Harlow 1990, p. 12.

92 Renn, Ortwin: Risikowahrnehmung der Kernenergie. Frankfurt/Main 1984, pp. 32, 195–203; cf. Renn, Ortwin: Wahrnehmung und Akzeptanz technischer Risiken. Vol. 3: Das Symbol Kernenergie – Einstellungen und ihre Determinanten. Jülich 1981, especially p. 162.

93 Weart, Spencer R.: Nuclear Fear. A History of Images. Cambridge 1988.

94 Weart, Spencer R.: The Rise of Nuclear Fear. Cambridge 2012, p. 219.

95 Hughes, Thomas P.: American Genesis. A Century of Invention and Technological Enthusiasm 1870–1970. New York 1989, pp. 444–453, 459–461.

96 Examples beyond the ones highlighted here include, among many others: Röthlein, Brigitte: Kernenergie, ein Thema der öffentlichen Meinung. Beschreibung und Analyse eines Kommunikationsprozesses. Diss., Munich 1979, pp. 147–150; Opp, Karl-Dieter et al.: Eine empirische Überprüfung des Modells 'rationalen Verhaltens', in: Opp, Karl-Dieter et al. (Eds.): Soziale Probleme und Protestverhalten. Eine empirische Konfrontation des Modells rationalen Verhaltens mit soziologischen und demographischen Hypothesen am Baispiel [sic, CG] von Atomkraftgegnern. Opladen 1984, pp. 47–198, especially p. 192; Gloede, F.:

Gesellschaftlicher Konflikt um die Kernenergie und Öffentlichkeitsbeteiligung als Verfahren. Thesen zum Handlungsbedarf, in: Blümel, Willi (Ed.): Die Öffentlichkeitsbeteiligung bei der Genehmigung von umweltrelevanten Großvorhaben. Zwischenergebnisse eines gemeinsamen Forschungsprojektes. Vol. 2. Speyer 1988, pp. 77–83; van der Pligt, Joop: Nuclear Energy and the Public. Oxford 1992, pp. 18–38; Joppke, Christian: Mobilizing against Nuclear Energy. A Comparison of Germany and the United States. Berkeley 1993, pp. 91, 95; Watts, Nicholas: 'Deconstructing Chernobyl'. The Meaning and Legacy of Chernobyl for European Citizens, in: Mez, Lutz, Lars Gerhold, Gerhard de Haan (Eds.): Atomkraft als Risiko. Analysen und Konsequenzen nach Tschernobyl. Frankfurt/Main 2010, pp. 33–73, here p. 34; Böttiger, Helmut: Kernenergie. Gefahr oder Nutzen? Petersberg 2011, pp. 5–9, 45–46; Sovacool et al., National, pp. 4–8, 234–235; cf., with a view to protests against nuclear weapons: Cordle, Daniel: Late Cold War Literature and Culture. The Nuclear 1980s. London 2017; Uekötter, Atomare, pp. 9, 17, 19, 70, 116, 253, 294.

97 Wagner, Peter: Contesting Policies and Redefining the State: Energy Policy-Making and the Anti-Nuclear Movement in West Germany, in: Flam, Helena (Ed.): States and Anti-Nuclear Movements. Edinburgh 1994, pp. 264–295, here pp. 267–268; questions regarding participation were also taken up by the contributors in: Renn, Ortwin, Thomas Webler, Peter Wiedemann (Eds.): Fairness and Competence in Citizen Participation. Evaluating Models for Environmental Discourse. Dordrecht 1995.

98 Cf. Schmoldt, Hubertus: Die Auseinandersetzung um die Zukunft der Kernenergie in Deutschland aus gewerkschaftlicher Sicht, in: Grawe, Joachim, Jean-Paul Picaper (Eds.): Streit ums Atom. Deutsche, Franzosen und die Zukunft der Kernenergie. Munich 2000, pp. 115–137, here p. 123.

99 The concept of 'Sociotechnical Imaginaries' was developed along similar lines: Jasanoff, Sheila, Sang-Hyun Kim: Containing the Atom. Sociotechnical Imaginaries and Nuclear Power in the United States and South Korea, in: Minerva 47 (2009), pp. 119–146.

100 Hecht, Gabrielle: The Radiance of France. Nuclear Power and National Identity after World War II. Cambridge ^{2}2009 (Original 1998), especially pp. 330, 334, 338–339; for a rather flippant version of the argument: Pons, Frédéric: Die Kernenergie in Frankreich. Aktivitäten im Herzen des nationalen Bewußtseins, in: Grawe, Joachim, Jean-Paul Picaper (Eds.): Streit ums Atom. Deutsche, Franzosen und die Zukunft der Kernenergie. Munich 2000, pp. 105–114. Highlighting interconnections between the country's nuclear history and wider, societal conflicts: Kalmbach, Karena: Tschernobyl und Frankreich. Die Debatte um die Auswirkungen des Reaktorunfalls im Kontext der französischen Atompolitik und Elitenkultur. Frankfurt/Main 2011; for the French perception: Jordan, Katrin: 'Die Wolke, die an der Grenze haltmachte'. Der Reaktorunfall von Tschernobyl 1986 im französischen Fernsehen, in: Themenportal Europäische Geschichte (2014), www.europa.clio-online.de/2014/Article=677 (23.1.2023).

101 Welsh, Mobilising, pp. 50–52; cf. Welsh, Nuclear Nation, pp. 2–4; Welsh, NIMBY, pp. 18–22; for a similar observation on the international level: Presas i Puig, Albert: A Comparative Study of European Nuclear Energy Programs, in: Presas i Puig, Albert (Ed.): A Comparative Study of European Nuclear Energy Programs. Berlin 2011, pp. 7–25, here pp. 15–16; for a specific British example: Laucht, Christoph: Atoms for the People. The Atomic Scientists' Association, the British State and Nuclear Education in the Atom Train Exhibition, 1947–1948, in: The British Journal for the History of Science 45 (2012), 4, pp. 591–608.

102 Flood, Nuclear Dream, pp. 39–42; cf. Gowing, Independence, especially Vol. 2; Elliott, Dave: Großbritannien. Aufstieg und Niedergang der Atomindustrie, in:

Mez, Lutz (Ed.): Der Atomkonflikt. Berichte zur internationalen Atomindustrie, Atompolitik und Anti-Atom-Bewegung. Mit einer Einleitung von Robert Jungk. Reinbek 1981, pp. 147–159, here pp. 147–151; Hall, Nuclear, p. 99; Hill, Atomic Empire, pp. 1–6, 21–49; Hogg, British, p. 77; Edgerton, David: The Rise and Fall of the British Nation. A Twentieth-Century History. Milton Keynes 2019, pp. 170, 394. A formally institutionalised German equivalent could be the Deutsches Atomforum e. V. (cf. 'Nuclear Debates').

103 Welsh, Mobilising, especially pp. 3–12, 20–23, 28–32, 163.

104 Kupper, Patrick: Atomenergie und gespaltene Gesellschaft. Die Geschichte des gescheiterten Projekts Kernkraftwerk Kaiseraugst. Zurich 2003, p. 16; Kupper, Patrick: From Prophecies of the Future to Incarnations of the Past. Cultures of Nuclear Technology, in: Nowotny, Helga (Ed.). Cultures of Technology and the Quest for Innovation. Oxford 2006, pp. 155–166; cf., for a similar argument: Saladin, Peter: Kernenergie und Verfassungsstaat, in: Roßnagel, Alexander (Ed.): Recht und Technik im Spannungsfeld der Kernenergiekontroverse. Opladen 1984, pp. 35–54; as an example for the USA: Harvey, Kyle: American Anti-Nuclear Activism, 1975–1990. The Challenge of Peace. Basingstoke 2014, p. 3.

105 Metzler, Gabriele: Demokratisierung durch Experten? Aspekte politischer Planung in der Bundesrepublik, in: Haupt, Heinz-Gerhard, Jörg Requate (Eds.): Aufbruch in die Zukunft. Die 1960er Jahre zwischen Planungseuphorie und kulturellem Wandel. DDR, CSSR und Bundesrepublik Deutschland im Vergleich. Weilerwist 2004, pp. 267–287, here p. 271.

106 Nehring, Holger: National Internationalists. British and West German Protest against Nuclear Weapons, the Politics of Transnational Communications and the Social History of the Cold War 1957–1964, in: Contemporary European History 14 (2005), 4, pp. 559–582, p. 578.

107 Weisker, Albrecht: Powered by Emotion? Affektive Aspekte in der westdeutschen Kernenergiegeschichte zwischen Technikvertrauen und Apokalypseangst, in: Brüggemeier, Franz-Josef, Jens Ivo Engels (Eds.): Natur- und Umweltschutz nach 1945. Konzepte, Konflikte, Kompetenzen. Frankfurt/Main 2005, pp. 203–221, here pp. 211–212; cf. Weisker, Albrecht: Expertenvertrauen gegen Zukunftsangst. Zur Risikowahrnehmung der Kernenergie, in: Frevert, Ute (Ed.): Vertrauen. Historische Annäherungen. Göttingen 2003, pp. 394–421, here pp. 407–408, cf. pp. 416–417.

108 Dannenbaum, Atom-Staat, p. 270.

109 Engels, Naturpolitik, pp. 328, 348–349.

110 Cf. note 49.

111 Schüring, Michael: 'Bekennen gegen den Atomstaat'. Die evangelischen Kirchen in der Bundesrepublik Deutschland und die Konflikte um die Atomenergie 1970–1990. Göttingen 2015, p. 8; cf. Kroll, Thomas: Protestantismus und Kernenergie. Die Debatte in der Evangelischen Kirche der Bundesrepublik Deutschland in den 1970er und frühen 1980er Jahren, in: Erhardt, Hendrik, Thomas Kroll (Eds.): Energie in der modernen Gesellschaft. Zeithistorische Perspektiven. Göttingen 2012, pp. 93–115; Schüring, Michael: 'Bekennen gegen den Atomstaat'. Historische und religiöse Codierungen im kirchlichen Protest gegen die Atomenergie in den 70er und 80er Jahren, in: Ostheimer, Jochen, Markus Vogt (Eds.): Die Moral der Energiewende. Risikowahrnehmung im Wandel am Beispiel der Atomenergie. Stuttgart 2014, pp. 230–243.

112 Mende, Silke, Birgit Metzger: Eco-Pacifism. The Environmental Movement as a Source for the Peace Movement, in: Becker-Schaum, Christoph et al. (Eds.): The Nuclear Crisis. The Arms Race, Cold War Anxiety, and the German Peace Movement of the 1980s. New York 2016, pp. 119–137, here pp. 122–124.

113 Tompkins, Better Active, pp. 30, 238; he emphasises, however, that the protesters' lack of a common concept of democracy meant that the anti-nuclear movement cannot be said to have strived *for* a specific new democratic model, while they certainly rejected the Atom State, Tompkins, Better Active, pp. 22, 24–25; cf. Tompkins, Grassroots.
114 Müller et al., Nuclear, especially pp. 2, 15; Müller, Wolfgang C., Paul W. Thurner: Understanding Policy Reversals and Policy Stability, in: Müller, Wolfgang C., Paul W. Thurner (Eds.): The Politics of Nuclear Energy in Western Europe. Oxford 2017, pp. 21–46, especially pp. 24, 27; for the question's persistence: Müller, Wolfgang C., Paul W. Thurner: Comparative Policy Indicators on Nuclear Energy, in: Müller, Wolfgang C., Paul W. Thurner (Eds.): The Politics of Nuclear Energy in Western Europe. Oxford 2017, pp. 47–64, here p. 60; Müller, Wolfgang C., Paul W. Thurner, Christian Schulze: Conclusion. Explaining Nuclear Policy Reversals, in: Müller, Wolfgang C., Paul W. Thurner (Eds.): The Politics of Nuclear Energy in Western Europe. Oxford 2017, pp. 286–324; for special explanatory problems in Britain's nuclear history: Müller, Wolfgang C., Paul W. Thurner, Christian Schulze: Appendix. Other European Countries – An Analytical Reconstruction, in: Müller, Wolfgang C., Paul W. Thurner (Eds.): The Politics of Nuclear Energy in Western Europe. Oxford 2017, pp. 324–360, especially pp. 352–356.
115 Gaumer, Wackersdorf.
116 The term utopia is used, in this study, to indicate positive visions of the future, as it is widely used in this way instead of the formally correct eutopia, cf. Woschech, Anke: Zwischen Luftschloss und Prognose. Der Terminus der 'technischen Utopie' im Fokus der Technikgeschichte, in: Neumeister, Katharina, Peggy Renger-Berka, Christian Schwarke (Eds.): Technik und Transzendenz. Zum Verhältnis von Technik, Religion und Gesellschaft. Stuttgart 2012, pp. 147–161; Metzner-Szigeth, Andreas: Technikzukünfte in Eu- und Dystopien. Fragestellungen und Forschungsansätze, in: Böhn, Andreas, Andreas Metzner-Szigeth (Eds.): Wissenschaftskommunikation, Utopien und Technikzukünfte. Karlsruhe 2018, pp. 17–52, here p. 18.
117 Augustine, Dolores L.: Taking on Technocracy. Nuclear Power in Germany, 1945 to the Present. Oxford 2018.
118 Kirchhof, Astrid Mignon: Pathways into and Out of Nuclear Power in Western Europe. An Introduction, in: Kirchhof, Astrid Mignon (Ed.): Pathways into and Out of Nuclear Power in Western Europe. Austria, Denmark, Federal Republic of Germany, Italy, and Sweden. Munich 2020, pp. 7–35, here p. 20.
119 Kirchhof, Pathways into and Out of Nuclear Power in Western Europe, p. 32.
120 Kirchhof, Pathways into and Out of Nuclear Power in Western Europe, p. 22; Kitschelt, Political Opportunity Structures.
121 Kaijser et al., Nuclear Energy, pp. 7–9; Kaijser et al., Future, pp. 280–283.
122 Oberloskamp, Ambiguities, p. 431; cf. Oberloskamp, Eva: Energie, Ökologie, Markt. Bundesdeutsche und britische Neuansätze in den 1970er und 1980er Jahren. Frankfurt/Main 2025.
123 Frickel-Pohl, Natalie: Die Anti-Atomkraft-Bewegung am Oberrhein. Chancen und Grenzen der grenzüberschreitenden Zusammenarbeit im 'Dreyeckland', in: Lakeberg, Beata Dorota, Hans-Christian Pust (Eds.), Atom. Strom. Protest. 50 Jahre Wyhl und anderswo. Ostfildern 2023, pp. 202–215, here p. 213.
124 Hager, Carol: Protest gegen Atomkraftwerke. Eine Schule der Demokratie?, in: Lakeberg, Beata Dorota, Hans-Christian Pust (Eds.), Atom. Strom. Protest. 50 Jahre Wyhl und anderswo. Ostfildern 2023, pp. 130–141, here pp. 131, 133.
125 Van der Pligt, Joop, J. R. Eiser, R. Spears: Public Attitudes to Nuclear Energy, in: Energy Policy 12 (1984), pp. 302–306, here p. 305.

126 Charnley-Parry, Ioan, John Whitton, Pieter Cools, Anne Bergmans: Integrated Comparative Report for In-Depth Understanding of the Mechanisms for Effective Interaction with Civil Society, www.honest2020.eu/sites/all/themes/Porto_sub/downloads/deliverables/D4.4.pdf (23.1.2023), p. 55.

127 Cf., arguing for the combination: Kocka, Jürgen: Comparison and Beyond, in: History and Theory 42 (2003), pp. 39–44, as a comment on: Werner, Michael, Bénédicte Zimmermann: Vergleich, Transfer, Verflechtung. Der Ansatz der *Histoire croisée* und die Herausforderung des Transnationalen, in: Geschichte und Gesellschaft 28 (2002), 4, pp. 607–636; furthermore: Josephson, Paul, Jan-Henrik Meyer, Arne Kaijser: Nuclear-Society Relations from the Dawn of the Nuclear Age, in: Kaijser, Arne et al. (Eds.): Engaging the Atom. The History of Nuclear Energy and Society in Europe from the 1950s to the Present. Morgantown 2021, pp. 27–51, here p. 28.

128 Cf. Haupt, Heinz-Gerhard, Jürgen Kocka: Historischer Vergleich. Methoden, Aufgaben, Probleme. Eine Einleitung, in: Haupt, Heinz-Gerhard, Jürgen Kocka (Eds.): Geschichte und Vergleich. Ansätze und Ergebnisse international vergleichender Geschichtsschreibung. Frankfurt/Main 1996, pp. 9–46.

129 Radkau, Joachim: Kernenergie – Großtechnik zwischen Staat, Wirtschaft und Öffentlichkeit, in: Wengenroth, Ulrich (Ed.): Technik und Wirtschaft. Düsseldorf 1993, pp. 346–366, here p. 361; Rucht, Dieter: The Impact of Anti-Nuclear Power Movements in International Comparison, in: Bauer, Martin (Ed.): Resistance to New Technology. Nuclear Power, Information Technology, and Biotechnology. Cambridge 1995, pp. 277–291, here p. 282; Rootes, Christopher: Environmental Protest in Britain 1988–1997, in: Seel, Benjamin, Matthew Paterson, Brian Doherty (Eds.): Direct Action in British Environmentalism. London 2000, pp. 25–61, here p. 34; Carrera, Diana Gallego, Jürgen Hampel: Die Situation der Kernenergie nach Fukushima, in: Mohr, Roland, Hannes Utikal (Eds.): Zukunft. Chemie. Perspektiven auf die Welt von morgen. Frankfurt/Main 2013, pp. 172–177.

130 Patterson, Nuclear, p. 112; Campbell, Collapse, pp. 164–166; Barnes, Michael: The Hinkley Point Public Inquiries. A Report by Michael Barnes QC to the Secretaries of State for Energy and the Environment. Vol. 1. London 1990, pp. 55, 97; Joppke, Mobilizing, pp. 142–144, 166; Rüdig, Wolfgang: Maintaining a Low Profile. The Anti-Nuclear Movement and the British State, in: Flam, Helena (Ed.): States and Anti-Nuclear Movements. Edinburgh 1994, pp. 70–100, here pp. 91–94; Engels, Naturpolitik, pp. 345–346; Wood, Janet: Nuclear Power. London 2007, pp. 117–119; Cooke, Stephanie: In Mortal Hands. A Cautionary History of the Nuclear Age. New York 2009, p. 127; Altenburg, Cornelia: Wandel und Persistenz in der Energiepolitik: Die 1970er Jahre und die Enquete-Kommission 'Zukünftige Kernenergie-Politik', in: Erhardt, Hendrik, Thomas Kroll (Eds.): Energie in der modernen Gesellschaft. Zeithistorische Perspektiven. Göttingen 2012, pp. 245–263, here p. 260; Hill, Atomic Empire, pp. 339–345. Joachim Radkau, considering orders instead of constructions, pointed to 1976 for Germany: Radkau, Joachim: Fragen an die Geschichte der Kernenergie – Perspektivenwandel im Zuge der Zeit (1975–1986), in: Hohensee, Jens, Michael Salewski (Eds.): Energie – Politik – Geschichte. Nationale und internationale Energiepolitik seit 1945. Stuttgart 1993, pp. 101–126, here p. 111; cf. Burn, Nuclear, pp. 25, 82–83; Radkau, Kernkraft-Kontroverse, pp. 324–325.

131 Rubio-Varas, Mar: The Changing Economic Context Influencing Nuclear Decisions, in: Kaijser, Arne et al. (Eds.): Engaging the Atom. The History of Nuclear Energy and Society in Europe from the 1950s to the Present. Morgantown 2021, pp. 52–80, here p. 53.

132 Häfner, Daniel: Eine Anti-Atom-Bewegung in der DDR?!, in: Lakeberg, Beata Dorota, Hans-Christian Pust (Eds.), Atom. Strom. Protest. 50 Jahre Wyhl und anderswo. Ostfildern 2023, pp. 88–99, here p. 88; Bösch, Frank: Zeitenwende 1979. Als die Welt von heute begann. Munich [5]2019, p. 349.
133 Cf., for a recent overview over discussions about technology life-cycles, with a focus on obsolescence: Weber, Heike: Made to Break? Lebensdauer, Reparierbarkeit und Obsoleszenz in der Geschichte des Massenkonsums von Technik, in: Krebs, Stefan, Gabriele Schabacher, Heike Weber (Eds.): Kulturen des Reparierens. Dinge – Wissen – Praktiken. Bielefeld 2018, pp. 49–83.
134 Used as a tool to criticise periodicity in art history by Wilhelm Pinder in the 1920s, and part of an ongoing discussion about the characteristics of time, moved forward, not least, by Ernst Bloch, cf. Schwartz, Frederic J.: Ernst Bloch and Wilhelm Pinder. Out of Sync, in: Grey Room 3 (2001), pp. 54–89.
135 Haupt et al., Historischer Vergleich, p. 25.
136 For a differentiation regarding this aspect, refer to the concept of 'political opportunity structures', discussed in 'The National, the Spectacular, and the Complex Atom: State of Research'.
137 For this analysis, such a transnational plane of debate can therefore be postulated, although, as Dick van Lente pointed out, it would be more correct to assume a transnational frame of reference providing for exchange between the national debates than one transnational debate: van Lente, Introduction, pp. 1–3. Especially so as he was, with Dolores Augustine, of the opinion that the debates about nuclear energy helped to establish a 'world opinion' as a force of history: Augustine, Conclusion, p. 245.
138 Hahn, Otto: Geleitwort, in: Löwenthal, Gerhard, Josef Hausen: Wir werden durch Atome leben. Mit einem Geleitwort von Nobelpreisträger Professor Dr. Otto Hahn und einem Vorwort von Bundesminister für Atomfragen Franz-Josef Strauß. Berlin 1956, pp. 9–12, here pp. 10–11; Bagge, Erich, Kurt Diebner: Zur Entwicklung der Kernenergieverwertung in Deutschland, in: Bagge, Erich, Kurt Diebner, Kenneth Jay: Von der Uranspaltung bis Calder Hall. Hamburg 1957. Beinhaltet eine Übersetzung von Jay, Kenneth: Calder Hall. The Story of Britain's First Atomic Power Station. London 1956, pp. 7–80, here pp. 78–79; Radkau, Joachim: Kernenergie-Entwicklung in der Bundesrepublik: ein Lernprozeß? Die ungeplante Durchsetzung des Leichtwasserreaktors und die Krise der gesellschaftlichen Kontrolle über die Atomwirtschaft, in: Geschichte und Gesellschaft 4 (1978), 2, pp. 195–222, here p. 201; Eckert, Michael: Primacy Doomed to Failure. Heisenberg's Role as Scientific Adviser for Nuclear Policy in the FRG, in: Historical Studies in the Physical and Biological Sciences 21 (1990), pp. 29–58, here pp. 41, 53; Bleidick, Dietmar: Pionier der deutschen Kernenergie. Geschichte des VAK Kahl 1958–2010. Kahl am Main 2010, here pp. 16–17.
139 To Joachim Radkau, this implied that Germany was no Atom State: Radkau, Zukunft, pp. 165–166; on a side note, it should be remarked that the different wartime experiences of both countries' populations are taken to be one element behind the different developments regarding its protest movements: Nehring, Holger: Angst, Gewalterfahrungen und das Ende des Pazifismus. Die britischen und westdeutschen Proteste gegen Atomwaffen, 1957–1964, in: Greiner, Bernd et al. (Eds.): Angst im Kalten Krieg. Hamburg 2009, pp. 436–464, here pp. 463–464.
140 Arnold, Windscale, p. xxii; C. N. Hill emphasised that the Russian Obninsk reactor was experimental and points out that, by 1968, Britain produced 150% as much electricity from nuclear power than all the other states that had initiated like programmes: Hill, Atomic Empire, pp. 1–6.

141 Foley, Gerald, Ariene van Buren: Introduction, in: Foley, Gerald, Ariane van Buren (Eds.): Nuclear or Not? Choices for our Energy Future. A Royal Institution Forum. London 1978, pp. 1–10, here p. 1; Watts, Nicholas S. J.: Mobilisierungspotential und gesellschaftspolitische Bedeutung der neuen sozialen Bewegungen. Ein Vergleich der Länder der Europäischen Gemeinschaft, in: Roth, Roland, Dieter Rucht (Eds.): Neue soziale Bewegungen in der Bundesrepublik Deutschland. Bonn 1987, pp. 47–67, here p. 55; Rüdig, Anti-Nuclear; cf. Radkau, Joachim: Das RWE zwischen Kernenergie und Diversifizierung 1968–1988. Kernkraftwerke und die Politisierung der Energieträger, Umweltpolitik und erste Ansätze der Diversifizierung, in: Schweer, Dieter, Wolf Thieme (Eds.): RWE – 'Der gläserne Riese'. Ein Konzern wird transparent. Wiesbaden 1998, pp. 221–244, here p. 229.

142 Cf. Oberloskamp, Energie; Rucht observed a higher number of environmental protest events in Britain in the late 1980s: Rucht, Dieter: Umweltproteste in der Bundesrepublik Deutschland. Eine vergleichende Perspektive, in: Politische Vierteljahresschrift 39 (2007), pp. 518–539, here p. 532.

143 For this assessment and the following one regarding Germany, cf. Breach, Windscale, p. 80; Watts, Mobilisierungspotential, p. 66; Brown, G. I.: Invisible Rays. The History of Radioactivity. Phoenix Mill 2002, pp. 210–211; Laucht, Christoph: Zurück in die nukleare Zukunft: Das Beispiel Großbritannien, in: Greiner, Bernd, Tim B. Müller, Klaas Voß (Eds.): Erbe des Kalten Krieges. Hamburg 2013, pp. 263–273, here pp. 263, 269–271; Radkau, Zukunft, p. 163; Butler, Stuart, Robert Bud: United Kingdom Short Country Report (2018), in: History of Nuclear Energy and Society (HoNESt) Consortium Deliverable 3.6, www.honest2020.eu/sites/default/files/deliverables_24/UK.pdf (23.1.2023), pp. 4–6; Espluga, Josep et al.: Risky or Beneficial? Exploring Perceptions of Nuclear Energy over Time in a Cross-Country Perspective, in: Kaijser, Arne et al. (Eds.): Engaging the Atom. The History of Nuclear Energy and Society in Europe from the 1950s to the Present. Morgantown 2021, pp. 147–169, here p. 147; Kalmbach, Afterlives, p. 197; Oberloskamp, Ambiguities, pp. 418–419; Uekötter, Atomare, pp. 101, 270–271, 298.

144 Although there was some critique of the protest's parliamentarisation and the politics employed by Green politicians later on: Paul, Reimar: Die Anti-AKW-Bewegung: Wie sie wurde was sie ist. Einleitung, in: Paul, Reimar (Ed.): . . . und auch nicht anderswo! Die Geschichte der Anti-AKW-Bewegung. Göttingen 1997, pp. 11–32, here pp. 24–26; Ehmke, Wolfgang: Bewegte Zeiten. Von Wyhl bis zum Tag X, in: Paul, Reimar (Ed.): . . . und auch nicht anderswo! Die Geschichte der Anti-AKW-Bewegung. Göttingen 1997, pp. 34–45, here pp. 39–40; cf., for the French roots of the idea: Milder, Grassroots Activism, p. 199.

145 Rüdig, Low Profile, pp. 70–71; Rucht, Dieter, Jochen Roose: The German Environmental Movement at a Crossroads?, in: Rootes, Christopher (Ed.): Environmental Movements. Local, National and Global. London 2002, pp. 59–80, here pp. 73–74; cf. Camilleri, State, pp. 89–90. This argument, of course, cannot explain changing outlooks on nuclear power at sites like Hinkley Point, cf. 'Rational Atoms'.

146 Cf. Kaijser, Arne, Jan-Henrik Meyer: Nuclear Installations at European Borders. Transboundary Collaboration and Conflict, in: Kaijser, Arne et al. (Eds.): Engaging the Atom. The History of Nuclear Energy and Society in Europe from the 1950s to the Present. Morgantown 2021, pp. 254–277, here p. 254.

147 Rüdig, Anti-Nuclear, pp. 181–195.

148 Flood, Michael: Torness. Keep It Green. London 1979, pp. 31–33; Radkau, Aufstieg und Krise, p. 423; Rüdig, Low Profile, pp. 70–71.

149 Duve, Tamara, Michael Sontheimer: Der Atom-Skandal oder Das Atom probt den Selbstmord, in: Traube, Klaus (Ed.): Der Atom-Skandal. Alkem, Nukem

und die Konsequenzen. Reinbek 1988, pp. 13–41; Mez, Lutz: Transnuklear oder Urangate in Hanau, in: Komitee für Grundrechte und Demokratie (Ed.): Jahrbuch 1987. Sensbachtal 1988, pp. 49–56; Kernenergie auf dem Prüfstand. Sichere Energiequelle oder nicht beherrschbares Risiko? Bericht des 2. Untersuchungsausschusses der 11. Wahlperiode des Deutschen Bundestages ('Transnuklear/Atomskandal'). Ed. by Deutscher Bundestag, Referat Öffentlichkeitsarbeit. 2 Vols. Bonn 1991; Radkau, RWE zwischen Kernenergie und Diversifizierung, p. 234; Arndt, Auswirkungen, pp. 75–77; Radkau et al., Aufstieg und Fall, p. 343; cf. 'Atomic Trust(s)'.

150 Laucht, Zurück, pp. 263, 270; Oberloskamp, Ambiguities, p. 436.

151 Cf. Heßler, Kreative Stadt, pp. 13, 29.

152 Arguing for this potential of locales in principle: Schlögel, Karl: Im Raume lesen wir die Zeit. Über Zivilisationsgeschichte und Geopolitik. Munich 2003, p. 10.

153 Hughes, British, p. 497; Hogg, British, pp. 4–6; cf., for a more general perspective: Hocke, Massenmedien, pp. 15–16. For the main focus points in the history of nuclear energy in Germany according to the magazine *Der Spiegel*: Stange, Susanne: Die Atomenergie im Urteil der Zeitschrift 'Der Spiegel' 1955–1985. Magister thesis, Kiel 1990: Stange, Susanne: Die Auseinandersetzung um die Atomenergie im Urteil der Zeitschrift 'DER SPIEGEL', in: Hohensee, Jens, Michael Salewski (Eds.): Energie – Politik – Geschichte. Nationale und internationale Energiepolitik seit 1945. Stuttgart 1993, pp. 127–152.

154 Hogg et al., Introduction, p. 493; Hogg, British, p. 4.

155 Engels, Naturpolitik, p. 348; cf. Tompkins, Europe of Struggles, p. 125.

156 Nehring, Internationalists.

157 Brown, Kate: Plutopia. Nuclear Families, Atomic Cities, and the Great Soviet and American Plutonium Disasters. Oxford 2013; for a project highlighting the perspectives of young people in 'Atomic Cities': McQuerry, Maureen (Ed.): Nuclear Legacy. Students of Two Atomic Cities. Columbus 2000.

158 Hecht, Being Nuclear.

159 One of them was Ian Welsh, who described the workings of the protesters and their transnational interrelations, not least with the Wyhl-protests, rejecting analyses claiming it had been a failure: Welsh, Mobilising, pp. 155–180, 197; cf. Oberloskamp, Ambiguities.

160 The other was Würgassen, where an accident delayed operations until 1975: Barthelt, Klaus, Klaus Montanus: Begeisterter Aufbruch. Die Entwicklung der Kernenergie in der Bundesrepublik Deutschland bis Mitte der siebziger Jahre, in: Hohensee, Jens, Michael Salewski (Eds.): Energie – Politik – Geschichte. Nationale und internationale Energiepolitik seit 1945. Stuttgart 1993, pp. 89–100, here pp. 95–96; Radkau, Joachim: Das RWE zwischen Braunkohle und Atomeuphorie 1945–1968. Wiederaufbau und Verbundwirtschaft, Einstieg in die Kernkraft und Quantensprung im Braunkohlentagebau, in: Schweer, Dieter, Wolf Thieme (Eds.): RWE – 'Der gläserne Riese'. Ein Konzern wird transparent. Wiesbaden 1998, pp. 173–196, here pp. 193–194; Weisker, Albrecht: Systemwettstreit oder Konvergenz durch Sachzwänge? Die Ausbaupläne der Kernenergie in der Bundesrepublik und der DDR in den 1960er Jahren, in: Haupt, Heinz-Gerhard, Jörg Requate (Eds.): Aufbruch in die Zukunft. Die 1960er Jahre zwischen Planungseuphorie und kulturellem Wandel. DDR, CSSR und Bundesrepublik Deutschland im Vergleich. Weilerwist 2004, pp. 185–206, here pp. 190–192.

161 For a discussion of the relation of examples, cases and the general in historical analyses: Pohlig, Matthias: Vom Besonderen zum Allgemeinen? Die Fallstudie als geschichtstheoretisches Problem, in: Historische Zeitschrift 295 (2012),

pp. 297–319; for the ability to analyse broader issues through local perspectives: note 17; for the wider nuclear debates being based in the local ones: note 45.

162 Cf. Uekötter, Atomare, pp. 78, 108.

163 Cf. Goerke, Dietrich: Untersuchung und Ausarbeitung von Maßnahmen zur Öffentlichkeitsarbeit bei Erörterungsterminen über den geplanten Bau kerntechnischer Anlagen. Part I and II, Vechta 1975, pp. 128–139, 173; Kitschelt, Kernenergiepolitik, pp. 192–193, 232; Freudenburg, William R., Rodney K. Baxter: Host Community Attitudes towards Nuclear Power Plants. A Reassessment, in: Social Science Quarterly 65 (1984), 4, pp. 1129–1136; Rüdig, Anti-Nuclear, pp. 9, 266–276; Hasenöhrl, Ute: Zivilgesellschaft und Protest. Eine Geschichte der Naturschutz- und Umweltbewegung in Bayern 1945–1980. Göttingen 2011, pp. 404–405, 458–460, 465–467; Venables, Dan, Nick F. Pidgeon, Karen A. Parkhill, Karen L. Henwood, Peter Simmons: Living with Nuclear Power. Sense of Place, Proximity, and Risk Perceptions in Local Host Communities, in: Journal of Environmental Psychology 32 (2012), 4, pp. 371–383; Tompkins, Better Active, pp. 35–36; for studies pointing to the inverse phenomenon: Goerke, Untersuchung Part I, p. 139 and Part II, p. 14; van der Pligt, Nuclear, p. 11.

164 Cf. Goerke, Untersuchung Part I and II, pp. 25–38; Wüstenhagen, Hans-Helmut: Bürger gegen Kernkraftwerke. Wyhl – der Anfang? Reinbek 1975, p. 64; Rucht, Wyhl, p. 144; Rüdig, Anti-Nuclear, pp. 131–132; Löser, Georg: Die Badisch-Elsässischen Bürgerinitiativen. Grenzüberschreitende Zusammenarbeit gegen Atomkraft und für Alternativen am südlichen Oberrhein und anderswo, in: Borchers, Uli (Ed.): Die AKW-Protestbewegung von Wyhl bis Brokdorf. Hamburg 2011, pp. 19–75, here p. 33; Milder, Fish, pp. 178–189; Tompkins, Better Active, p. 30; Milder, Example, pp. 167–168.

165 Cf. Goerke, Untersuchung Part I and II, pp. 95, 121, 127; Rüdig, Anti-Nuclear, pp. 150–151; Renner, Jens: Der Bauplatz muss wieder zur Wiese werden. 1976–1986. Zehn Jahre Kampf gegen das AKW Brokdorf, in: Borchers, Uli (Ed.): Die AKW-Protestbewegung von Wyhl bis Brokdorf. Hamburg 2011, pp. 87–110, here pp. 87–88; Biehl, Karsten, Eberhard Hollweg: Brokdorf – ein zweites Wyhl? BRD 1976, a television documentary available as an appendix to Paul, Chronologie.

166 Cf. Goerke, Untersuchung Part I and II, pp. 42–52; Hasenöhrl, Zivilgesellschaft, pp. 419–420; Opp, Karl-Dieter, Peter Hartmann: Die Zielsetzung und Konzeption eines Forschungsprojekts zur Konfrontierung soziologischer Hypothesen mit dem Modell 'rationalen Verhaltens', in: Opp, Karl-Dieter et al. (Eds.): Soziale Probleme und Protestverhalten. Eine empirische Konfrontation des Modells rationalen Verhaltens mit soziologischen und demographischen Hypothesen am Baispiel [sic, CG] von Atomkraftgegnern. Opladen 1984, pp. 13–29; Kretschmer, Winfried: Wackersdorf. Wiederaufarbeitung im Widerstreit, in: Linse, Ulrich et al. (Eds.): Von der Bittschrift zur Platzbesetzung. Konflikte um technische Großprojekte. Laufenburg, Walchensee, Wyhl, Wackersdorf. Berlin 1988, pp. 165–218, here p. 189; Rüdig, Anti-Nuclear, pp. 156–157, 163–164; Gaumer, Wackersdorf, pp. 50, 95–97, 148–149, 164–165, 302; contributions by Dave Banks and Jill Perry in: Davies, Sellafield, pp. 279–288 and 289–295.

167 Peters, Wolfgang: Das Kernkraftwerk Biblis. Räumliche Distanz als Parameter bei der Meinungsbildung, in: Geographische Rundschau 32 (1980), 2, pp. 71–77; Renn, Wahrnehmung, p. 15; Lieb, Matthias: Aktivismus in der 'Verschnarchte[n] Provinzmetropole'? Umwelt- und Anti-AKW-Bewegungen in Mainz, Wiesbaden und der Rhein-Main-Region, in: Paulus, Julia (Ed.): 'Bewegte Dörfer'. Neue soziale Bewegungen in der Provinz 1970–1990. Paderborn 2018, pp. 211–234, here pp. 228–230; Goerke, Untersuchung Part I and II, pp. 139, 149; Rüdig, Anti-Nuclear, p. 269; Venables et al., Sense of Place.

168 Goerke, Untersuchung Part I and II, p. 173; Glasson, John et al.: Power Station Impacts. The Socio-Economic Effects of Power Stations on Their Localities. Oxford 1979, especially pp. xxiii, 69–71, 114; Kitschelt, Kernenergiepolitik, p. 232; Renn, Wahrnehmung, pp. 17–19; Camilleri, State, pp. 75–76; van der Pligt et al., Public; van der Pligt, Joop, J.R. Eiser, R. Spears: Construction of a Nuclear Power Station in One's Locality. Attitudes and Salience, in: Basic and Applied Social Psychology 7 (1986), pp. 1–15; Rüdig, Anti-Nuclear, pp. 163–164; Hasenöhrl, Zivilgesellschaft, pp. 458–460.

169 For Biblis: Lieb, Aktivismus, pp. 228–229; questioning the explanation's reach: Peters, Biblis; Biehl et al., Brokdorf; cf. Hasenöhrl, Zivilgesellschaft, pp. 419–420; van der Pligt, Nuclear, pp. 70–73; Gleitsmann et al., Fortschrittsfeinde, pp. 143–145; Kretschmer, Wackersdorf, p. 189; Rüdig, Anti-Nuclear, pp. 163–164; Breach, Windscale, pp. 89–91; contribution by Jill Perry in: Davies, Sellafield, pp. 289–295, here p. 293.

170 Ester, P., C. Mindell, J. van der Linden, Joop van der Pligt: The Influence of Living Near a Nuclear Power Plant on Beliefs about Nuclear Energy, in: Zeitschrift für Umweltpolitik 6 (1983), pp. 349–362; van der Pligt, Joop, J. R. Eiser, R. Spears: Attitudes to Nuclear Energy. Familiarity and Salience, in: Environment and Behavior 18 (1986), pp. 75–93; Rüdig, Anti-Nuclear, pp. 9, 273–275; van der Pligt, Nuclear, p. 69; Eiser et al., Nuclear Neighbourhoods, pp. 42–43, 66; Pidgeon, Nick F.: Living with Nuclear Power in Britain. A Mixed-Methods Study. Cambridge 2010, www.eprg.group.cam.ac.uk/wp-content/uploads/2009/10/Nick-Pidgeon-v2.pdf (23.1.2023); Venables et al., Sense of Place.

171 Pointing to studies by Wolfang Peters: Rüdig, Anti-Nuclear, pp. 273–274; van der Pligt, Nuclear, p. 70; Pidgeon, Nick F., K. L. Henwood, K. A. Parkhill, D. Venables, P. Simmons: Living with Nuclear Power in Britain. A Mixed Methods Study. Cardiff 2008; Parkhill, K. A. et al.: From the Familiar to the Extraordinary. Local Residents' Perceptions of Risk When Living with Nuclear Power in the UK, in: Transactions/Institute of British Geographers 35 (2010), 1, pp. 39–58; Venables et al., Sense of Place.

172 Special cases were places like Richland and Ozerk in the USA and the Soviet Union, dedicated to the production of plutonium, which profited while their surroundings suffered: Brown, Plutopia, especially pp. 133–267, 335.

173 Cf. for details regarding follow-up industries and assistance to hosting communities by plant operators: Dosky, Doris von: Untersuchungen zu Auswirkungen der Kernkraftwerke Obrigheim, Stade, Goesgen. Diss, Darmstadt 1987, pp. 46–66, 82–84; furthermore: Glasson, John et al.: Power Station Impacts. Socio-Economic Impact Assessment. Proposed Hinkley Point C Power Station. Final Report. Headington 1987.

174 Glasson et al., Impact Assessment, pp. 71–73; cf. Heßler, Kreative Stadt, p. 163.

175 Pahl, Harald: Gemeinde Biblis ist stolz auf ihre 'umweltfreundliche Industrie'. Wohlstand durch 'Atommeiler', in: *Neue Rhein Zeitung Wesel*, 14.2.1976; Verwunderung über Kernkraftwerkslob, in: *Aichacher Zeitung*, 17.10.1977.

176 os: Wenn Block C gebaut würde . . . Bibliser Haushaltsvolumen hat sich reduziert, in: *Darmstädter Echo (DE)*, 5.12.1978.

177 os: 'In Zukunft kleinere Brötchen backen'. Gemeindevertretung verabschiedet Nachtrag und erhöht Hebesätze, in: *DE*, 11.12.1981.

178 Glasson et al., Effects of Power Stations, especially pp. xxiii, 69–71, 114.

179 ch: Bei sofortigem Ausstieg aus Kernkraft: Industrie kündigt Werksschließung an. PreussenElektra: Stade bleibt günstiger Standort, in: *Stader Tageblatt*, 21.7.1986.

180 Bürgerinitiative gegen Atomenergie Osnabrück, Fortschritt, pp. 18–20.

181 Welsh, NIMBY, pp. 17–18.

182 A British one, pointing to zones of 0–5, 6–10, and 11–15 miles, is presented by Rüdig, Anti-Nuclear, p. 274.
183 I use Nuclear Debates with capital letters when referring to the thusly named chapter in this book. Otherwise, nuclear debates refer to the debates about the 'Peaceful Atom' in general.
184 For modifications of the concept of what was 'local' during the nuclear debates, cf. Harvey, Activism, pp. 124–128, 141–142.
185 Rüdig, Anti-Nuclear, pp. 273–274; van der Pligt, Nuclear, p. 70.
186 Cf. Pidgeon, Living.
187 Radkau, Joachim: 'Angepasste Technik' in der Vergangenheit. Der regionale Ansatz in der Technikgeschichte, in: König, Wolfgang (Ed.): Technikgeschichte. Stuttgart 2010, pp. 235–242.
188 For this observation in the case of Biblis, cf. Lieb, Aktivismus, p. 230; cf., with a view to German nuclear power debates overall: Tompkins, Better Active, pp. 35–36.
189 Cf. Heineberg, Heinz: Großbritannien. Raumstrukturen, Entwicklungsprozesse, Raumplanung. Gotha 1997, pp. 26, 28, 30–31, 33.
190 In fact, five main types of sources were considered relevant for a reasonably realistic (re)construction of local, public debates, which, in the democracies under consideration here, can be accepted as the foundation for the formation and expression of political will locally: contemporary witnesses; company archives; archives of citizens' initiatives; archives of local or regional authorities, namely parliamentary bodies; and local and regional media. The first three were ruled out early on: In the case of contemporary witnesses, test interviews revealed that memories regarding the period under consideration were often transformed by later events, to a point at which relying on them would have been impossible independently of corroborative materials. This made information gained using this method alone unusable for the analysis, while on the other hand implying that those interviews were quite ineffective to scrutinise other source materials. Archives of companies or citizens' initiatives were excluded because the wealth of material produced for public consumption would have contorted the analysis of the debates' dynamics. It would have created the impression that there were phases of intense debate whenever one of the groups generated more output to achieve one, as opposed to the flow of debate that took place in a wider public, beyond the most active voices on either side.
191 A similar situation was observed with a view to Shoreham Nuclear Power Plant by McCaffrey, David P.: The Politics of Nuclear Power. A History of the Shoreham Nuclear Power Plant. Dordrecht 1991, p. 23.
192 Minutes, Meeting, Williton Rural District Council, 15.12.1970, in: Somerset Heritage Centre (SHC), D/R/wil 2/2/18, item 2295.
193 Openness, in: *Northavon Gazette*, 5.5.1975.
194 In an exemplary, quantitative analysis, Rucht found that, in national newspapers of the early 2000s, only about 10% of protest events were covered. Nonetheless, he argued, this coverage – which was higher in local newspapers – was "the political relevant 'social construction' of protest events": Rucht, Dieter: Soziale Bewegungen der 1960er und 70er Jahre in der Bundesrepublik, in: Hermle, Siegfried (Ed.): Umbrüche. Der deutsche Protestantismus und die sozialen Bewegungen in den 1960er und 70er Jahren. Göttingen 2007, pp. 91–107, here pp. 93–94, quote p. 94. With a view to protest events in Freiburg, Peter Hocke found that about 38% were covered in local media, compared to about 5% in national media. Nonetheless, he argued that the reports were largely representative of the protests that happened, insofar as more protests resulted in more reports. Also, he pointed out that the rate of reported protests about environmental topics

grew after spectacular events like Chernobyl: Hocke, Massenmedien, pp. 151, 180–181, 266–268, 277–278. For the stronger 'filtering-function' of media's attention logics with national media, focusing them on spectacular events, cf. Overhoff, Klaus: Die Politisierung des Themas Kernenergie. Regensburg 1984, p. 1; Milder, Fish, pp. 190–232; furthermore: Mazur, Allan: Media Influences on Public Attitudes toward Nuclear Power, in: Freudenburg, William R., Eugene A. Rosa (Eds.): Public Reactions to Nuclear Power. Are There Critical Masses? Boulder 1984, pp. 97–114, here pp. 101–107.

195 Cf. van Buiren, Shirley (Ed.): Kernenergie und Medien. Protokolle, Referate, Analysen, Themenmatrix, Pressespiegel eines BMFT-Seminars in Zusammenarbeit mit dem Battelle-Institut e. V., Frankfurt am Main. Villingen 1980, especially pp. 12–20; Roth, Bewegungsnetzwerke, p. 431. For crass examples of accusations of one-sidedness: Hermann, Armin: Wie die Wissenschaft ihre Unschuld verlor. Macht und Mißbrauch der Forscher. Stuttgart 1982, pp. 226, 228; Kepplinger, Hans Mathias: Vom Hoffnungsträger zum Angstfaktor, in: Grawe, Joachim, Jean-Paul Picaper (Eds.): Streit ums Atom. Deutsche, Franzosen und die Zukunft der Kernenergie. Munich 2000, pp. 81–103.

196 Cf. 'The Potential of Local Perspectives'.

197 Welsh, Mobilising, p. 211; cf., with a view to wider protest events: Hocke, Massenmedien, pp. 59, 155–157.

198 Cf. Hocke, Massenmedien, pp. 277–278. That should not imply that any mention on the national level of a power plant under consideration here automatically became part of the local or regional public debate.

199 Regarding the national level, this was done in several studies, not least by Hans Mathias Kepplinger, who argued that, because of the internal logics of mass media, the national press became critical of nuclear power before the public did: Kepplinger, Hans Mathias: Die Kernenergie in den Massenmedien oder Die Rhetorik der Angst, in: Krämer, Hermann (Ed.): Kernenergie heute. Die großen Themen. Bonn 1988, pp. 71–103, here pp. 78, 87; Kepplinger, Hans Mathias: Kernenergie. Die Kernenergie in der Presse. Eine Analyse zum Einfluß subjektiver Faktoren auf die Konstruktion von Realität. Erlangen 1989, pp. 19–20; for said logics, cf. van Lente, Introduction, pp. 7–8; for a contrarian position regarding the influence of media on opinions, cf. Mez, Lutz: Zur Rolle der Medien in der deutschen Energiepolitik, in: Koch-Baumgarten, Sigrid, Lutz Mez (Eds.): Medien und Policy. Neue Machtkonstellationen in ausgewählten Politikfeldern. Frankfurt/Main 2007, pp. 85–100, here p. 99; Iwaniec, Mario: Die Betrachtung der Risiken der Kernenergie in ausgewählten deutschen Printmedien. Berlin 2010.

200 Cf. Herring, Energy, p. 92.

201 Even though, at the time, letters to the editor seemed less important than news reports or even interviews, which were mostly reserved to the local elites: van Buiren, Kernenergie-Kontroverse, pp. 114–115.

202 Sometimes, letter-writers remained anonymous by choice or because their names were only published in the form of abbreviations. These letters are considered in the analysis nonetheless, as their contribution to the debates was evident and could be reacted to. Also, in the case of members of the local power plant staff, political parties or institutions, or citizens' initiatives taking part in the debate in the form of letters to the editor without identifying themselves as holders of said positions, they are considered to be writing as private individuals if it is reasonable to assume that they were not generally known for their institutional affiliation.

203 On a side note, newspapers do not consist of text alone, but also contain pictures as often telling sources. Unfortunately, in practice, these were usually only accessible in very low quality, so that, in this analysis, they are only taken up in indirect form and described, where necessary, not printed.

204 For the general argument, cf. Requate, Jörg: Öffentlichkeit und Medien als Gegenstände historischer Analyse, in: Geschichte und Gesellschaft 25 (1999), 1, pp. 5–32, here pp. 8–9, 11–16; Negrine, Ralph: Politics and the Mass Media in Britain. London [2]2000; Bösch, Frank: Mediengeschichte. Vom asiatischen Buchdruck zum Computer. Frankfurt/Main [2]2019, pp. 7, 13–24.

205 Van Buiren et al., Bürgerinitiativen, p. X; for a similar diagnosis regarding British power plant sites: Glasson et al., Effects of Power Stations, pp. 165–169; for the situation in Switzerland: Saxer, Ulrich, Heinz Gantenbein, Martin Gollmer, Walter Hättenschwiler, Michael Schanne: Massenmedien und Kernenergie. Bern 1986, pp. 3–30; cf. Schulz, Winfried: Die Konstruktion von Realität in den Nachrichtenmedien. Analyse der aktuellen Berichterstattung. Freiburg 1976, pp. 28, 115–120.

206 Cf. Renn, Wahrnehmung, p. 93; Stöber, Rudolf: Neue Medien. Geschichte. Von Gutenberg bis Apple und Google. Medieninnovation und Evolution. Bremen 2012, pp. 435–436, 443–444, 446, 453; for a comprehensive discussion of the development of research into the persuasive powers of mass media, cf. Schenk, Michael: Medienwirkungsforschung. Tübingen [3]2007; furthermore: contributions in Wober, J. Mallory (Ed.): Television and Nuclear Power. Making the Public Mind. Norwood 1992; Kepplinger, Hans Mathias: Individual and Institutional Impacts Upon Press Coverage of Sciences. The Case of Nuclear Power and Genetic Engineering in Germany, in: Bauer, Martin (Ed.): Resistance to New Technology. Nuclear Power, Information Technology, and Biotechnology. Cambridge 1995, pp. 357–377, here p. 374.

207 Gorman, Lyn, David McLean: Media and Society in the Twentieth Century. A Historical Introduction. Padstow 2004, p. 18; Götter, Christian: Die Macht der Wirkungsannahmen. Medienarbeit des britischen und deutschen Militärs in der ersten Hälfte des 20. Jahrhunderts. Berlin 2016, pp. 2, 7–11.

208 Cf. Stamm, Karl-Heinz: Alternative Öffentlichkeit. Die Erfahrungsproduktion neuer sozialer Bewegungen. Frankfurt/Main 1988, pp. 164–167.

209 Kliment, Kernkraftprotest, pp. 405–408; cf. Spears, R., Joop van der Pligt, J. R. Eiser: Sources of Evaluation of Nuclear and Renewable Energy Contained in the Local Press, in: Journal of Environmental Psychology 7 (1987), pp. 31–43.

210 Cf. Mazur, Media, p. 111; Overhoff, Politisierung; Eiser et al., Nuclear Neighbourhoods, pp. 107–109; Bösch, Frank: Kommunikative Netzwerke. Zur globalen Formierung sozialer Bewegungen am Beispiel der Anti-Atomkraftproteste, in: Mittag, Jürgen, Helke Stadtland (Eds.): Theoretische Ansätze und Konzepte der Forschung über soziale Bewegungen in der Geschichtswissenschaft. Essen 2014, pp. 149–166, here p. 152.

211 Cf. the concept of "Relational Coproductionist Participation": Chilvers, Jason, Matthew Kearnes: Remaking Participation in Science and Democracy, in: Science, Technology and Human Values 45 (2020), 3, pp. 347–380, especially p. 353.

212 Stamm, Alternative Öffentlichkeit, pp. 160–161; cf. Rucht, Soziale Bewegungen, p. 94.

213 As quoted by Lepore, Jill: The Atomic Origins of Climate Science. How Arguments about Nuclear Weapons Shaped the Debate over Global Warming, in: *The New Yorker*. American Chronicles, 30.1.2017.

214 For the principles of historical discourse analysis: Landwehr, Diskursanalyse, especially pp. 49, 52–55, 96–98, 163, 165–166, 168–170; for the importance of working from empirical observations instead of preconceived assumptions: Dewey, John: The Public and Its Problems. Denver 1927, pp. 4, 8–9, 12, 21; for the reconstruction of a foreign (in this case: temporarily removed) society's culture: Geertz, Dichte Beschreibung, especially pp. 9–10, 15, 20–21, 31–32; for an

implementation in a cultural history of technology: Gugerli, David: Redeströme. Zur Elektrifizierung der Schweiz 1880–1914. Zurich 1996.

215 In sum, in the German cases, there are more than 4,000 relevant articles in the collections.

216 In general, quotes refer to the newspaper issues in question, not the shelf marks of archival collections because, first, these were not available in all cases and second, they make finding a specific article more cumbersome than looking it up in the complete collections of the newspapers in question. The signatures of the collections of newspaper cuttings are given when the specific sequence or selection of articles within these are relevant for the interpretation, or when the paper could not be identified.

217 This concerns documents from the Archiv für Christlich-Soziale Politik, the Laka Documentation and Research Centre on Nuclear Energy, and the National Archives, Dwight D. Eisenhower Presidential Library.

218 Large parts of the relevant collection of the Gemeindearchiv Biblis (GAB) did not have standardised shelf marks.

219 Part of the collection Hessisches Staatsarchiv Darmstadt (HStAD), H2 Bergstraße.

220 Cf. Peters, Biblis; Schweer, Dieter, Wolf Thieme (Eds.): RWE – 'Der gläserne Riese'. Ein Konzern wird transparent. Wiesbaden 1998; Lieb, Aktivismus, covers mostly external opponents of Biblis C; Götter, Christian: Kernenergie – eine Vertrauensfrage. Vertrauen als Katalysator für die Kernenergiedebatte am Beispiel der Standorte Biblis und Stade, in: Fiedler, Gudrun (Ed.): Stader Jahrbuch 2016. Stade 2016, pp. 193–211; Götter, Christian: Emotionen als Argument – Die Debatte um die Kernenergie im Biblis der 1970er Jahre, in: Smeddinck, Ulrich (Ed.): Emotionen bei der Realisierung eines Endlagers. Interdisziplinäre Beiträge. Berlin 2018, pp. 31–48.

221 In the collection Stadtarchiv Lingen (StadtA LIN), RP.

222 StadtA LIN, AP, Verw/Umw, respectively.

223 In the collections StadtA LIN OStDir 82–87 and 93.

224 StadtA LIN Erwerbung 670.

225 As part of Niedersächsisches Landesarchiv, Abteilung Oldenburg (NLA OL), Rep 410.

226 Emsländischer Heimatbund (Ed.): Emsland im Werden. Vom Handwerk bis zur Kernkraft. Sögel 1986; the search for a nuclear waste storage site in the Emsland was covered by Tiggemann, Achillesferse.

227 Within the Stadtarchiv Stade (StadtA STD) Bestand 10.

228 Mainly in the collections StadtA STD Zeitungsausschnittsammlung and Pressearchiv, but also Stadtangelegenheiten IV; the original creator of the collections within the city administration could not be determined in every case.

229 Namely StadtA STD I – Stadtdirektor acc. 1994/001 Nr. 39–45.

230 These are part of the collection Niedersächsisches Landesarchiv, Abteilung Stade (NLA STD) Dep. 28.

231 Mainly part of NLA STD Rep. 180.

232 Cf. Frerichs, Klaus, Gerd Mettjes (Eds.): Neue Energie!? Abschied vom Kernkraftwerk Stade. Bremen 2004.

233 In SHC C/CC/M/2/22–29.

234 In SHC D/R/will 2/2/15–19.

235 As part of London's National Archives (NA) EG 4.

236 Files on the topic are collected under the SHC A\BPR shelf mark.

237 County council to store Hinkley papers, in: *Bridgwater Mercury*, 8.8.1989.

238 Partially in SHC C/FB/50/1 and C/PHH/201.

239 Part of the collection NA EG 4.

240 Including, for example, the report of the first Public Inquiry, NA HLG 79/1172.

241 Cf. van der Pligt et al., Public; van der Pligt et al., Familiarity and Salience; van der Pligt, Nuclear, especially pp. 69–73; Pidgeon et al., Living; Venables et al., Sense of Place.
242 Cf. Pocock, Nuclear, pp. 93–102, 175–176, 232–234; sadly, however, it turned out to be impossible to gain access to a dissertation on the later public inquiry into Hinkley Point's third power plant in the context of the British electricity industry's privatisation: Roberts, Susan Jane: The Hinkley Point 'C' Public Inquiry and the Privatisation of the UK Electricity Supply Industry. Diss., Open University 1991.
243 In Gloucestershire Archives (GlouA) P237a/PC/1.
244 In GlouA DA38/100 and DC94/1.
245 Under the shelf mark GlouA D10125.
246 In GlouA D5277/5.
247 For example: NA COU 1/1395; NA HLG 79/1570; NA POWE 14/1177; NA POWE 14/1248–1250.
248 Namely NA HLG 120/1044.
249 NA HLG 120/230.
250 Bradshaw, Jane, Joyce Tibbenham (Eds.): The Millennium Book of Oldbury on Severn. Clevedon 2001; Pidgeon et al., Living; Parkhill et al., Familiar; Venables et al., Sense of Place.
251 Especially John Gray Centre (JGC) East Lothian Scrapbooks 38–54.
252 In JGC LH/SD.
253 In JGC ELDC/2/2/116.
254 In JGC ELDC/5/3/4.
255 Mostly in NA POWE 74.
256 Especially: Flood, Torness; Oberloskamp, Ambiguities.

1 Nuclear Debates

"Times and attitudes are certainly changing," the editors of the *Thornbury Gazette* observed in June 1989, "so that we may now see the rise of the Greens as a serious political force, instead of being dismissed as cranks. They have already established themselves with real power in Germany."[1] It was not uncommon during the 1980s to compare the United Kingdom of Great Britain and Northern Ireland's (UK) Greens to their counterpart in the Federal Republic of Germany (FRG). Six years earlier, Colin Hunt, the secretary of Somerset's Ecology Party, had already pointed out that "[t]he Green Party, our sister party in West Germany, is poised to enter the Federal Parliament in Bonn after their General Election" and formulated the equivalent goal for his party in the upcoming British general elections of 1983.[2] Germany's Greens served as a benchmark for ecologically minded people in Britain, from an inside perspective as well as in outside observers' eyes. The level, synchronous comparison seemed to dwarf the British side – a phenomenon perpetuated until today by various analyses of both countries' anti-nuclear movements.[3]

The "Short Country Report", presenting results of the research into Britain's nuclear history undertaken as part of the European research project "History of Nuclear Energy and Society" (HoNESt), described Britain as a place where "debates about nuclear power have been infrequent, taken place in relatively small sections of society and have focused on individual topics of concern such as cost-management, reactor choice or pollution".[4] The equivalent report on the FRG emphasised that here "[t]he controversy was carried out at all societal levels and integrated not only small groups of experts and stakeholders, but numerous intermediary social groups and actors".[5] This impression of an organised, broad, and critical debate about nuclear power in Germany, compared to which the one in the UK lacked in scope and coherence, is very much present when it comes to the nations' nuclear histories and consistent with a diagnosis of a British public less politically active in general.[6] Over time, various explanations were put forward for this seemingly obvious difference in the degree the atom was splitting both societies, ranging from the model of political opportunity structures to more specific differences between Germany and Great Britain. These include

DOI: 10.4324/9781003583677-2

different building phases (more drawn out in Britain, with especially few activities during the 1970s), a greater degree of secrecy in the UK (due to its military nuclear programme), the effects of transnational networks of protest (reinforcing the FRG's anti-nuclear movement since Wyhl), differences in reactor technologies used, and the more co-operative attitude of British environmental groups and state actors. The local perspectives taken up here not only help to identify other important factors behind this obvious difference, but more importantly, help to question whether it really was that obvious a difference – or only an apparent one – brought about by applying a misleading, seemingly "neutral" temporal perspective when comparing the two countries' nuclear histories.[7]

In order to elaborate this, in this chapter, I *first* argue that differing conditions of debate were formative for the obviously different development of the countries' nuclear histories.[8] John Dewey describes the conditions of debate crucial for democratic societies as broad, free, and informed.[9] The term is employed to describe the ways in which administrative or political bodies, industries, associations, scientists, and, very importantly, private citizens publicly engaged with a given topic. The concept considers the amount of activity these actors developed to publicly state their positions, as well as the way they interacted with and regarded one another, including their respective concession of legitimate participation in public discussions in the first place. It is about who took part in the debates, which arguments they employed regarding which issues, and about their demeanour or style. With this focus on the scope and limits of the topics and actors as well as the relationships between them, the conditions of debate loosely lean upon two of the limiting factors of discourse according to Michel Foucault, or, more specifically, two of the "rules" making up what he called an "archive", namely "[t]he limits and forms of the *sayable*" and "the limits and forms of *appropriation*".[10] Understood thus, conditions of debate have been broached by existing research.[11] With a few exceptions however, such as Ian Welsh's analysis of a British "social scientific movement",[12] they, and especially the consequences of the way they played out in each country's early debates, have not received the systematic attention they deserve considering their importance for the course of the history of nuclear energy in both democratic nations. By turning to the local perspective, the conditions of debate can be rendered visible. In particular, the comparison of local discussions allows the use of an inclined temporal plane instead of aligning two monolithic, national nuclear histories along an apparently neutral, calendrical timeline. It allows the comparison of two public debates that were not simultaneous but in phase regarding their position relative to their countries' nuclear histories: the debates in two locales at the beginning of both countries' commercial nuclear power programmes – when, it should be pointed out, the nuclear aspect of the projects still tended to play a secondary role only – and their development from there on.[13]

Once the analysis is set up in this way, it becomes clear that communities asked to live with early commercial nuclear power plants in both countries entered their public controversies against the background of widely differing conditions of debate, indicating their differing democratic cultures. In Britain, it was quite open, allowing for conflicts of interest and opinion to be publicly stated, with various parties acting more or less on an equal footing.[14] In the German case, a much stronger hierarchical element came to play in the early nuclear debates. Reacting to experiences with protests elsewhere, the upper tier of political, administrative and industry groups successfully attempted to dominate other participants and their opinions by framing themselves as responsible experts,[15] able, willing, and destined to educate everybody else, especially private individuals, to make them see things from the right perspective and overcome their "information deficit".[16]

These differences, I *secondly* argue, unintentionally led to the diametrically opposed courses of the two countries' nuclear histories, resulting in the prevalent impression of small, hardly controversial nuclear debates in Britain, and escalating, all-encompassing ones splitting German society. Initially, however, it had been the other way round: Britain had started with a wide, broad and controversial debate about the first commercial nuclear power plants developed.[17] This resulted in the critics of nuclear energy being able to win a number of concessions, compromises, and assurances from atomic power's proponents that could openly be debated at a public inquiry.[18] In its wake, an equally broad, albeit not complete, acceptance of the new technology ensued, lasting for decades. In Germany, on the other hand, initially there was hardly any debate at all. The electricity industry, aided by local, regional, and state authorities as well as lobbying associations, managed to steer the few public questions and remarks there were to topics of its choice, and thereby to frame the debate in a rather patronising way: It seemed not to be a conflict of political opinions and interests, but a question of the education of an allegedly under-informed population by responsible experts. This suppression of critical perspectives resulted in smooth initial licensing processes. It also laid the groundwork for festering conflicts that had not been worked through, for their expansion and, in the end, violent eruption, including a tendency to search for ways to protest outside the established pathways.[19]

What is more – and this is this chapter's *third* argument – the differing course of the nuclear debates in Britain and Germany, which resulted from their conditions of debate, had, in turn, consequences for the latter. In Germany, governments and the nuclear industry attempted to present a more participative decision-making process, aimed at calming the debates. In Britain, on the other hand, the calm that had resulted from the early escalation led to the impression that governments could act rather freely when it came to nuclear power. This, however, was not the case, especially in view of the information that came to light about nuclear power in the 1970s (cf. "Rational Atoms") – as the administration of Margaret Thatcher was to discover when it aimed to expand nuclear power as a weapon in its conflict with the miners' union during the 1980s.[20]

The specific debates analysed here are those concerning Hinkley Point in Somerset, and Biblis in Hesse. These sites are ideally suited for investigating the influence of and transformations undergone by the British and German conditions of debate during their anchoring to the conflict about the "Peaceful Atom". Both saw the construction of one of the earliest, though not the first, commercial nuclear power plants, rendering visible the early conditions of debate without evoking the suspicion that they could be exceptional because they had no precursors. What is more important, both saw the construction of a second power station soon after the first, and in both cases, a further expansion was debated, too. Therefore, the process by which the two countries came to switch places when it came to their acceptance of and resistance to nuclear energy, and the consequences these developments had on their conditions of debate, can also be sketched out. In both cases, I take up the nuclear debates at the point where the plans to build their first reactors became public, follow them through the public inquiry and official hearing, respectively, and conclude with an overview over their further development from thereon.

To best illustrate the differences between, the immense impact of, and the transformations that came about in the two countries' conditions of debate, the sub-chapters on Hinkley Point and Biblis are both primarily structured along the chronological timeline. They start with a section on the very different early phases of the public debates, then highlight the public inquiry and official hearing, respectively, holding up a mirror to each country's initial conditions of debate. Each sub-chapter ends with a section presenting the inverted character of the public debates in the following years. In the case of Hinkley Point, to illustrate the scope of the debate, the sections are further structured by elements of its content. This way of further breaking down the debate highlights the various contact points between the conditions of debate and the other issues that became anchored to atomic energy analysed in later chapters. In the case of Biblis, to emphasise the contrast to the British situation and highlight the debate's unfolding and escalation in the third temporal section, after the official hearing, where this sub-chapter's focus lies, the chronological order of events is allowed to rule supreme. The first section, however, covering the non-debate before the official hearing, is further structured according to the participants of the debate. This serves to emphasise the dominance of officials and industry representatives here – in marked contrast to the situation in Hinkley Point. The latter, however, being the older one, is taken up first.

Hinkley Point – A Wide and Open Debate

In its 1955 White Paper "A Programme of Nuclear Power", the British government envisaged building 12 nuclear power stations with a combined generating capacity of up to two gigawatts of electricity over the next decade.[21] To do so was considered "vital . . . if we are to keep our position as a leading industrial nation",[22] because "[n]uclear energy is the energy of the future"[23] and "marks the beginning of a new era."[24] The power stations were to "be designed and built by private industry for the Electricity Authorities",[25] namely, in England and Wales, the Central Electricity Authority (CEA), founded in April 1955 and reformed into the Central Electricity Generating Board (CEGB) in 1957.

One of the first nuclear power stations was to be built at Hinkley Point in Somerset, in the English West Country, on the southern coast of the Bristol Channel. It was the second nuclear power station around the Severn Estuary, after Berkeley, and a third station soon followed at Oldbury. The site, whose name was only established during the public debate about the project,[26] was deemed suitable by the CEA after more than one year of surveys and tests,[27] because it fulfilled four criteria: It lay close to an area where electricity was needed and which was so far removed from coal pits that investment in nuclear energy seemed feasible – which, at the time, was only possible as long as plutonium, produced as a waste product, was calculated as a sellable resource;[28] solid rock formations could support the heavy buildings; cooling water was available in large quantities; and the area was sparsely populated.[29] The latter had been formulated as a precautionary measure for early nuclear power stations in the government's programme, "even though they will be of inherently safe design".[30] Beyond that, the site offered room for further expansion,[31] which the CEA publicly envisioned making use of about five years after erecting the first power station.[32]

The CEA needed the consent of the Minister of Fuel and Power for formal permission to build and operate a nuclear power station at Hinkley Point, for which it applied on 7 December 1956.[33] As the local planning authority, Somerset County Council was also heard, although it was not competent to approve or deny a nuclear power plant. It was tasked, however, with gathering the opinions of other bodies it would have consulted if it had been responsible.[34] In fact, selected members and officers of the council had already been involved in confidential, preliminary discussions with the CEA since 1955.[35] At the end of 1956 then, the CEA contacted the owners of the land it needed and announced a press conference at County Hall in Taunton, Somerset's administrative centre, for 28 November 1956.[36] With these steps, the public debate about the plans began.

On the one hand, a special subcommittee of the county council's planning committee was set up to assess the CEA's plans in conjunction with interested bodies and authorities. These included Bridgwater Borough Council, Bridgwater Rural District Council, Burnham-on-Sea Urban District Council, Friends of Quantock, the National Parks Commission, the Somerset River Board, Taunton Borough Council, Weston-super-Mare Borough Council, and Williton Rural District Council, in whose area the site lay.[37] On the other hand, the public became directly involved, with the *Bridgwater Mercury*, a local newspaper widely read in the area around Hinkley Point, beating the CEA's press conference by one day.

On 27 November 1956, the *Mercury* published an article about the authority's plans for a nuclear power station at "Hinkley Point, at Stolford, an outlandish spot on the coast of the Bristol Channel, in the parish of Stogursey". The newspaper had sent its reporter Ken Powell for an eyewitness report on the area. Almost poetically, he described a scene where, "[w]hipped by a strong North-Westerly gale the sea was lapping at the coastal defences, the sky was black and ominous, and the only living creatures in sight were the gulls at the water's edge. It was hard to imagine that within the next decade this wild, deserted spot might well be a hive of activity supplying energy to all parts of the South-West corner of England."[38] Powell here foreshadowed the two main strands of the public debate that was to unfold during the following months. He described a solitary, rural landscape that many people would try to preserve,[39] while, just like the White Paper, framing the planned power station as a materialisation of progress and economic opportunity.

Indeed, the article, in combination with the CEA's press conference, marked the beginning of quite a remarkable public debate. Over the next few months, until a public inquiry in May 1957, the nuclear power station invoked numerous letters to the local authorities concerned,[40] various, sometimes passionate, public statements by private citizens from the local area, the surrounding region, and even faraway parts of England, and saw local, regional, and national authorities and associations weighing in. The participants hotly debated four broad themes, each consisting of several smaller topics, which often merged into each other at the fringes. The first of these themes comprised the planned power station's implications for the countryside, with the underlying topics of its landscape, environment (although the term was not used at the time)[41] and rural character. It was often discussed in close connection with the second theme, which covered the project's economic prospects and included the local area's development on the one hand, and national progress on the other. Both of these most widely debated themes were not specific to nuclear power stations but common to various industrial or infrastructure projects in rural surroundings.[42] As such, it had been expected that they would be brought up by the county council's officers.[43] The other two themes, however, were specific to nuclear power projects or would become so in the years ahead, albeit receiving limited attention in the mid-1950s. The first of them comprised the implications of artificial radioactivity, subsuming the topics of nuclear waste, radioactive emissions, and the relationship with the experts who devised the technology. The final theme was the conditions of debate itself, namely the question as to who could legitimately hold a debate, and in what form, especially when it came to the decision about a public inquiry. When this inquiry did take place, it was welcomed by most participants and played an important role in the re-pacification of a society that, at least in part, had been split by the CEA's proposal – although, when it came to solving the controversies in question, it did not add to the compromises already reached with the local authorities before the public debate began.

Conflicts from the Start

The most controversial theme was the countryside. The three oft-interwoven topics it consisted of had already been broached by Ken Powell's report: the seemingly untouched, natural landscape, the rural way of life, and the environmental challenges, particularly when it came to water.[44] The theme was of special concern to private citizens from the area and beyond, who took an active role in the public debate from the start.

Somerset's landscape was central to one of the first letters to the editor published by the *Bridgwater Mercury* on 1 January 1957. It stemmed from Dorothy Irving-Bell, who lived in Bristol, some 40 kilometres to the north-east of Hinkley Point, but owned land and a caravan locally, near Stogursey.[45] She was one of the most active correspondents with the authorities, also broaching the topic with other people and motivating at least one to take up his pen himself.[46] Irving-Bell was cross with Powell's description of the site, which, she insisted, was not "an outlandish spot" but would be turned into one "if the proposed station is built". Up until then, she argued, it had been "one of the very few stretches of unspoilt coast in all England. It is unique. Here Coleridge was inspired to write . . . To those who know and love the place, it is a joy in all winds and weathers, and so it should remain."[47] To her, the CEA's plans came close to the desecration of a heritage site for a bygone age of pre-industrial England, revered and immortalised by romantic poets like Samuel Taylor Coleridge, who had lived in and written about Somerset in poems such as *This Lime-Tree Bower My Prison*. Later, she equated the project to "a sinister destruction of our country".[48] Besides Irving-Bell, one of the landowners directly concerned, Colin Mackenzie from Stolford,[49] just about two kilometres east of the proposed building site, took a similar, albeit less poetical, line of argumentation by emphasising that the power station would be "a monstrous eyesore of vast proportions" and that "the inevitable horrors of modern buildings" could not be hidden from sight. Therefore, "the beauty of" one of the "few areas of England as are still unspoiled" would be ruined.[50] Other local people concurred. To Mrs K. Margaret Page from Fiddington, five kilometres inland from Hinkley Point, even a well-designed power station would not fit into the area visually.[51] J. Richardson from Combwich, just to the east of Fiddington, called the CEA's plans "the desecration of . . . beautiful countryside" through "the projected atomic monstrosity" that would "spoil for all time" a wide area that was the people's "heritage".[52] Writing anonymously, a self-styled "newcomer to Bridgwater" was equally convinced that the area's amenities would suffer because "engineers and planners, quite reasonable people in themselves, will disregard all aesthetic considerations when their technical enthusiasm takes control", recommending as a consequence "to fight the project tooth and nail".[53] It was not just the locals who protested, however. David Jones[54] from Bath, some 50 kilometres to the east of the site, thought it imperative to preserve this "beautiful tract of country" and claimed that the power station would, in the end, not only "spoil a magnificent stretch of coastline" but also devalue the nearby Quantock Hills,[55] Britain's first officially designated area of outstanding natural beauty, just a few kilometres west of Hinkley Point.[56] Even if the power plant turned out to be "as beautiful as a cathedral", he insisted, "it would still be out of place on this coastline!"[57] Renowned artist Sir Alfred James Munnings from Dedham near the east coast of England finally took up Irving-Bell's view that the project would lead to "a desecration" of "Coleridge's country!!"[58] Worries about the local landscape, in other words, elicited sometimes heated public protests from private citizens living next to the site in the local area, but also in a wider region and even farther away.

The local councils took up a more moderate position. The county council, whose planning committee had set up a special subcommittee to consider the CEA's proposal,[59] at least in part shared the opinion that the local landscape was, "a heritage of beauty which it is our duty to try and preserve"[60] as the committee's chairman, William George Hervey Jolliffe, the Lord Hylton, put it. Nonetheless, it did not reject the power station because of this. It considered the "national interest"[61] and the fact that "there are some who think these buildings are worth seeing on their own merits and, if well designed, will not be an objectionable feature on the landscape".[62] Expecting public resistance, however the planners suggested a compromise:[63] The power station should be moved a few dozen metres west, where it could be hidden behind ridges and trees, and architects and landscape consultants should be employed.[64] The CEA was unwilling to comply with the first demand because of its costs and the necessary characteristics of the reactors' rock foundations.[65] They were resolved, though, to make the design of the power station "as unobtrusive as possible" and promised that it would be "screened by trees on the south side".[66] Gerald Wills, Bridgwater's Member of Parliament, believed in the CEA's ability to minimise the impact on the "countryside" in this way,[67] and Stogursey Parish Council, in whose area the site lay, was willing to accept the original site given these promises.[68] Other councils, including the county council, at least resolved not to oppose it.[69] Several private citizens from the area and farther away took up a similar position. Walter John Carter from Combwich was convinced that "surely the best landscape architects" would be able to design the station in a way that "would not spoil any view now enjoyed" and preserve the Quantocks.[70] Similarly, A. Hancock from Kendal, in the northwest of England, called for a balanced judgement and spoke out against rashly "condemning progress".[71] A reader self-identifying as "Home Guard" from Stogursey argued that nuclear power stations, one day, would be seen to be just as normal and necessary as the "railways and motor cars" that once had been rejected, too. Therefore, the reader argued, the best way was to try "to alleviate the nuisance and shorten the transition period" and to "prevent unnecessary changes". "Home Guard" explicitly expected the local councils to strive "to prevent the erection of unsightly houses, etc. . . . If the development schemes are carefully organised, the surrounding district should benefit by improved amenities, and villages like Fiddington get the electricity for which they have appealed in vain for years."[72] Regarding the site's optics, in other words, a compromise, conceived with a view to early resistance, seemed to gain public backing.

Somerset's coastal area was not only discussed as a beautiful landscape. It was also appreciated as a recreational space. This touches upon the topic of the area's rural way of life – a subject of interest mostly to local people and institutions, and part of Irving-Bell's early letter to the *Mercury*. She had explicitly defended Hinkley Point as a place of retreat for "those who love quiet. As the pressure of modern life increases it is increasingly important to preserve quiet and unpopulated places."[73] The power station did not only endanger a haven for those stressed out by "modern life", but also the day-to-day life of some local residents. This, at least, was a central concern of the local councils. They were especially worried about the impact that many construction workers and, later, the settlement of the station's employees and their families would have.[74] Yet when the CEA stated that it wanted to house construction workers on-site and expected the station's staff to settle dispersed over the area, "mingling in with the present community",[75] Stogursey Parish Council accepted the burden as bearable. Part of its reason was that road improvements, financed by the CEA, would benefit the local population, and little agricultural land would be used.[76] Relentless in his critique of the impact of the expected 2,000 construction workers, however, was Mackenzie, who argued that they would urbanise a rural area that was reserved for agricultural use in the county council's development plans.[77] This fact, in the end, was given as one of the county council's reasons to seek a public inquiry.[78] It was also stressed by Page, who opposed the "transformation of a purely agricultural area",[79] although her arguments mostly rested on environmental grounds, which formed the third topic within the theme of the countryside.

Page, who was worried about the pollution of rivers and freshwater sources,[80] was not the only one pointing out potential environmental problems. These also troubled Mackenzie, who, as a local landowner, was directly affected.[81] The county council's planning committee, too, had attached great importance to the issue from the beginning, especially regarding the preservation of the Bridgwater Bay Nature Reserve and the local water situation.[82] Therefore, it demanded, among other things, that a meteorologist should "advise on the effect warmed water will have on the incidence of fog as affecting navigation"[83] – an aspect also highlighted by Newton A. Rowe from Otterhampton, about four kilometres to the southeast of Hinkley Point[84] – and expected, with Stogursey Parish Council, an assurance that local beaches would remain accessible for bathing.[85] The CEA was willing to comply, stating that it "would be perfectly safe" to use the seawater for cooling and return it to the Bristol Channel afterwards.[86] Fog would not be caused, and bathing would remain possible.[87] Regarding freshwater supplies, the authority entered into negotiations with Williton Rural District Council, which was responsible for water supply. Consequently, the Clatworthy Reservoir, planned for 1959, was to be expanded for the benefit of all its users.[88]

In sum, the planned power station's effects on the countryside were rejected as negative by various citizens living in the local area, the wider region and even farther afield. At the same time, the local councils and other citizens involved in the public debate had developed a position that was not opposed to the CEA's project, if its effects on the local landscape, the rural way of life and the environment would be mitigated or at least controlled.[89] The CEA, in turn, had been open to some concessions.

It should be mentioned that some local contributors to the debate, and people with connections to the CEA or local business associations, did not consider concerns about the countryside valid. The CEA's representatives in early confidential talks had been bewildered that anybody would argue against a power station of national importance because of local amenities.[90] During the public phase of the debate, the former conservative parliamentarian and parliamentary secretary to the Minister of Transport under Winston Churchill's premiership, Sir Joseph Gurney Braithwaite, deemed large-scale protests over "a deserted stretch of coastline" unlikely.[91] John S. Giles from Stogursey assented, as Hinkley Point's "drab and ill-favoured coastline" was beautiful "only in their [the critics', CG] imagination".[92] A reader of the *Bridgwater Mercury* self-identifying as "Local Bricklayer" emphasised that it lacked in beauty compared to the Quantocks,[93] and B. J. Webb from Bridgwater called the proposed building site "an area of little charm" that "offers little except perhaps to the naturalist or ornithologist".[94] It was not the primary goal of these contributors to the debate, however, to devalue the landscape. Rather, they argued that its protection did not outweigh the station's possible economic gains, regarding which they postulated a kind of zero-sum argument: Webb was convinced that the power station would bring with it a higher "standard of living", worth the sacrifice of this coastline.[95] This position was shared by P. Hutchings from Stogumber,[96] and the Bridgwater and District Trades Council, which passed a resolution stating that there were other, more beautiful stretches of coastline, while the region's "future prosperity lies in cheaper electricity produced by atomic power stations".[97]

The theme of economic prospects broached by these contributions was also raised for its own sake. From a national perspective, it constituted the power station's *raison d'être*. In November 1956, one of the local landowners interviewed by Powell had already stated – albeit without enthusiasm – that he was willing to cooperate with the CEA, as the power plant seemed "inevitable". His assessment probably referred to the "national necessity" of nuclear power stations, proclaimed by the government,[98] and widely accepted by local councils.[99] A. C. Thirtle, controller for the CEA's southwestern division, took up this position more enthusiastically. To him, as the *Mercury*'s editorial writer "Nomad" reported, "nuclear power is the fuel of the future". It seemed destined to become ever cheaper and to "usher in a new age of clean air in which smoke, smog and fumes will be banished from cities and countryside".[100] Besides this longed-for environmental improvement – air quality had been an important topic during the early 1950s, culminating, not least, in the Clean Air Act of 1956 –[101] he expected that electrical heating and "new and improved types of appliances for use in home and factory" would make life more agreeable for all. The people of Somerset, he concluded, had reason to be proud of being part of the vanguard of the "development of the peaceful use of atomic power".[102] His position was shared by the popular mathematician, historian, and public intellectual Jacob Bronowski, who, at that point in time, worked for the National Coal Board.[103] Taking part in a BBC radio show on Hinkley Point, he thought that Britain, at that moment leading the world in civil nuclear power, had to expand its capacity to generate electricity "to become a first-class industrial power".[104]

Others brought the economic argument closer to home by focusing on Somerset, where, to the *Mercury*, regular blackouts and several settlements still not connected to the grid served as a case for additional supplies of electricity.[105] Beyond that, Walter A. Lawrence from Combwich welcomed the economic opportunities for local workers and businesses promised by the station. He, a self-declared "lover of Somerset" and "also a lover of progress", encouraged his fellow citizens to "respect the modern wonder of this age brought about by the brains and hands of mankind" and "make clear the way for the construction of this wonder power house . . . Work, prosperity and a chance for thousands of working men, that's my motto. Spare a moment and think of the materials that will be required to complete this atomic miracle, . . . all of which are produced in traditional British high quality in our own industrial village."[106] "Local Bricklayer" concurred. To him, the plant offered a way out of rising unemployment and a chance for a "golden era for Bridgwater".[107] The Bridgwater Chamber of Commerce, Shipping and Agriculture entertained similar hopes. While two members from Stolford were critical of the project, the chamber's majority expected to find business opportunities for "local industries and services" and profits for the town of Bridgwater.[108] To further these, the chamber had prepared a list of all the products manufactured in the vicinity,[109] which had been presented to the CEA. The authority, in turn, had promised that local firms would be kept in mind[110] and pointed out that the station would offer training and job opportunities for local people.[111]

Reciprocal to the situation regarding the countryside, some of the station's opponents sharply disagreed with these positive expectations. Early on, Mackenzie had pointed out that the project could bring trouble with "the labour position on farms".[112] More precisely, some farmers and other businesses[113] were afraid of being unable to compete for semi-skilled workers in the face of over 2,000 well-paying jobs during the construction period.[114] Consequently, Somerset's National Farmers' Union (NFU), asked the Minister of Power, Lord Percy Herbert Mills, to ensure that the project would not lead to a shortage of workers for local farms.[115] Local councils at least made the CEA promise to keep the question in mind.[116] Furthermore, the councils were concerned about the implications a large influx of workers might have on housing, education, policing, and more. They were able, however, to get the CEA to agree to cover some of the additional costs expected.[117] Finally, Jones highlighted a potential negative economic impact of the station on places that profited from the view they offered of the proposed construction site, like Burnham-on-Sea.[118] Naturally, this aspect was a concern for local authorities such as Burnham's Urban District Council.[119]

The CEA's proposal, in other words, had given rise to quite intensive debates about the closely connected themes of countryside and economy that could, in the end, have been expected for any kind of industrial project proposed for a rural area. In addition, the CEA's plan provoked debates about features that were distinctive for a nuclear power station. They fell broadly under the headings of nuclear waste, radioactive emissions, and the relationship with experts.

Nuclear waste was the first of these questions that was publicly broached – albeit in a rather ironic way. Lionel P. Mantle from Holford wrote to the *Mercury* in December 1956, suggesting such waste could be disposed of in Holford Glen, a small valley next to his home, where "all and sundry" used to deposit "every conceivable form of refuse." "Maybe", he speculated, nuclear waste dumped there "would kill off some of the offenders". The prospect was commented by the *Mercury*'s editor with a succinct "[b]ut not, we sincerely hope, the residents?"[120] Nuclear waste, in other words, while mentioned as a problem of nuclear power generation, was not taken seriously. While this approach could have been a coping strategy for existing fears (cf. "Emotional Atoms"),[121] it was not uncommon for the 1950s, when the dominant opinion in Britain (and beyond) was that nuclear waste was something that would be taken care of by the responsible experts (cf. "Rational Atoms"). In the words of the nuclear power programme: "The disposal of radioactive waste products should not present a major difficulty."[122] It was to be recycled, with only small amounts of waste remaining to be stored or discharged, and "harmless and comparable in effect to the natural background radioactivity which is always present".[123] Few were asking critical questions when it came to this, as debates about radioactivity were dominated by physicists who heralded its future possibilities,[124] discarding risks pointed out by some biologists or physicians,[125] arguing, with Pascual Jordan, that potential consequences like cancer would soon be overcome by medical research.[126] Correspondingly, the topic of nuclear waste did not have an observable impact in the debate about Hinkley Point, although Weston-super-Mare's Borough Council addressed this worry, too.[127] Rather, the CEA's assurance to local authorities that "[w]aste material would be disposed of to the Atomic Energy Commission"[128] (AEC) – in all probability a mix-up of the United States AEC with the United Kingdom Atomic Energy Authority (AEA) – seemed to settle the affair.[129]

It was Page who, in her letter to the editor in January 1957, pointed out a second problem specific to nuclear power. She wondered about "the effects of radioactive fall-out" and the consequences of a regional concentration of multiple nuclear power plants, given that one was already under construction at Berkeley.[130] Two months later, Rowe implicitly stated that this question could not be answered, as it would take many years to be able to gauge the long-term effects of nuclear power,[131] a point of view shared by Jones.[132] Rowe underscored these uncertainties by implying that the water reserves of the Quantocks could be endangered, establishing a concrete environmental impact of the new technology. In his view, which was based upon the national level of debate in the form of an article in the *Daily Telegraph* presenting radioactive emissions as a reason for siting nuclear power stations in sparsely populated areas, the local population was turned into guinea pigs "in the name of officialdom and progress", a development he wanted to be stopped.[133] R. W. Barnes from Bridgwater, who referred to an article in *The Times*, shared Rowe's verdict.[134] In addition to the consequences of a nuclear power station operating normally, a "Bridgwater District Resident" and "father of two children" pointed out the dangers of "a major accident", the possibility of which he saw indicated by the stations' remote location. He also emphasised that nuclear power stations would render Britain undefendable in wartime.[135] Both aspects were taken up by Rowe,[136] who, from April 1957 onward, called for a coal-fired plant to be built instead of an atomic one. Such a station, he added, to reach those interested in economic benefits, "would provide work for local men, instead of merely for a scientific priesthood".[137] His demand was endorsed by Jones.[138] Summing up his position on nuclear energy and implicitly answering Page's question, Rowe wrote: "Atomic energy, as it stands, is a cancer."[139]

In view of the various worries about radioactivity that were aired by local people as well as those from a wider region, the CEA conceded during a press conference on 22 March 1957 that nuclear power was not without "hazards". But they also emphasised the extensive safeguards that were taken, the ten-year record of safely operating reactors that the AEA had enjoyed, and the low chance of "only a million to one of anything ever going wrong".[140] Although the statement failed to dissuade critics like Rowe or Jones, the county council was under the impression "that this seemed to have allayed public apprehension on that score".[141] Proponents of the scheme were willing to trust the CEA's experts unquestioningly. Local Member of Parliament Gerald Wills believed their assurances that water and fish would not be negatively affected by radioactivity and refused to forgo progress because of potential consequences of a hypothetical war.[142] The Nature Conservancy, tasked by the government with the protection of national parks,[143] also accepted the CEA's statement,[144] and "Local Bricklayer" was explicitly willing to trust in its experts: "As to the fears of radiation, radio-active gas, contaminated water supplies, etc. . . ., surely we can leave the safety precautions and disposal of waste matter to the brains of the scientists who are clever enough to devise such a plant."[145]

The role of experts, on whom these supporters of the power plant in the end rested their confidence, formed the third topic within the theme of radioactivity-related questions. In later years, it would move to the very centre of debates about nuclear power. In the early debates about Hinkley Point, however, it was rather uncontroversial, as the example of nuclear waste had indicated. The British government and parliament operated based on trust in the experts (cf. "Rational Atoms", and "Atomic Trust(s)").[146] "Local Bricklayer" wanted to do the same, and the *Mercury* recommended this course regarding the open economic questions: These should "be left to experts to explain", given the complexities of higher capital costs and lower fuel expenses when compared to coal-firing stations.[147]

This widespread practice, however, was challenged by Page. In her January letter, she had emphasised that, so far, all information about the project came from the CEA, none from "independent scientists".[148] With this observation, she broached the question as to whether the CEA's experts could be trusted or whether they should be double-checked. By chance, a form of double-checking did indeed occur in the form of an assessment on the CEA's proposal by the Western Counties' Laboratory, which had previously given advice to the county council concerning a phosphorus factory.[149] Nonetheless, "Bridgwater District Resident" later called into doubt the expertise available with reference to the transnational scientific discourse about nuclear energy. He referred to a report in a newspaper from New Zealand, wherein Dr Harry R. Lillie stated that maximum permissible doses for radioactivity, supposed to guarantee the safety of human health over decades, were based upon short-time trials with animals. Herein, "Resident" saw proof for the shaky foundation of nuclear experts' knowledge. He was convinced that the public should be informed about this – and astonished that no English newspaper had written about it.[150] Like him, A. M. Hogwood from Bridgwater thought it necessary to challenge the CEA's experts, albeit for different reasons: To him, it seemed strange to build a nuclear power plant next to an area like the Bristol Channel, where tidal power could be exploited (cf. "Nuclear Futures"). His explanation for the one-sided concentration upon nuclear power was that "this is the Era of the Expert, the scientific expert who likes messing about with expensive, credit-consuming uranium, and regards H^2O [sic, CG] as a very old-fashioned formula".[151] In addition, former Somerset resident J. F. Hillier,[152] now from West Buckland, about 60 kilometres west of Hinkley Point, pointed out that the experts often abstained from presenting alternatives to their proposals, while local authorities tended to be dominated by people strongly interested in economic opportunities.[153]

There was, to sum it up in the Nature Conservancy's words, "considerable public interest in the proposed Hankley[154] Point Nuclear Power Station".[155] Various private citizens, both local and from further afield, challenged the CEA's plans for a nuclear power plant at Hinkley Point using arguments that were specific to nuclear power, namely nuclear waste, the consequences of radioactive emissions during normal operation and as a consequence of accidents, and the widespread delegation of decisions to and trust in experts. Although not all the arguments presented were made in earnest, owing to the contemporary principle of delegating certain questions to experts, the points of nuclear waste and the effects of radioactivity had at least provoked the CEA to try and mitigate the critique.

The authority's limited willingness to consider its opponents was in turn part of the final theme of the discussion about the Hinkley Point power station. It was a debate about the British conditions of debate, civilians' participation in this debate, and the way it was conducted, especially by the CEA. Like the theme of the consequences of radioactivity, it became a core element of many disputes about nuclear power later on (cf. "Nuclear Democracy"), although it was only brought up by a few people during the technology's early days. Nonetheless, brought up it was.

At the start of the debate, the county council had already encouraged a broad discussion, when its planning committee had emphasised that it would, as usual,[156] listen to other local bodies and authorities. Beyond that, it had pointed out, a public inquiry could be scheduled by the minister if "a sufficient volume of local opposition" arose.[157] The council favoured such an inquiry for two reasons. On the one hand, only people owning or leasing land within 300 yards of the site were formally seen as relevant "local opposition" and, therefore, asked to object in writing until 19 January 1957.[158] As the *Mercury* quickly pointed out, this "takes no account of any other persons in the area which would be dominated by the power station".[159] This question moved the county council, which had received various letters from objectors and concluded that people were not sure about the process itself. Therefore, it saw a public inquiry as a way for people not formally able to oppose the project to be heard in an official setting.[160] A number of people, including Irving-Bell, had explicitly argued for such an inquiry.[161] On the other hand, the council hoped that a sufficient quantity of public protest would strengthen its own position in negotiations with the CEA.[162] The latter, unsurprisingly, was keen to avoid a wider public inquiry, considering it a hurdle that would unduly delay its plans,[163] although it had resolved, in late 1956, to "give the local authorities every possible facility for expressing their view at the earliest stage" and to "give serious consideration to any points which appeared to be important".[164]

The question regarding whether people not formally considered as "local" should be heard gained some public traction after a letter to the *Bridgwater Mercury* by David Jones from Bath had stated that, to him, the local opposition seemed weak, while there were various "people living far from the area" who were concerned about the CEA's plans. He was convinced that "single-handed" he could "rally enough opposition from outside to have the power station moved to a more suitable site, where it would really benefit Somerset even more!"[165] This statement provoked Lawrence to label Jones part of a group of "outsiders" who probably had "not read the details of the programme" and to whom the local supporters of the project should pay no heed.[166] Jones' position was clear: "I suppose none of the supporters of the Hankley project would care to deny that people who do not live in the area have any right to an opinion on the subject?"[167] Apparently, nobody seemed to want to do so until after the public inquiry, when Lawrence somewhat gleefully came back to this question by pointing out that the power station was welcomed by those living close by, while the opposition of people from farther away "thank goodness . . . does not matter".[168]

That, as already indicated, was not the view of the county council, whose planning committee, in March 1957, recommended holding a public inquiry, as 19 objectors "from all parts of the country" had come forward and would be able to "submit their views" there.[169] Reacting to this proposal, the CEA once more argued that a public inquiry would not be necessary, as only a handful of people had protested directly to the authority, with only half of their statements "considered to be true objections. And of that number only two have come from local people",[170] whereas the public in general, including groups like the National Union of Agricultural Workers and local authorities such as Stogursey Parish Council, supported the station.[171] This attitude now led to a veritable outcry by the power plant's opponents,[172] who had tried from the start to organise their resistance. In January 1957, Mackenzie had already resolved "to fight it [the power station, CG] tooth and nail", and had used a letter to the *Bridgwater Mercury* to search for others willing to join forces.[173] "Bridgwater District Resident", who was convinced "that only strong public opinion will deter the Government", stated his readiness to "give my full support to any public body to oppose the setting up of an atomic power station at Hinkley Point".[174] Jones was soon able to announce the founding of the "Quantock Seaboard Defence Committee" which aimed at opposing the station to preserve "an unspoiled stretch of coastline".[175] In addition, Irving-Bell called for women to make "their feelings known to the Minister of Power" regarding the "men, so blind to beauty, so lacking in imagination and so careless of the feelings of their fellow men that they can contemplate such vandalism [the construction of the power plant, CG]".[176]

On a side note, it should be remarked that, in doing so, Irving-Bell provided an early example of two divisions along gender lines that appeared in the debates time and again.[177] On the one hand, she placed women on the side of the critics of nuclear power, which was very much a fact in the majority of the cases examined here.[178] Nonetheless, there were contrarian examples, especially in Britain, right up to Prime Minister Margaret Thatcher.[179] Second, she linked female criticism of nuclear energy with an emotional perspective. In doing so, too, she was not alone.[180] However, especially in Germany, this connection was for a long time pointed out by proponents of nuclear power to devalue precisely this female criticism (cf. "Emotional Atoms"). In fact, it would be wrong to limit female criticism to questions of emotion. Rather, notably in the early debates, women provided a broad spectrum of arguments against the technology. Irving-Bell herself had prominently presented the issue of landscape protection and added further ones later. She was not alone. The already mentioned Page had been a pioneer in pointing to radiation as a potentially problematic consequence of nuclear power plants, thereby citing an argument that was later to take centre stage in the debates.

To return to the veritable outcry in reaction to the CEA's position, Rowe even accused the authority of trying to dupe Somerset's people by pushing its plans through in as short a time as possible.[181] Jones restated that limiting objections to people within a 300-yard radius was not fitting for nuclear power stations, and that there was a "growing opposition . . . whether it be 'official' or not, and sooner or later the fact will have to be faced and not shelved".[182] He also criticised Bridgwater's Member of Parliament for pointing out that he had only received ten letters regarding the power station.[183] Assuming that he had received none at all supporting the project, he called on all opponents to write to Wills.[184] Therefore, while the absolute number of people objecting to the power plant was small, they covered a broad variety of aspects and were willing and able to make their voices heard, in the end playing a role in forcing the CEA into a number of concessions, including a public inquiry. It was not without foundation, then, that the *Mercury*, reporting on the latter, chose to refer to "Stormy protests" in its headline.[185]

The Public Inquiry

In April 1957, the Minister of Power announced that a public inquiry was to be held in Bridgwater on 14 May, expected to last for two days.[186] The *Bridgwater Mercury* welcomed the decision because there was "a clear divergence of public opinion about the project", and it was therefore "right that the authorities should give people the opportunity publicly to voice their objections so that these may be given proper consideration in the appropriate quarter".[187] From the ranks of the project's critics, Page publicly welcomed the inquiry,[188] and Rowe argued that, possibly, the *Mercury* had played a part in effectuating it because it had offered a stage for all positions in the debate.[189] Certainly, the fact that a public inquiry was to be held was of great importance to the people rejecting the power station,[190] who had demanded, in Irving-Bell's words, "a chance of expressing their views"[191] that went beyond letters to authorities, councils, and editors.[192] Although, formally, the public inquiry served only as a tool for the minister's decision-making process, its form, similar to a legal trial, offered the power plant's opponents the chance to publicly cross-examine the CEA's experts. Although opponents were in a structurally weaker position than the CEA, they would, at least symbolically, meet the authority on equal footing, as its experts would have to argue the authority's position in front of inspectors, who would also hear the opponents' arguments and weigh both sides in their final report to the minister.[193] Indeed, during the public inquiry, no completely new arguments were presented before Howard Walker Grimmitt, the Chief Engineering Inspector of the Ministry of Power, and H. R. Wardill, Senior Inspector of the Ministry of Housing and Local Government.[194] Now though, the CEA had to deal directly with the views brought forward, and the compromises and assertions, which had so far mostly been talked about with the local councils, prominently came to the fore.

At the beginning, the CEA's counsel, S. B. R. Cooke, summed up their known arguments for their proposal to build a nuclear power plant at Hinkley Point, with the added weight of an expanded nuclear programme: The government had, in March, trebled the output of the nuclear power stations to be constructed by 1965. Therefore, Hinkley Point, now envisaged to become the third of 19 plants, was to achieve an output of 500 MW, instead of the 300 MW planned before. To go ahead was, in the words of the CEA's deputy chairman, Sir Josiah Eccles, to fulfil the "demands of the nation" and of "vital necessity".[195]

Irving-Bell once again argued for the protection of a landscape that once had inspired the leading English romanticists and now served as a retreat, especially for the "elderly", while questioning the CEA's assertion that local maritime life would not be adversely affected. Her concerns regarding the landscape were backed by Cyril Williams, who presented a petition of the Quantock Sea Board Defence Committee for its protection, which had been signed by 600 people, while N. Crouch of Wick, a hamlet halfway between Hinkley Point and Stogursey, concentrated upon the intrusion into a valuable agricultural area.[196]

Accordingly, the CEA's representatives sought to allay concerns regarding the countryside from the outset. Sir Josiah pledged to "do as little damage as possible" to the amenities, while Cooke promised that the station's "design would be discussed with the County Council" and emphasised the authority's "intention to appoint a landscape consultant who would help with the fitting of the building into the countryside". After all, the CEA "endeavour[ed] to be good neighbours", and further industrialisation was not to be expected. Regarding the freshwater supply, the authority pointed out that Williton Rural District Council and Taunton Borough Council had agreed to improve and extend the planned Clatworthy reservoir,[197] "which would benefit all people relying on that supply". Regarding the sea, the CEA stressed that the water would not exceed radiation limits for drinking water and it would be "possible to swim in it in perfect safety". Although, as station development engineer Donald Clark admitted, it "would be warmed". Finally, Cooke emphasised that the site lay outside any kind of protected area, that the station would not generate "smoke, grit, noise or smell", and that neither the Nature Conservancy nor the National Parks Commission raised objections.[198]

Regarding the economic consequences, Burnham Urban Council and Geoffrey Knowles[199] from Weston-super-Mare, who spoke for the tourism trade of Weston, feared severe losses "if the area was to be ruined", while the regional branch of the NFU and the County Landowners Association pointed out local worries about the labour situation. Concerning the latter, the CEA was unable to do more than promise to highlight these worries to its contractors, while Clark affirmed that the power station would not be visible from Burnham. Beyond this, improvements to the rural infrastructure, especially roads, criticised by Rowe because of the potential destruction of "historical places" in Cannington, were explicitly welcomed by Cannington Parish Council and Bridgwater Rural District Council. The construction workers' settlement, which had been controversially debated by some councils, was to be erected on-site, while the permanent employees would be settled in cooperation "with the local housing authority" with the goal "to absorb them in the community." It was a plan that had the explicit support of the county council. As the CEA also "envisaged training schemes for young people interested in atomic energy", and the Somerset Federation of Women's Institutes welcomed the electrification of rural settlements, the economic questions appeared, in the end, to be settled to almost everybody's satisfaction.[200]

Doubts regarding radioactivity, which were expressed by Burnham Urban Council and Knowles, were immediately discarded, and countered with an extensive explanation as to why there could be no problems: "There is no danger of explosion . . . If there is any danger the only one we can think of is a remote one. There may be for a short time a slight increase in the discharge of radioactivity, but . . . [i]t is inherently safe . . . In every part of the world there is a background of radioactivity and, as far as this station is concerned, it will increase local activity by only five per cent . . .", and "[i]f there was any slight increase in radio-activity beyond the normal, the only danger would be if it went on for a long time so it is a good thing to have this station in a comparatively sparsely populated area", later on adding "and any emission will do no harm to animal or plant life". Clark pointed out that even if the power station were hit by a plane from a nearby bombing range, "[t]here would be steam escaping but no nuclear danger." Whenever radioactivity escaped, it would happen slowly so that it would be possible to react. Frank Reginald Farmer, the AEA's chief safety officer, added that experiments had shown a power station's liquid discharges would not affect fish in the area. "[S]pent fuel", finally, amounting to "one lorry load" per day, was to be shipped to the AEA and would, therefore, pose no problem in Somerset.[201]

In the end, the CEA had done little more than officially and publicly pointing out the compromises reached with the local authorities beforehand and repeating the assurances already given.[202] Viewed with scepticism, therefore, the public inquiry had been little more than a performance. Nevertheless, the inspectors were of the impression that the CEA's "explanations appeared to satisfy some of the objectors. They were impressed when Eccles said that the Authority desired to become part of the community", and by the fact that he had given the CEA's assurances in person. In the inspectors' eyes, worries regarding the station's safety seemed to have been allayed, and it had become obvious that most opponents lived outside the area, while many local people tended to favour the project.[203] One example was Stogursey Parish Council, representing the community closest to Hinkley Point. Its chairman, Percival Chidgy, emphasised "that the feeling of people attending the annual parish meeting was that nothing but good would come from it".[204] In the aftermath of half a year of "stormy protests", then, which had been enabled by the rather open British conditions of debate, these very protests appeared to be settled because they had been played out.[205]

The End of the Debate (for the Time Being)

The end of the public inquiry, by and large, marked the end of the highly controversial phase of the wide-ranging public debate about nuclear power at Hinkley Point. Certainly, the debate did not come to a complete stop. Staunch opponents of the project, including Jones,[206] Rowe[207] and Enid Williams from St Audries in the Quantocks,[208] remained critical of the nuclear reactor. Nevertheless, the debate settled down remarkably: Where there had been 28 letters to the editor in the six months up to the public inquiry, it took about four years for the same number to appear afterwards.[209] At least in part, this slowdown was due to some critics deeming their cause lost. Eventually, this group even included Rowe, who at first had resumed his protest with six further letters to the editor of the *Mercury* after the Windscale Fire (discussed later). He explained that he was motivated by the death, ahead of the public inquiry, of his friend Mackenzie, who had been determined "to fight the project tooth and nail".[210] Finally, however, Rowe saw no sense in further arguing against a power station that was already under construction.[211]

These examples should not distract from the number of people who were quite satisfied with the fact that there had been a public inquiry, where everybody who wanted to contribute to the debate had been heard, and with the compromises and assurances offered by the CEA. Some of these were explicitly taken up by Lord Mills in early summer, when he, as Minister of Power, approved the CEA's plans.[212] To some, the compromises reached were just that – an ambivalent outcome. B. Yeudole, from Bridgwater, although disappointed with the station's design and the landscaping when both were put forward, admitted that at least the CEA's promises in that regard had "partially allayed" widespread "fears of a grand extension of Subtopia" – and that "[t]he buildings themselves appear remarkably smooth and un-fussy".[213] Others were more positive, to all appearances enthralled by the technological sublime (cf. "Emotional Atoms"): Lawrence felt "sure that when the giant reactors of the world's largest atomic power station tower majestically against the skyline all those now opposed to it will gaze at it in admiration and will marvel at its success."[214] In a similar vein, the *Bridgwater Mercury*'s journalists soon praised the power station as the area's new "landmark"[215] and a technological marvel,[216] largely in tune with the industry's high gloss brochures.[217] Indeed, some of the area's residents seemed to have been appeased especially by the power station's design: When the plant was to be coloured blue instead of the original grey in 1984, they complained that it would no longer merge with the background of the Bristol Channel.[218]

The development of the power station's economic consequences was even more positively regarded. The NFU's Somerset branch soon conceded that it was satisfied that the CEA would manage to avoid trouble regarding the local labour situation,[219] and, in December 1957, Lawrence pointed out that the power station had already exerted a positive influence on the area, as the first roads had been completed.[220] In the following year, the *Mercury* observed that Hinkley Point had stimulated an economic upturn in the Bridgwater region,[221] which was also recognised by Stogursey Parish Council.[222]

In addition, the issue of the conditions of debate itself did not give rise to much critique after the public inquiry, although some fundamental questions raised in its course remained unresolved. The aspect of the difference between the people formally represented in the decision-making process and those potentially affected by the station, shortly taken up again by Lawrence,[223] saw next to no further discussion. This held true even when it was once more highlighted in 1960 by a governmental declaration about limits to the development of the residential population within a five-mile radius around the power station.[224] Even when Rowe mentioned his impression that the public inquiry had been conducted as a mere formality, he nonetheless conceded that the CEA did respond to its critics there, claiming that better outcomes could have been realised if the power station's opponents had had access to their own experts to question the authority's claims.[225]

The trust in experts thus addressed remained entwined with other parts of the theme of radioactivity, as the assurances given by the CEA at the public inquiry regarding radioactive emissions and their consequences rested heavily on this very trust. It was soon to be tested by two events that brought the consequences of dealing with radioactive materials back out in the open: the Windscale Fire in the autumn of 1957 and the government's decision to modify Hinkley Point for use in the production of military-grade plutonium in 1958.

The Windscale Fire refers to Britain's worst nuclear accident so far. In October 1957, one of its piles producing weapons-grade plutonium caught fire after attempts to release the accumulated Wigner energy from the graphite core.[226] Reacting to the (governmentally doctored) news about the transnationally debated accident,[227] some local critics once again took up their pens. First and foremost, Rowe argued that the fire proved the possibility of nuclear power plants endangering their surroundings,[228] thereby pioneering a way of thinking that was to find a widespread following transnationally after the Three Mile Island and Chernobyl disasters. A few days later, in reference to an article in the *Daily Telegraph*, he added that, owing to their constantly expanding dimensions, all nuclear power plants now under construction were experimental ones, turning the surrounding population into guinea pigs for a "cynical grab for material benefits" that put lives at stake, and not only those "of the present generation".[229] Jones joined him in this critique,[230] and "Resident" explicitly thanked Rowe "for his public-spirited common-sense contributions on this devilish project at Hinkley Point", wondering why only a few people raised their voices, although "[s]urely there must be thousands like him who feel as we do". As he thought it was too late to stop Hinkley Point now, he considered moving, together with his company of 40 employees.[231]

Despite this once more intense critique, in general, contributors to the debate remained positive about the power plant and the safety assurances given ahead of and during the public inquiry. A reader called "Onward" pointed out that, although there had now been an accident, "surely we can put our faith in our British scientists" when it comes to dealing with radioactivity.[232] A more practical approach was taken by the NFU. Alarmed by the losses that farmers around Windscale had to endure because of the accident there, it asked the Ministry of Power for an assurance that nothing similar could happen in Hinkley Point. The aim was to establish an argument for calls for compensation in case something did happen in the future.[233] Once this goal had been reached,[234] they seemed willing to go back to the quiet acceptance of the power plant they had exhibited before (cf. "Rational Atoms").

This, in fact, was what happened in general. The debate died down once again. It is fair to assume that most people continued to put their trust in the CEA's capability to deal with the challenges posed by nuclear power – which was advertised by Hinkley Point's "resident engineer", D. M. Longman. Talking to the *Mercury*, he reassured its readers that a "mishap" like Windscale was out of the question, as Windscale, unlike Hinkley Point, was "primarily concerned with plutonium" and used normal air to cool the reactor, which was returned to the atmosphere through filters, whereas in Hinkley Point a closed circuit of CO_2 was employed. Therefore, even if there were a leak, "no harmful effects" would result from it.[235]

In view of the importance attached to the differences between Windscale and Hinkley Point in that regard, it is not surprising that the public debate flared up anew when, in the summer of 1958, the government highlighted the always present dual-use character of nuclear power by declaring that Hinkley Point would be modified to be able to produce plutonium for military purposes.[236] Once more, Rowe criticised the project and lamented the fact that there were no experts present at the public inquiry representing the opponents of Hinkley Point.[237] The newly formed Campaign for Nuclear Disarmament had also now joined in the critical debate.[238] Founded in 1958 to argue for the renunciation of atomic weapons by Britain against the background of a nuclear arms race that was gathering pace, the group quickly turned into a prominent voice in matters of nuclear weapons and nuclear war. It also became a forum for discussing the acceptability of illegal actions in the context of protests, which were later to occupy the anti-nuclear movements transnationally. It remained as such, with ups and downs, through the period under consideration here, frequently joining forces with groups opposing the "Peaceful Atom", especially on matters obviously connecting it to nuclear weapons, like reprocessing. Thus, their speaking up on this occasion in Hinkley Point was quite prototypical for what was to become a given from the mid-1970s onwards.[239] Furthermore, Stogursey Parish Council also protested against the plutonium plans,[240] and Bridgwater Borough Council called the decision "thoroughly dishonest" in view of the promises made in the beginning, besides highlighting the fact that this made the station more similar to Windscale.[241] This connection was also drawn by Herbert W. Haydon from Weston-super-Mare, who called for alternative ways to generate electricity, like a tidal barrage.[242] Others agreed,[243] and over 100 citizens of Stogursey and Combwich initiated a petition.[244] The debate died down once more, however, with the prime minister emphasising that the modifications were of a purely precautionary nature and would, probably, never be put to use.[245]

Beyond these two events, the theme of radioactivity was hardly discussed at all in public in the years following the inquiry, just like the countryside, economy, and the conditions of debate themselves, while the power plant's emissions became a regular topic for its Local Liaison Committee (LLC), which first convened in April 1961.[246] LLCs had been introduced as a standard for British nuclear power plants as a lesson drawn from the Windscale Fire. Combining representatives "from local authorities, Government Departments and interested organisations" and members of the supply industry,[247] they served as a component of emergency planning, but also to provide information about the station and environmental monitoring.[248] The latter practice, measuring radioactive pollution around nuclear reactors, had been established at Windscale in 1947.[249] In the case of Hinkley Point, monitoring was called for in 1958,[250] and, after preliminary measurements, began "in earnest" in 1960, covering an area up to 20 miles around the site.[251] The LLC served as central distribution channel,[252] with reports regularly disseminated locally – but not, of course, in the national media.[253] The committee thus "aimed to demonstrate to the public that the power station operated as a completely safe plant."[254] This was one of the main purposes of the LLCs overall,[255] for which they were implemented "very early in the detailed planning stage of the first stations".[256] Consequently, they played an important role in debates about participation and trust. They were also typical of the British conditions of debate in that they institutionalised public questions and information exchange. Certainly, the issues ascribed to them were not necessarily confined to the committees. Occasionally, they were again taken up in public, as were questions regarding the transport of used fuel elements,[257] and incidents in the power plant.[258] And while shortcomings in this regard were held against the technology and its operators, the rather transparent information about the area's radioactive pollution clearly contributed to its acceptability.[259] Overall, then, the broad and public controversy from the months before the public inquiry was over. Summing up the situation when the power station was officially opened in 1966, one year after it had taken up operations, one of the *Mercury*'s commentators put it this way:

> [O]n reflection, it can be said that the pessimistic predictions way back in 1956, and at the public enquiry the following May, have not been fulfilled. It was said that . . . there might be risks of explosion, and dangers from radioactivity. That the whole area would be turned upside down; historic places would be damaged by road widening; the fish would leave when the warm water flowed in; quiet and undisturbed villages would be spoiled . . . We know now that most of these fears were completely unfounded, and that neither the sea nor the air have been polluted. And in the light of what has – or has not happened, over the years, social workers must be relieved that there was little need for concern about the effect of the sudden appearance of so much that was alien to a rural area.[260]

In fact, the controversial debate had settled down so much that plans for a second nuclear power station at Hinkley Point and its subsequent construction received hardly any public attention at all,[261] even though the antecedent talks stretched over some years and the CEGB initiated quite an active public relations strategy. This included an exhibition on the project in Taunton in October 1967 and opening its deliberations with local authorities for the media.[262] Consequentially, when, later, that power station's radioactive waste received some public attention, the most remarkable statement made in a debate in the Bridgwater Rural District Council came from councillor Bernard Smith. It was not remarkable because he was dissatisfied with the way the council was treated in the affair, but rather because he reflected that there should have been a public inquiry on Hinkley Point B, since the one on the first power plant had allowed for "certain guarantees" from the CEA to be acquired.[263] In a way, it seemed to have dawned on him that the public inquiry, which had been held as a result of the wide and open public debate that the first power plant's announcement had evoked, had unquestioningly contributed to the plant's widespread acceptance afterwards.

A change back to an open and critical debate manifested by the end of the 1970s,[264] especially so once the Conservative government of Margaret Thatcher, soon notorious for its aversion to discussion,[265] attempted to force an accelerated nuclear power building programme. As early as 1977, in the wake of an accident resulting in one of its reactors being shut down for many months,[266] Hinkley Point had already been chosen by a group affiliated with the well-connected environmental group Friends of the Earth as a place to protest against breeder reactors and a plutonium economy.[267] Friends of the Earth had been founded by David Brower and others in the USA in 1969 in the context of the Sierra Club, with the explicit aim of establishing it as a transnational voice for environmental protection.[268] Soon developing into an anti-nuclear voice,[269] willing to take action for its goals, including civil disobedience,[270] it gained sections in Britain, Sweden, France, Germany, and about a dozen other countries during the first half of the 1970s.[271] By the mid-1970s, there were 94 local groups in Britain alone.[272] A protest organised against Hinkley Point in 1977 by one of these groups, while most likely triggered by an open day at the power stations, which guaranteed publicity for the critics, signified the interconnections between Hinkley Point and the growing protests against atomic energy on a (trans-)national level.[273] These went along with a rise in environmental topics playing a role in party politics, as exemplified by the foundation of the new Ecology Party in 1975.[274] In addition, some local and regional politicians, many with a Labour affiliation,[275] had begun to view Hinkley Point more critically, and to point out accidents as well as structural problems, including the treatment of nuclear waste. In a few later cases, there were even attempts to fabricate or at least magnify sensational events.[276] The increase in activities by critics went along with other developments that served to highlight the points raised by them. These included attempts to locate an underground nuclear waste storage facility in the region,[277] worries about the incidence of leukaemia around the power station, and deficiencies in the emergency plans, which were taken up in the wake of the nuclear catastrophe of Chernobyl.[278] Combined with the Thatcher government's growing assertiveness about plans to erect a third power station at Hinkley Point based upon the widely criticised American pressurised water reactor (PWR) technology,[279] the debate about nuclear power turned into an ongoing issue around Hinkley Point. In this context, various people and institutions started to affiliate themselves with anti-nuclear networks. For example, a Somerset Anti-Nuclear Alliance was formed in 1979, combining, among others, activists from Friends of the Earth, the Taunton Anti-Nuclear Group, and regional bodies of the Ecology Party.[280] When, in 1988, a public inquiry into Hinkley Point C began,[281] the centre of national debate shifted to Somerset. For 14 months, the issue nearly split the area in two. The story of this newly confrontational debate, however, is taken up in "Rational Atoms".

For the argument made here, it is important to keep in mind that, when a first commercial nuclear power plant was to be constructed at Hinkley Point in the middle of the 1950s, there had been a broad debate from the outset, both in terms of the issues dealt with and in terms of the groups involved. Strikingly, critical arguments, namely regarding the countryside, the consequences of radioactivity, and the way the debate was conducted, were pointed out by private citizens from the local area, with vocal support from others living in the region and the nation. Local authorities, while mostly interested in the project's effects on the countryside and the economy, took up a rather moderate position, arguing for an open discussion and, explicitly, a public inquiry to give the numerous citizens partaking in the debate a chance to be officially heard. The local citizens and associations who supported the project, on the other hand, concentrated mainly on its economic side and in part tried to exclude from the debate the voices of opponents from outside the local area as irrelevant, more or less in line with the CEA itself. In this regard, they had had to give in to those demanding a public inquiry. With the inquiry, and the way the compromises and assurances arranged beforehand were presented there, the debate was, to a large extent, settled. The power station seemed to become a widely accepted aspect of the area for more than two decades – until the 1970s. Then, critique rose once more, especially in view of a seeming onrush of negative news and new nuclear power projects, including storage sites for nuclear waste and, most prominently, an American-style PWR. It was a development from vocal opposition, via compromise-based acquiescence, back to vocal opposition once a pro-nuclear government attempted to exploit this very acquiescence.[282] This was quite different from what happened when one of the first commercial nuclear power plants of Germany was erected in Biblis at the end of the 1960s.

Biblis – "Education" for the Population

The Rheinisch-Westfälisches Elektrizitätswerk AG (RWE) had been sceptical of the economic viability of nuclear energy when Britain started its "Programme of Nuclear Power", only to establish light water reactors as the standard for the German industry with the experimental nuclear power plant Kahl a few years later.[283] The RWE was pushed by Heinrich Mandel, an engineer, physicist, and long-time champion of nuclear power who became a member of the company's executive board in 1967.[284] Aiming to secure its leading position in the German energy sector, the RWE was looking for a suitable site for a commercial nuclear power station with a capacity of about one gigawatt of electricity, somewhere between Frankfurt (Main) and Mannheim, by the end of the 1960s.[285] At first, there were rumours about a cooperation with the Badische Anilin- & Soda-Fabrik Aktiengesellschaft (BASF) and, potentially, the Großkraftwerk Mannheim AG (GKM). According to them, a power plant was to be built on Mannheim's Friesenheimer Insel, an island in the river Rhine just opposite the BASF works in Ludwigshafen, to simultaneously produce electricity and steam for the BASF.[286] The RWE, however, aimed solely at electricity for the wider Rhein-Main-Neckar region and an integration into the European super-grid.[287] In spring 1968, it therefore applied for a licence to build the world's largest nuclear power plant at the time in Biblis, a community of a few thousand people in the Hessian Rhine Valley, just opposite the outskirts of Worms.[288] Around 20 possible sites had been considered for some months, during which Biblis' mayor Josef Seib from the conservative Christian Democratic Union of Germany (CDU) actively tried to convince the company to come to the small town in the Rhine Valley's reed landscape, the so-called Ried.[289] At the time, public opposition was troubling the construction of a PreussenElektra nuclear power plant in North-Rhine-Westphalia's Würgassen, marking the beginning of the German controversy about atomic energy to Joachim Radkau.[290] To avoid a similar situation for Biblis, the RWE's PR department under Josef Buderath came up with an active communications strategy. It advertised nuclear power in a 25-kilometre radius around Biblis with the help of high gloss brochures and, on 21 June 1968, sent a remarkable open letter to the citizens of Biblis and the surrounding region. It was published in various newspapers all over the south-west of Germany.[291]

"We have to talk", the RWE opened. Then, after portraying its own dimensions and emphasising its responsibility for an area with nearly nine million inhabitants, the company stated its mission and pointed out its national importance in reliably delivering sufficient and cheap electricity to its customers, now and in the future. To be able to do so, the RWE was "obligated to use technological progress wherever possible", which meant "using nuclear energy, in the interest of all of you". After running two smaller nuclear power plants "to its full satisfaction", the RWE explained that it now wanted to build a "super-large nuclear power station". A suitable site for such a plant lay near Biblis. The exact timing, and even the final decision to build, were being withheld for now, but the application to Hesse's Ministry for Economic Affairs and Transport had been made,[292] and the responsible local authorities informed. Subsequently, the company assured the letter's readers that "[t]oday, a nuclear power plant runs as safely as every other kind of power plant. For one decade, we have collected our own experiences and utilised those from other countries . . . All precautions and security measures have already been calculated." Then, the RWE urged: "Please consider the opportunities arising out of this power station, for your community and its population, today as well as tomorrow, think about it. We know for sure that you have many questions. And it is your right to ask them. Rest assured that we will answer. Write to us or send your representatives. We are willing to always talk to you."[293]

With this letter, the RWE – successfully, as it seemed in hindsight,[294] – tried to pre-shape the expected public debate about its plans. It described nuclear power plants as safe and necessary installations, which were not only progressive but also an economic boon for the community hosting them, thereby denying the existence of any potential downsides of the project. At the same time, it painted itself as an obliging procurator of the public's interests – and as the responsible and experienced expert who would enlighten all those who did not know enough about nuclear energy to recognise its hazard-free benefits. Indeed, from here on until the official hearing in January 1970, which was an integral part of the German licensing process, the RWE's strategy seemed to work. There was next to no public debate about their project, especially regarding potential problems, and what little debate there was remained dominated by officials and proponents of the scheme, who focused mainly on the planned power station's economic benefits.[295]

The Absence of Debate

The ministry responsible for licensing the proposed power plant, the Hessian Ministry for Economic Affairs and Transport, with which the RWE had applied for the reactor in 1968,[296] publicly welcomed the project. Minister Rudi Arndt from the Social Democratic Party of Germany (SPD), when asked about the power plant in a live programme from the "Hessentag", the annual Hessian state festival, pointed out that it was "cheaper" to generate electricity from nuclear power than from conventional energy sources, while at the same time there were "various systems, guaranteeing the population's absolute safety".[297] Dr Carl Otto Lenz, the CDU's Member of Parliament for Bergstraße, the district to which Biblis belonged, concurred, and called for the technology's further development while emphasising its safety.[298] A known supporter of nuclear energy, he was a member of the Deutsches Atomforum e. V. (DAtF), the country's most prominent pro-nuclear lobbying group. It had been founded as an umbrella organisation containing four pro-nuclear lobbying groups in 1959. Two years later, its member groups dissolved to completely merge into the DAtF, which then also opened for new members.[299] It was well connected in political, industrial, and scientific circles and at the time chaired by Karl Winnacker, chairman of the Hoechst AG and formerly of the IG Farben, who was replaced by the RWE's Mandel in 1973.[300]

Within the chorus of voices endorsing the RWE's plans, the Bergstraße's district administrator and SPD member, Dr Ekkehard Lommel, turned out to be the most critical. He had four conditions for his endorsement of the project: The RWE had to uphold trustful relationships with the authorities in Biblis, the surrounding communities, and the district, "the citizens' rights and security" had to be "guaranteed", the "up to now unspoilt landscape" had to be treated with care, and Biblis and its neighbourhood had to profit economically. In practice, of course, these conditions rather resembled arguments for the project, given that Lommel expected them to be fulfilled and the population to accept the RWE's proposal. After all, as he put it, "[w]e have to live with modern technology's accomplishments".[301] On the other, unconditionally positive end of the spectrum of publicly stated positions stood Biblis' mayor. Seib, trusting in the experts from the RWE and the authorities to guarantee the project's safety, focused completely on its economic prospects. He looked forward to about 1,000 jobs that would come to his municipality during the phase of construction and hoped that Biblis would "profit taxwise", as it was put by Friedel Fabrie, a member of the editorial board of the *Wormser Zeitung*, which was based in Biblis' Rhineland-Palatinatian neighbour.[302] At a press conference on the occasion of Biblis' annual "Gurkenfest", a village festival centring upon the gherkins that had been Biblis' economic foundation for half a century, Seib gave an overview of the local authority's plans for the following year. In this context, he emphasised: "The proudest project though is not a project executed by the municipality, although it will be realised within its borders. It is the RWE's nuclear power plant . . . A plant of similar capacity does not exist anywhere in the world, neither in America nor in Russia."[303] The whole community, he pointed out, would benefit. Landowners could expect good prices and would be offered high-quality municipal land as compensation for what they sold to the RWE.[304] He added that, so far, only one person had criticised the plans,[305] admitting later that he had heard from five sceptics.[306]

Clearly, the picture Seib painted of a community welcoming nuclear power in near unison was one-sided. It was not structurally flawed, however. Together with the mayor, Biblis' administrators were convinced that the population regarded the RWE's plans as "entirely positive".[307] The chairmen of the party groups in its municipal council shared Seib's hopes for a significant improvement of the economic situation,[308] which was no small matter. During the first half of the 20th century, cucumber farming and canning had dominated Biblis' economy. The industry, however, had collapsed when cheaper produce from southern and eastern Europe entered a globalising market.[309] During the 1960s, the community had been stagnating, with few industrial enterprises within its borders and most of its inhabitants commuting to work in other towns and villages.[310] In September 1968, however, atomic energy already seemed to be bringing benefits for Biblis. Advertisements for flats construed as investment properties emphasised that they were situated in "Biblis, near the nuclear power station".[311] Biblis' council also had a positive perception of the power plant, welcoming it unanimously.[312] In September 1968, it resolved to sell the municipal land necessary for the project.[313] A few weeks later, about 80 private landowners followed suit after the RWE had agreed to pay 4 German marks per square metre for land that was, on average, estimated to be worth less than half that. In addition, as Seib had promised, an option to buy better land out of the municipality's possessions at favourable conditions was offered to those selling to the company.[314] By early 1969, the RWE had acquired the site it wanted.[315]

The positive reception of the project was not restricted to those directly involved economically, however. Biblis' Catholic priest Erich Löw soon registered rising interest in education among his "fearless" parishioners.[316] During the campaigns for the communal elections in October 1968, Biblis' political parties even fought about how much Seib had – or had not – contributed to securing the RWE's interest in the community. The mayor's opponents claimed that he had overstated his involvement, whereupon his party attempted to prove the mayor's active role by quoting a letter from Mandel to him on a leaflet.[317] In the end, Seib's CDU won the majority in Biblis, while the SPD was the overall winner in Hesse at large, a result convincing the mayor that his engagement for the nuclear power station had been crucial in his party's victory.[318]

More indirectly, the reception of the RWE's attempts to "educate" the local population about nuclear energy, together with the DAtF, also demonstrated the dominance of positive attitudes regarding the reactor project.[319] In October, a series of lectures was given in the region, propagating nuclear energy as safe and economic.[320] In a Biblis gymnasium, an exhibition was staged, accompanied by three days of "educational lectures".[321] From the ranks of the DAtF, a chemist and a physicist argued for nuclear energy and the RWE's project. They were supported by Seib and the Ministry for

Economic Affairs and Transport's permanent secretary, chartered engineer Heinz Wagner. Wagner, a renowned expert in reactor security and related environmental questions within the ministry, was not only head of the department responsible for the oversight over nuclear power stations, but also a member of the RWE's administrative advisory board.[322] As such, he advocated actively educating the population about nuclear power.[323] At the end of the well-attended event,[324] Hans-Joachim Noack observed for the *Süddeutsche Zeitung*, that "everybody was convinced that the parish stood at the threshold of an incommensurable boom".[325]

In view of these developments, newspapers describing the upcoming transformation of the "gherkin village" into an "atom city"[326] with the help of a "daring jump into a future full of golden opportunities"[327] do not seem to have been exaggerating all that much.[328] Even Noack's claim that Biblis was envied by neighbouring communities did not seem implausible in view of Gundremmingen, where, allegedly, tax revenues had tripled after a nuclear power plant had been built.[329] Other communities bordering on Biblis and belonging to the same district hoped that they, too, would profit. As the *Bürstädter Zeitung* put it: Once the nuclear power plant is operational, "our district's future will have begun".[330]

By and large, the positive perception of the RWE's project dominated the public debate in Biblis and the region. Especially once the first works were underway, fascination for the scientists' and engineers' achievements and the project's dimensions – in short, the technological sublime – came to the fore (cf. "Emotional Atoms"). As early as July 1968, the *Bürstädter Zeitung* proudly claimed to have been the first newspaper in the region to explain how a nuclear reactor worked and was calling the project "gigantic".[331] One year later, after the RWE had placed the order for the power plant with the Kraftwerk Union AG and Hochtief,[332] the *Mannheimer Morgen* wrote about the "super-project . . . for a better future, not only for the gherkin-metropolis of Biblis".[333] Later on, it delved into the project's technical data and pointed out the RWE's utopian vision of the "totally electrified household",[334] including the widespread use of electric cars and air conditioning in flats,[335] while the *Frankfurter Neue Presse* envisioned "a new landmark of the Bergstraße and the Rhine valley".[336]

Even most of the few critical voices that came to the fore were, in the end, not sceptical about nuclear power as such – on the contrary. The foremost example is Mannheim's mayor, physicist Dr Ludwig Ratzel, who had been concerned with atomic energy as a Member of the Federal Parliament one decade earlier.[337] In June 1968, he had pointed out that the Biblis project could not be realised soon, as studies and expert reports would first be needed. The reason for his observation, however, was that he worried about the project's implications for the chance of Mannheim's GKM getting their own nuclear power plant.[338] In the same vein, he criticised the Badenwerk's plans to build a nuclear power plant upriver from Mannheim on the banks of the Rhine that would heat up the river and thereby diminish its cooling capacity in Mannheim. Herein he was seconded by the city's newspaper, the *Mannheimer Morgen*. To deal with this problem, coordination beyond the federal states' borders was called for.[339] It should be noted that, at the time, a similar argument was also present on the national level, between Germany, France, and Switzerland.[340] It was, in other words, a rather prominent issue. As such, it was taken up by the *Darmstädter Echo*, the only newspaper in the region criticising nuclear power in principle now. This should not imply that the newspaper would have been all too vocal in its critique – or aiming especially at Biblis. In August 1968, it merely pointed out that, transnationally, nuclear power had so far been neither reliable nor able to produce electricity cheaply.[341] One year later, Karl-Heinz Baum, later well known for his work as a West Germany correspondent in the German Democratic Republic, observed that there were plans for five nuclear power plants within 40 kilometres of the Rhine's course. This, he argued, reflected an interest of the states in question to erect monuments for themselves instead of working together for a better electricity supply.[342]

Rather unsurprisingly, the issue of lacking coordination between the German federal states concerning cooling water warming the Rhine, possibly endangering fish or, via influenced currents, even shipping, was also brought forward by the river authorities commenting on the Biblis project.[343] The RWE, however, was of the opinion that there would be no trouble in this regard.[344] In fact, when its application files were publicly presented in Biblis during the official display period at the end of 1969, all potential problems were presented as having been solved. Low-activity waste would, for a time, be stored on-site before being shipped to a final storage facility. Used water would be cleaned, checked, and returned to the Rhine when its radioactivity was within the approved limits,[345] raising the river's radioactivity "only slightly". The steel containment would be able to withstand even the greatest conceivable accident.[346] In effect, the security measures would protect the surrounding population from any damage during normal operations and in case of accidents. Their appropriateness would be guaranteed by regular inspections.[347]

On the eve of the official hearing, then, all seemed well for the RWE's Biblis nuclear power plant. Problems that had been pointed out by a few journalists, by Mannheim's mayor or from within specialised authorities were considered to have been solved. Other critical voices, namely those of private individuals, had not been raised publicly,[348] and what little debate had taken place had been dominated by the industry, lobby groups, and authorities backing the expansion of nuclear power in general and the Biblis power plant in particular. An economic upturn for Biblis and its regional surroundings was expected, and trust in experts' safety guarantees was limitless. The same holds true for the official hearing, which was an integral part of the German licensing process for nuclear power plants. Here, however, for the first time in Biblis' nuclear history, critical opinions received some publicity, as they garnered the attention of the media. Although they were noticed, however, critics were not taken all too seriously by the journalists – and treated in a downright patronising way by the responsible officials.

The Official Hearing

In the middle of January 1970, the official hearing for the RWE's application was held in a gymnasium in Biblis,[349] in front of an audience of 150, which included mayor Seib and representatives of the municipality. It was extensively covered in regional newspapers and reported by the Süddeutscher Rundfunk, the regional broadcaster, besides being featured on the Hessenschau. The hearing was presided over by permanent secretary Wagner, who was joined by representatives of the Hessian Ministries for Economic Affairs, for Social Affairs, of the Interior, and for Agriculture, as well as experts from the Free University Berlin and the Technischer Überwachungsverein (TÜV, Technical Inspection Association).[350] Out of seven objections originally raised, only five were dealt with, as two had not been submitted in time,[351] hinting at the importance of formal criteria in Germany. The three objectors who took part in the hearing were the Rhein-Selz water supply board, the Schutzgemeinschaft Deutscher Wald e. V. (SDW), an association founded in 1947 and originally dedicated to reforestation projects before broadening its mission to the protection of and lobbyism for woodlands in Germany,[352] and Marga Müller, a housewife from Biblis.[353]

In the beginning, Günther H. Scheuten, one of the RWE's directors, once more explained the necessity of the planned reactor by pointing to the demand for electricity, which was doubling every decade. The Biblis site, he argued, was suitable because cooling water from the Rhine was available, there was an access point to the RWE's integrated grid nearby, and there was existing traffic infrastructure in place. The public, he added, had been informed early on and most of the objections raised concerned the federal parliament as a legislative body, not the officials implementing the existing law. He emphasised that the power plant would operate safely, basing his assessment not the least on experiences with the RWE's reactors in Kahl and Gundremmingen, and that it would not produce "dust, noise etc."[354] A representative for Siemens went further into the details of the safety systems, pointing out that they "virtually excluded all dangers for staff and population", while thorough checks by independent experts would rule out mistakes during the construction.[355] He emphasised that all safety systems had two backups, so as to exclude "every risk, human or technological". With a view to normal operations, he stated that the reactor would emit "a fraction" of the natural level of radioactivity.[356] This statement was backed up by physicist Karl Aurand from Berlin, who was to become a member of the Commission on Radiological Protection in 1974.[357] The point was also taken up by Fred Meyer, the director of the Kahl nuclear power plant, who would soon head the one in Biblis. Referring to his experiences, he saw no hazards for the population.[358] Also, the TÜV had no objections.[359]

The objections that had been raised during the official display period were then commented upon by the representatives from the authorities, the TÜV and the RWE.[360] The water supply board, which had announced recourse claims in case their wells were contaminated by radioactivity,[361] did not object in principle, and they were assuaged by the Ministry for Agriculture's representative.[362] The SDW had presented a wider array of worries. The Rhine valley's landscape, they argued, was endangered by the project, as were the fish, due to the discharge of used cooling water. The latter could also lead to intensified fogging, and radioactivity could pose problems analogous to dichlorodiphenyltrichloroethane (DDT), which had come to fame as an environmental problem thanks to Rachel Carson's book "Silent Spring".[363] Not least, it could lead to an increase in cases of cancer.[364] Müller's objections were even more wide-ranging. She "feared a negative impact" of the station's emissions "on the Ried population's health",[365] saw their lives endangered,[366] was worried about potential catastrophes, saw nuclear power plants as uneconomic and dependent on subsidies,[367] which should be cancelled,[368] and complained that prices for electricity would not go down.[369] With that, Müller had taken up the position at least partially held by the German supply industry for most of the years since the Second World War.[370] In her eyes, the public had been insufficiently informed, and she wanted to know whether she would receive subsidies if she decided to move because of the power station's hazards.[371]

The objectors received support from singular voices from the audience, particularly Dr Wilhelm Braunewell from the Weltbund zum Schutze des Lebens (cf. "Nuclear Futures"). It was pointed out, for example, that American studies showed negative health effects in the areas surrounding nuclear power stations. The experts and representatives of the various authorities nonchalantly rebuffed them. Aspects like rising rates of cancer, negative effects on the Rhine and the prohibitively high costs of nuclear energy were dismissed quite matter-of-factly, as was Müller's request for a subsidy to leave the region.[372] The woman from Biblis, however, who was the only prominent female voice in the early years of public debates about the local power plant, received especially patronising treatment. Wagner lectured her, pointing out that most of her objections had been "cleared up by the previous talks and explanations". He also reproached her for, in the words of the *Südhessische Post*, "not having visited the Atomforum's events, whereby she had neglected to inform herself".[373] Her rejection of the station was not, in other words, accepted as a different opinion on the matter, but framed as a wrong conclusion based upon an "information deficit".[374]

To the media covering the event, the hearing was "very factual",[375] with the RWE's application passing "with flying colours",[376] as it had demonstrated that everything would be done to exclude any "dangers for humans and animals".[377] On the other hand, "the power plant's opponents, besides their expert knowledge, ran out of air after some hours of discussion".[378] This impression was shared by Wagner, who opined that there were no further reservations in the ministry against the RWE's plans after the hearing.[379] His position was shared by the municipal and district authorities.[380]

After the hearing, the public debate seemed to be over. Quite in accordance with the RWE's publicity strategy, which was supported by the state government and local and regional authorities, it had never been broad or critical. Regarding the problems experienced by the energy industry in Würgassen, the impression springs to mind that the RWE, Seib and the government could have believed that they had overcome a potentially critical phase. In fact, however, this was not the case. Indeed, there had been few objections, and as a part of the hearing, the assembled experts from the RWE and the authorities had formally dismissed them, opening the way for the construction and later operation of the Biblis nuclear power plant. There loomed, however, two problems with the procedure as it was: Some of the objectors were dissatisfied not only with the hearing's outcome but also with the way their objections had been discarded flat out, so that they kept up their critique afterwards, and even escalated their activities.[381] The second problem was even more fundamental. The fact that a broad and critical public debate about the nuclear power plant had successfully been avoided meant that many people in the region had not really considered such a station's implications. In a way, they had been kept on the sidelines, forming a fertile ground for a later debate.[382] This did indeed start once the power plant rose ever more visibly on the bank of the Rhine – and, since it only began once construction works on the plant and even a second one had been taken up, the debate quickly resulted in quite fundamental positions being taken by the project's critics. They did not wish to be caught off guard again.

The Start of the Debate

Of the three objectors present at the Biblis hearing, most prominently, the SDW was unwilling to accept its outcome. They resolved to continue searching for an authority willing to see the validity of their arguments and able to overturn the unwelcome result. To that end, they wrote to the Hessian Minister for Economic Affairs and Technology, Heinz Herbert Karry, pointing out that heating up the Rhine could endanger the river's ecosystem and, via the food chain, human health. They suggested instructing the RWE to construct cooling towers to alleviate the problem. When Karry dismissed their arguments in August 1970, the group petitioned the Hessian State Parliament to intervene, emphasising that the power plant would undermine the goal of protecting air and water – a goal that the representatives had written into the current regional planning programme.[383] In fact, the call for cooling towers became part of the conditions imposed by the Ministry for Economic Affairs when the RWE, at the beginning of 1971, planned the construction of a second power plant in Biblis,[384] the application for which was granted after another, even less publicly active debate, with the hearing taking place in November 1971.[385] The Hessian

Minister for Agriculture and the Environment, Dr Werner Best, even retroactively imposed the erection of cooling towers for the first power station.[386] This decision, to be clear, was not brought about solely by the SDW's continued objections against the first Biblis power plant. It must be seen against the background of the broader debates concerning the Rhine's temperature in the region around Biblis, where, in 1970, up to five nuclear power plants were under discussion, and even in an international context. With the number of pondered projects shrinking in the second half of the year, however, and in view of efforts in (inter-)national coordination and measures being taken such as the construction of cooling towers, the issue receded.[387]

Still, this did not put an end to the fledgling public debate that had sprung up after the first hearing. On the contrary, the cooling towers were the next topic to be scrutinised by a public that became more active as the construction works in Biblis moved visibly forward. In an analogy to what was discussed in south-west Germany around Wyhl,[388] the SDW soon pointed out that too little was known about the consequences of large amounts of vapour on the local climate.[389] In addition, the CDU party group of the city of Lorsch, some five kilometres from Biblis and a part of the Bergstraße District, worried about ecological dangers for the Ried. Among those dangers were the Biblis nuclear power plants with their cooling towers, which seemed to imply changes in the area's climate and the possibility of various environmental hazards reinforcing one another.[390]

These other environmental hazards clearly included radioactivity, the effects of which had entered the public debate around Biblis soon after the first hearing had taken place. The main reason for this was the BASF's plan to build its own nuclear power plant in Ludwigshafen. The company had applied for a licence from the Ministry for Economic and Social Affairs of Rhineland-Palatinate in May 1969. After initially being broadly welcomed as a bold step forward for the "Peaceful Atom", the question of reactor safety moved to the fore in view of the densely populated surroundings and the proximity of the BASF's chemical installations. The Reaktorsicherheitskommission (RSK, Advisory Commission on Reactor Safety) took up the matter in the autumn of that year.[391] This body of governmentally appointed experts was initially assembled in 1958. It was rather closely linked with the Deutsche Atomkommission (cf. "Nuclear Democracy") and designed to advise the German government on matters of nuclear safety – often in secret.[392] Its position regarding the BASF reactor underwent a rather rapid change against the background of changing conceptions of nuclear accidents and their consequences. In November 1969, most of its members had opined that the BASF reactor could be built with safety installations similar to Biblis' design, while a minority argued for additional bursting protection to safeguard against a failure of the pressure vessel. In the following months, in view of requirements by the licensing authorities, the RSK changed its position. In January 1970, it demanded additional safety measures, including this bursting protection. Once the BASF had taken these up, in May 1970, the RSK accepted the company's plans.[393]

Meanwhile, unlike the first and second reactor in Biblis, where, as Karlheinz Eirich from the *Mannheimer Morgen* put it at the end of 1970, the RWE had been able to "prove" the plant's radiological safety,[394] the BASF's project was also debated more broadly and ambivalently in public. Unlike Ludwigshafen City Council, which accepted the plans in early 1970,[395] Mannheim objected. Once more, presumably, one important reason was the interest in the GKM's own nuclear power plant and the trouble regarding the Rhine's cooling capacity. Publicly, the city demanded better coordination between projects, expert reports on the power plant's effects on air and water, and additional security measures along the lines of the demands discussed in the RSK, including improved containment or an underground construction.[396] Other objectors with less interest in their own reactors – there were 49 overall – raised more fundamental questions regarding nuclear power's dangers. They pointed to the consequences of radioactive emissions for living organisms, especially regarding bioaccumulation. They also referred to a growing debate about atomic energy's radioactive dangers in the USA, highlighting rules regarding minimal distances between reactors and residential areas applied there. These, if directly transferred, would prohibit building nuclear power plants anywhere in Germany. Furthermore, the TÜV's lack of competence in the fields of biology, agriculture, or water management was deplored, and statistics about accidents in nuclear power plants and the unresolved question of nuclear waste were emphasised.[397]

In August, the BASF's project was put on hold for two years by Federal Minister for Education and Research Hans Leussink, after the American policy of avoiding the construction of nuclear reactors near larger cities had been highlighted by the RWE's Mandel in a bid, according to Radkau, to avoid competition for Biblis.[398] Publicly, Leussink explained that nuclear power plants could never achieve "absolutely safety" and that, because of this small but existent "residual risk" that radioactivity could escape, it seemed prudent to keep a distance to larger cities. At least, he argued, there should be more research into the matter.[399] The BASF reacted stating its readiness to undertake any research the minister deemed necessary, besides emphasising the studies already conducted and the measures planned, including a doubling of the containment's thickness and the installation of bursting protection, even though, as it pointed out, this had been deemed unnecessary by a number of experts.[400] The company was not the only party irritated by Leussink's scruples. Among others, Dr Hanns Neubauer, Rhineland-Planatinate's Minister for Economic Affairs and Transport, publicly criticised the decision given the facts that the population density around the planned BASF reactor was equal to that around the Obrigheim power plant, which had been approved; that it was to be of the same type as the one under construction in Biblis; that it was to be equipped with better security systems than any other German power plant in existence; and that it had been accepted by the RSK.[401] Nevertheless,

Leussink remained unmoved, thereby marking a turning point in the German national debate about nuclear power. From now on, it was harder for the atom's proponents to argue for the acceptability of nuclear power plants anywhere when the technology, at a safety level above the norm, was not acceptable at Ludwigshafen.[402]

As regards Biblis, the BASF debate contributed to critical perspectives taking hold there, too. During the months following Leussink's decision, a few readers of the *Mannheimer Morgen* started to controversially debate nuclear power in principle. They discussed the officially presented arguments or even went beyond these. For example, they actively linked the issue of nuclear energy to questions about the role played by experts in the licensing process, about the local value of international statistics, and about the deplored tendency of nuclear power's proponents to try to muzzle its critics.[403] Soon, these voices were backed up by groups like the Weltbund zum Schutze des Lebens, which added its – rather typically German – emphasis on potential genetic damage, cancers, and additional deaths to the building critique of atomic energy in general.[404]

The doubts regarding the safety of nuclear power soon began to be connected to the Biblis project. As early as September 1971, Erwin Bugert, chairman of Bergstraße's SPD, observed a growing number of questions about its safety being raised throughout southern Hesse.[405] In Darmstadt, about 30 kilometres from Biblis, the Interessengemeinschaft Umweltschutz Südhessen (IUS), headed by Klaus-Dieter Henninger,[406] organised an exhibition on the dangers of atomic energy in July 1972. It referred to the works of Ernest J. Sternglass who had worked on the effects of radioactivity on children during the 1960s, and had just published a book on low level radiation.[407] The IUS demanded a stop to Biblis' construction,[408] backed by about 1,000 signatures.[409] Even in Biblis itself, some people began to wonder why a reactor near their community of 9,000 people was deemed safe, while one next to a chemical plant was not. In general, however, the local population remained unafraid, as Biblis' priest, Löw, emphasised, and retained its trust in pro-nuclear experts and authorities.[410]

The situation was different in the region around Biblis. Half a year after the Darmstadt exhibition, on 18 January 1973, all the aspects debated during the previous years – the BASF power plant plans, nuclear power in principle and, with a view to the SDW, Biblis' own reactors – were brought together and explicitly connected to the Biblis nuclear power plants at a meeting of environmentalists in Hamm am Rhein, just about two kilometres away from the construction site, on the other side of the river.[411] The meeting was organised by Wolfgang Staab and the Aktionsgemeinschaft für Umweltschutz Rheinhessen-Süd e. V. (AGURS), originally founded to oppose a refinery at Worms' Rheindürkheim, on the banks of the Rhine more or less opposite Biblis,[412] as well as the District Farmers' Association Worms-Alzey.[413] The meeting was also backed by the SDW.[414] It took the form of a panel discussion that included three environmentalists, three scientists, one winegrower, and two

RWE representatives among others. It was chaired by Hans-Helmut Wüstenhagen, a chemist who, from 1973 to 1977, was head of the Bundesverband Bürgerinitiativen Umweltschutz before leaving in the wake of internal conflicts over his critical position regarding communist groups. This umbrella organisation for citizens' initiatives dedicated to environmental protection had been founded in 1972 by the German Friends of the Earth and the Weltbund zum Schutze des Lebens, with some initial support from the Federal Minister of the Interior, Hans-Dietrich Genscher, and his Free Democratic Party. Although not primarily anti-nuclear in the beginning, Wüstenhagen shifted its focus in that direction and soon it was perceived as a primary anti-nuclear voice in Germany. It came to include between three and four thousand groups, including initiatives from communities close to nuclear power plant sites.[415] Admittedly, it was not unrestrainedly appealing to the latter. In the case of Biblis, only five out of 35 groups in existence at the end of the decade were to be part of it.[416] One explanation for this fact tends to be that Wüstenhagen was rather critically received by parts of the anti-nuclear movement later on.[417] For now, however, he was not, and the meeting drew about 450 people.[418] During the event, meteorologist Dr Hans von Rudloff and Dr Reinhard Muth from Rhine-Hesse's Winegrowers' Union concentrated upon the potential effects of cooling towers, with von Rudloff focusing the rising likelihood of intense fog, rain, thunderstorms, and black ice, while Muth pointed out the potential degradation of wine grown in the area as a consequence.[419] In addition, the cooling water's thermal effect on the Rhine continued to be scrutinised, as did the potential of sabotage or large-scale accidents.[420] As a trigger for the latter, plane crashes were highlighted because of a nearby low flight path used by the North Atlantic Treaty Organization (NATO).[421] Besides these scenarios of large-scale releases of radioactivity, the consequences of low-level emissions over a longer period of time were discussed.[422] Regarding the latter, a Dr Trunko[423] pointed to the lack of knowledge about their effect on rates of leukaemia or genetic defects, especially seen long term.[424] Meanwhile a physics professor, Klumb,[425] referred to the transnational level of debate when he considered it odd that the construction of nuclear power plants seemed to be gathering pace in Germany, while reactor security was becoming an ever more important topic in the USA.[426] Dr Hermann Schmitt-Vockenhausen, Member of Parliament for Groß-Gerau, to the north of Biblis, who, according to the *Wormser Zeitung*, spoke for the SDW here, added that radioactive emissions would inevitably damage air and water, and that existing limits were too lax.[427]

Beyond these explicitly discussed issues, the conditions of debate were taken up as a subliminal topic. This is particularly visible in the tendency to delegate questions to experts whose answers were then to be trusted, and in the exclusion of large parts of the population in the power plant's vicinity. Regarding the first aspect, the *Wormser Zeitung* reported that von Rudloff received intense applause for pointing out that it would have been better to undertake meteorological studies before starting construction on Biblis, not afterwards. "He could well understand that, in view of this kind of procedure, the population considers itself insufficiently informed and at the mercy of an almighty industry."[428] This was deemed especially true for those living (just) outside Biblis' borders. A farmer from Worms-Ibersheim, on the Rhine's banks opposite the power plant in Rhineland-Palatinate, complained that it had been sited at Biblis' municipal border, closer to Ibersheim and Hamm than to Biblis itself.[429] Reacting to this, Fabrie from the *Wormser Zeitung* opined: "A public debate, especially on the left bank of the Rhine, where people are closer to Biblis' 'breeder' than the gherkin-metropolis' citizens themselves, is indispensable".[430]

But that was not what the RWE had in mind. Replying to the critics, Dr Kuhnt, one of its employees,[431] simply pointed to people's experiences so far and studies currently underway, asking "the people present to trust the security measures". In addition, the RWE's representatives, who, in the words of the *Wormser Zeitung*, were partially "put on the spot by pointed questions from a student of physics from Heidelberg", "questioned the scientific merits of their opponents". The power plants' opponents, however, were not satisfied, especially when Seib called them "fanatical like Jehovah's witnesses",[432] drawing criticism from Wüstenhagen and Staab.[433]

In the end, the meeting in Hamm, which can be considered the first public debate about the Biblis nuclear power plants not controlled by its proponents, could not make up for the lost opportunity to openly discuss the reactors and the different opinions about them before construction began. It was not at all suited to settling the conflict, nor to hedging it in once more. On the contrary, participants regretted not having taken note of the power plant before it had come into existence. All that seemed possible now was to demand improved security measures and to fight further expansions.[434] With options limited in this way, the event in Hamm turned into the starting point of a regional public debate, which quickly became wider, broader, and more critical because there could be "no rapprochement reached between the two opposite positions",[435] no compromise negotiated after the fact. The way the debate had been conducted up to this point had precluded widespread acceptance. Now began, as Fabrie put it, the time for watchful citizens' initiatives.[436]

Indeed, these groups started to become very active. Long before the oft-mentioned events in Wyhl, they initiated a public debate about the Biblis power plants that grew over the following years, in terms of both its topics and its participants. It could not be reined in again because of the way it had been initiated: in a moment of shock, when citizens of a democracy realised that they had not been able to bring their opinions to bear in a question dear to them – one that had already been settled when they woke up to it. They would not be caught napping again. First, the AGURS used 247 signatures, collected in Hamm,[437] to write to the Hessian Prime Minister Albert Osswald and the Hessian and Rhineland-Palatinatian Ministers responsible for the environment, Werner Best and Otto Meyer. They demanded a stop to Biblis' construction until "independent studies" "proved 'without a doubt that the nuclear power plant's operation was completely harmless' for humans, animals and vegetation, especially agriculture and winegrowing",[438] while continuing to highlight the effects "of the waste heat for the environment".[439]

As expected, the demand was dismissed. Instead of giving in, the RWE and Hesse's Minister for Economic Affairs, Karry, once more tried to reassure the activists by claiming that, despite the critics not having participated in the initial debate, it had still resulted in a compromise solution because the authorities had acted on the people's behalf. Explicitly, they pointed out that environmental aspects had played an important role during the licensing process. The cooling towers, for example, would only be used for a few weeks per year without negatively affecting the climate or grapes, as a study currently underway would show.[440] The argument puzzled Fabrie: How could the cooling towers' harmlessness be accepted before the study into their meteorological impact had been finished? Behaviour like this, to him, was counterproductive to building "trust in the licensing authority". The same held true for the fact that Best argued that the RWE's Biblis power plants were environmentally friendly, while opposing – for environmental reasons, including heating up the river Main, diminishing air quality, or emitting dangerous radiation – one planned by the RWE in Großwelzheim in Bavaria, just opposite the Hessian town of Seligenstadt.[441] Once more, in other words, the people in the Biblis region wondered why a member of the government would think the reactor next to them acceptable, but had doubts about another one, when the only visible difference was geographical.

Obviously, the now attentive citizens would not let the RWE and the authorities assume the role of patriarchs with unquestionable supremacy once more, or accept being pushed aside as uninformed or living out an environmental "mania".[442] Quite like the SDW after the first hearing in Biblis, the AGURS were not prepared to trust the official statements and keep quiet. Instead, they wrote another letter, this time addressed to Osswald and the President of the State Parliament, Georg Buch.[443] It demanded a vetting of all decisions taken by Karry and his ministry regarding the RWE because permanent secretary Wagner was a member of the company's administrative advisory board.[444] To the AGURS, this was "a blatant case of partiality" and "a scandal". They expected members of the administration to "decide in complete freedom and independence, like a judge, between collective goods and private economic interests".[445] As was to be expected, their demands were not met by the state authorities.[446] The group then wrote to Genscher, who had overseen the birth of environmental politics on the federal level.[447] It was a clear demonstration of the escalatory dynamic that resulted from the way the debate had been started after the decisions had been made – and it was not the only one.

Besides calling for ever higher authorities to intervene, groups such as the IUS and the AGURS, as well as individual citizens, ramped up their regional activities and started to coordinate themselves.[448] People began to make their views known publicly, writing to the *Wormser Zeitung*, recapitulating and extending the argumentation against the power plants. They pointed out, for example, that, by law, emergency plans had to cover a radius of 10 kilometres around the Biblis reactors, including places like Ibersheim, Rheindürkheim, Hamm, Eich, and Nordheim and Wattenheim, which had recently merged with Biblis.[449] This was seen as proof that the consequences of the power plant were not limited to Biblis. Based on this, people demanded a halt to the construction until all questions from the entire region had been answered to their satisfaction: "We, the ones directly affected, have a right to full and neutral information."[450] Again, the initial absence of a wide and open debate and a compromise built upon it was visible behind the demand. As a participant in a later debate in Rheindürkheim put it: "We can only bark after the fact".[451]

The critique even started to resonate in Biblis itself. Here, the SPD in Wattenheim decided to have a closer look into the matter and, if need be, join the environmentalists.[452] A new citizens' initiative from within the municipality, the Aktionsgemeinschaft für Umweltschutz südliches Ried, originally founded to fight logging in a nearby forest,[453] also turned its attention to the power plants. It began collecting signatures demanding a halt to the construction until more safety measures had been taken.[454] While mayor Seib pointed out that the group was unsuccessful in Biblis proper,[455] it repeated and expanded the publicised arguments against the reactors. It asked, for example, why the Hessian fire insurance agency excluded payment for damages resulting from radioactivity, and concluded that this could only mean that there was a real risk for those to occur.[456]

While the critics formed a network around Biblis and even within the municipality itself, nuclear power's proponents attempted to rein in the debate once more and to re-establish the dominance of their perspective – which did not quite work out beyond Biblis proper. In May 1973, Wagner was invited to a meeting of the municipal council to explain the licensing process, the monitoring of the power plants, their safety and their emissions, and the workings of insurances in view of Germany's Atomic Energy Act, the 1959 "Gesetz über die friedliche Verwendung der Kernenergie und den Schutz gegen ihre Gefahren (Atomgesetz)".[457] Afterwards, council chairman Hans Bart thanked Wagner "for his contribution to correcting lots of concerns".[458] Outside Biblis, however, the *Wormser Zeitung* emphasised that he had not known about the NATO low flight path, drawing into doubt the credibility of his assurances that plane crashes had been taken into account during the containment's construction.[459] Just one month later, the CDU's youth organisation staged a panel discussion with representatives from the authorities, the media, and environmental groups. Here, topics like radioactive emissions were debated in detail between environmentalists like Wüstenhagen and Staab on the one hand and representatives from the Ministry for Economic Affairs and the power plant's director Meyer on the other, in the end overstraining many in the mostly nuclear-critical audience of about 150.[460] Meanwhile, the RWE initiated so-called power plant talks (Kraftwerksgespräche), a series of talks between representatives from the power plant and the company at large and from the municipality, aimed at answering the latter's questions.[461] They were to be repeated every few months and were seemingly able to satisfactorily answer the questions of those who were invited.[462] With a view to a wider audience, the company staged open days at the site of the two power plants in October, which drew about 15,000 people. In addition, an information centre was provided for visitors.[463]

Thus, while many people in Biblis itself, particularly those with direct access to the RWE's local representatives who had participated in or, at least, been primary "targets" of the initial debate – or rather, monologue – still supported the power plant, the belatedly awakened critical voices from the region could not be won over. Debates about the reactors continued to be driven by letters to the regional newspapers,[464] and the citizens' initiatives remained active, putting permanent pressure on the RWE and the authorities. Staab, for example, backed by the IUS, demanded that the meteorological study into the cooling towers be handed over to the environmentalists, not just made accessible to them in the Wiesbaden ministry.[465] When this demand was declined, he did go to Wiesbaden, only to start accusing the ministry of trying to keep a lid on the story, as he was not allowed to take photographs.[466] In the summer of 1974, when the first reactor in Biblis went critical on 16 July,[467] the citizens' initiatives changed their demands. Instead of a halt to construction, they now demanded postponing the start of commercial operations until more studies regarding the reactor's safety became available, new ones into the emergency cooling system were undertaken, 60 centimetres of concrete were added to the containment, and the emergency plans were published in full.[468] In 1975, shortly after the national debate about nuclear power heated up due to the Wyhl protests,[469] the regional one around Biblis escalated, too.

The trigger was an announcement by the RWE in April, during the opening ceremony for the first power plant – which Seib declared a "festive day in Biblis' life and history"[470] – that they intended to construct two further reactors.[471] In fact, a third reactor had been on the cards from the very beginning.[472] Time and again, director Meyer had pointed out that a Biblis C would be built. However, he had mostly given the impression that this was a question for a more or less distant future. In March 1974, for example, he had stated that a third reactor would not start operation that decade.[473] In September, he had claimed that a third reactor would be built, "but not before the eighties",[474] drawing a comment by district administrator Lommel that "two reactors are enough".[475] Now, though, the RWE said they were planning to build and operate two further reactors. Construction would start in 1976, Biblis C become operational in 1981, and Biblis D in 1983. To the company, the only potentially problematic aspect of the plan was "aesthetics".[476] That assumption turned out to be quite wrong.

For one, the local and regional citizens and citizens' initiatives opposed the expansion.[477] Groups like the AGURS and the District Farmers' Association Worms-Alzey started off a series of protests in the region with a demonstration of 300 people in Worms.[478] In Darmstadt, the IUS, displaying a perspective on its role relative to the general public not dissimilar from its opponents', organised an event for "the population's education", directed against the "publicity drive of the electricity industry" which, they pointed out, ignored scientific dissent, especially regarding the effects of radioactivity.[479] Letters to the editor continued to appear in regional newspapers, protesting against the expansion. Staab accused the RWE, which had once more spoken to the region's citizens in an open letter, of not addressing the people of Worms separately, although they were "the most affected".[480] In addition, marking a qualitative change to the debate, 26 local authorities in the region around Biblis resolved to oppose the new reactors.[481] Groß-Rohrheim's Municipal Council was one of the first, with only one councillor abstaining.[482] Worms' City Council followed suit,[483] as did Lampertheim Council, arguing primarily with the heightened demand for groundwater and further disruptions of the landscape by additional overhead lines.[484] Soon, the Ried's SPD took a stand against the RWE's plans,[485] with the result that Seib, who was still convinced that, as he had put it earlier, "so far, the nuclear power plant's coming had only positive consequences",[486] accused the party of being in league with communists.[487] This did not, however, prevent a political coup by Biblis' own SPD in June 1975. To the applause of 30 citizens present,[488] the party achieved a one-vote majority in the council, thereby instructing Seib to ask the Bergstraße District and the Hessian state authorities to suspend the licensing procedures until experiences regarding the first two plants' effects on the population and the environment had been gathered over several years.[489]

From now on, the regional and local public debate about the Biblis nuclear power plants did not stop until, and even beyond, the end of the 1980s, while the RWE's expansion plans were upheld.[490] The topics changed, as did positions and demands, but, while Biblis never became a stronghold of protest,[491] the quiet acceptance aimed for by the RWE and the state, district, and local authorities at the end of the 1960s, when the first gigawatt reactor was to mark the breakthrough of nuclear power in the field of commercial electricity generation, had been broken. It could not be re-established, despite some attempts in that direction, for instance, as part of a national political initiative called the Bürgerdialog Kernenergie (Citizens' Dialogue on Nuclear Energy, cf. "Emotional Atoms"). In fact, one could go as far as calling the early acceptance an illusion, based upon the successful management of the public debate. To avoid conflicts like those in Würgassen, the RWE had reframed the debate as being solely about the economic benefits of a technology they painted as well managed by the responsible experts who did not need to discuss with the public but were able to "educate" them. It is quite probable that, as a result, many people were either not motivated to or even discouraged from partaking in the debate themselves. Criticism was effectively outmanoeuvred. This strategy, however, by averting a wide and open debate in the beginning, also forewent the chance to form compromises with those who held different opinions to build a foundation upon which possible future problems could have been resolved. Instead, the public in a wider region, sometimes living closer to the power plant than the local citizens of Biblis itself, came to feel they had been duped, resulting in heightened mistrust in the RWE and the authorities. Now, organised groups emerged where individuals had not been able to stand up to the "educational campaigns", and a regional anti-nuclear network formed, ultimately reaching into Biblis proper. The people who had not been actively involved in approving the power plants beforehand were now motivated by their inability to inject themselves, their perspectives, and their demands into the decisions after the fact, to retrospectively question all licences granted, to gather all potential critical perspectives, and to form alliances bent on never being caught off guard again. They settled in for a drawn-out conflict that neither the RWE nor the authorities could avert with their intensified attempts to once more frame themselves as being able to answer all questions. It had become clear that the issue was not one of right or wrong information but of opinions.

Nuclear Histories Shaped by the Conditions of Debate

The close-up, local perspective on the nuclear debates in Hinkley Point and Biblis revealed that, indeed, the conditions of debate prevalent in both cases did have a dominant influence on the way their nuclear histories and their disputes about the "Peaceful Atom" developed, on the degree the issue split both societies. In Hinkley Point, the rather uncontrolled initial phase after the public announcement of the CEA's plans gave rise to a wide and open debate, accepted by the authorities.[492] Individual people and associations from the local and regional plane of debate actively weighed in – and even some from farther beyond – including Irving-Bell, Mackenzie, and Jones. This strong showing contributed to compromises regarding several criticised aspects of the power plant, which were formally aired during the public inquiry, in what could be called a prime example of a "public technology". This meeting between the project's proponents and critics, under the auspices of inspectors situated outside both camps, had been forced on the rather unwilling nuclear industry. Yet on this foundation, a rather serene local nuclear history could develop. Even changing circumstances and deviations from the compromise were, in most cases, accepted – including a second power station. This quiet often tends to be considered "normal" in Britain's nuclear history, blanking out its quite tumultuous historical foundation. It only came to an end when, against the background of a more active conflict about nuclear energy on the (trans-) national plane, the government attempted to push through further nuclear projects during the 1980s, overstraining the compromises originally reached. The open and critical character of the British conditions of debate thus resurfaced.

In Biblis, on the other hand, the number of private individuals actively voicing their view was negligible – with the remarkable exception of Marga Müller. In any case, critical voices were brushed aside by industry and authorities, not least Seib, working hand in hand, dominating the public debates, limiting, in a way, nuclear power's "public technology" dimension, most prominently visible at the official hearing, and aiming for acceptance and quiet. This, in fact, did work quite well, up to the point when people living in the region picked up on (trans-)national critiques of nuclear power, belatedly realising what had happened locally. Rather shocked by this realisation, they resolved not to be ignored or ignorant again. They consequently mounted a protest that was, in principle, open-ended and, in practice, escalating. Its specific goals were adapted accordingly whenever the circumstances changed, as the reactions to new studies, the construction progress, and the changing plans about further power stations in Biblis demonstrated. And while it must be admitted that radioactivity as such played a subordinate role in both early debates, the debates about nuclear power were nonetheless clearly shaped by the way early critics and their arguments appeared and were treated in Britain and Germany. They were shaped by a difference in the interaction between decision-makers and less powerful actors, visible in the conditions of debate – with a rather exceptional conflict developing in the German case.

This chapter's analysis has also shown that the way the nuclear debates developed in Hinkley Point and Biblis, under the influence of the conditions of debate, did bring these conditions themselves up for debate and, in the end, shaped them. In Hinkley Point, the reason for the return to a livelier dispute was the Thatcher government's assumption that the quiet acceptance of the first two nuclear plants predestined the place for further atomic projects aimed at weakening the coal miners with whom it was at loggerheads.[493] This renewed dispute was also infused with more resolve on the side of the critics, who had seen their acquiescence after the first compromise being used for additional measures that had not been covered originally – or even went against the assurances received. In Biblis, the most important development regarding the conditions of debate was the awakening of the critics, first in the region, later, in part, even locally. This had been brought about by the nuclear developments locally, as well as on the (trans-)national level. In view of the problems this entailed for atomic power's expansion, slight changes in the conditions of debate were initiated by the "Peaceful Atom's" proponents, visible especially in the initialisation of power plant talks, which were to provide a space for exchange. While the British LLCs became a nationwide standard, however, the power plant talks and the Informationskreis Kernkraftwerk (Nuclear Power Plant Information Circle), which was founded later (cf. "Emotional Atoms"), did not. Also, they were initially mostly restricted to the local level, leaving aside the wider region and thus most critics. This indicated that nuclear power's German proponents were not willing to change their concept of a debate. It had to be a monologue to which critics should listen without being listened to themselves, not even when they were afforded the opportunity to speak in forums like the power plant talks or, nationally, in the Bürgerdialog Kernenergie.

There were also clear indicators for the conditions of debate analysed here being national rather than local phenomena. In both cases, the debates were not geographically limited to the local, or even regional, area. There were contributors from the outside who sometimes weighed in very actively, as Munnings did in the case of Hinkley Point.[494] Others influenced it unintentionally, like Minister Leussink in the case of Biblis, or were taken into consideration by local people, as the debates about nuclear power in the USA were in both cases (cf. "Rational Atoms").

This assessment is supported by the role the conditions of debate played in the other locales under consideration here, where, despite the individual situations, national characteristics stood out. When the plans for a nuclear power plant in Oldbury were announced in 1959, another atomic power station was already under construction in nearby Berkeley, so the debates here started out from compromises already established, which could to a large degree be extended. Oldbury's integration into the LLC that had been established because of Berkeley is one example. Nevertheless, critical perspectives on nuclear power were also taken up and openly discussed. However, most of these subsided quickly, resulting in a quiet phase – like the one at Hinkley Point – that extended into the late 1970s.[495] Then, again like Hinkley Point, Oldbury came under mounting pressure, especially so when worrying numbers of leukaemia cases were registered in the region (cf. "Emotional Atoms"). Even in Torness, where the discussion began only in the 1970s, the British conditions of debate were visible, albeit in a temporally peculiar form. After a rather smooth initial debate and a public inquiry, a remarkably broad, open, and critical debate unfolded as the start to construction of the power plant began to drag on, so that several beliefs held locally began to be questioned. In the end, it was not least the implementation of an LLC that contributed to a broader acceptance of the station (cf. "Atomic Trust(s)" and "Nuclear Democracy").

In the other German cases, meanwhile, parallels to the development in Biblis were obvious. In Stade, the initial none-debate even remained in place throughout most of the 1970s, during which time critics of nuclear power were to a large degree deflected to the nearby Brokdorf power plant site. At the end of the decade, however, an unexpected breach of trust by the Stade power plant's operators shook the city and the surrounding region, marking the start of a broad conflict about the local reactor. It continued all through the period under consideration here and could not be reined in again, although numerous attempts were made by the technology's proponents. These, however, were limited like the ones in the case of Biblis, in that they were mostly aimed at generating the impression of critics being given their say – without being listened to. This became especially clear during a hearing in the wake of the Chernobyl catastrophe in 1986 (cf. "Atomic Trust(s)"). The case of Lingen, not unlike the case of Torness, differed somewhat because the debate here did not begin until the end of the 1970s. In addition, a smaller demonstration power plant was already in existence, and several other nuclear projects were debated in the region. Consequently, when the plans for a new Lingen power plant were publicised, quite a wide and open debate sprung up. Nonetheless, as in the case of Biblis, actors from the industry and the authorities were very much willing to frame it in a manner like the one employed by the RWE, with critics being seen as lacking information that could only be provided by the "official" experts. As the atom's proponents were not open to trying a different approach, to accepting different opinions as such, instead of framing them as "wrong", the conflict could not be settled – because Lingen's critics, aware of the experiences at other nuclear power plant sites, were as prepared to permanently be on guard as the critics in Biblis had been since the Hamm meeting (cf. "Rational Atoms").

In sum, the similarities between the individual locales analysed in the two nations were so striking that it seems fair to see, in Hinkley Point and Biblis, not two isolated cases of, but quite prototypical examples for, the British and German conditions of debate, respectively.

The observable differences, on the other hand, can mostly be attributed to the changes the debates underwent on the (trans-)national planes. They were, in other words, not least a consequence of the development of the conditions of debate, following their anchoring to the "Peaceful Atom", which turned them into a topic of debate themselves. This observation indicates that the local debates were not one-directionally determined by the national ones and their "global" timeline. What happened in Hinkley Point or Biblis did have an influence beyond the local, or even regional level. The way the debates developed was also, in each case, clearly determined mostly by what was happening locally, with the respective publication of construction plans, formalised debates, construction periods, and ongoing operations shaping the development of the local debates. Indeed, these local time phases also had no small influence on the way the power plants were regarded emotionally.

Notes

1 Our Point of View, in: *Thornbury Gazette (TG)*, 9.6.1989.
2 Hunt, Colin (Secretary, Somerset Ecology Party, Taunton): Green Party Seeks Votes, in: *Bridgwater Mercury (BM)*, 15.3.1983.
3 Cf. Bösch, Zeitenwende, p. 294; implicit: Uekötter, Atomare, pp. 232–233.
4 Butler et al., United Kingdom, pp. 4–5.
5 Kirchhof, Astrid Mignon, Helmuth Trischler: Federal Republic of Germany Short Country Report (2018), in: HoNESt Consortium Deliverable 3.6, www.honest2020.eu/sites/default/files/deliverables_24/FRG.pdf (23.1.2023).
6 Jefferys, Kevin: Politics and the People. A History of British Democracy since 1918. London 2007, pp. 6, 121, 155.

7 The focus on differences is not to imply that there were no similarities between the countries. There were, including the fact that, in confrontations between citizens and the nuclear industry, the latter had way more leverage. Cf. 'Coming to Terms with Conflicts Anchored to the Atom: Methods'.
8 For the central role of the national in these, cf. Kirchhof, Pathways, pp. 9, 22.
9 Dewey, Public, pp. 207–209.
10 Foucault, Michel: Politics and the Study of Discourse, in: Burchell, Graham, Colin Gordon, Peter Miller (Eds.): The Foucault Effect. Studies in Governmentality: With Two Lectures by and an Interview with Michel Foucault. Hemel Hempstead 1991, pp. 53–72, here pp. 59–60, emphases in the original; cf., for limits on topics and participants in the discourse, Foucault, Michel: Die Ordnung des Diskurses. Frankfurt/Main 1992 (Original 1972), pp. 16, 26; furthermore Foucault, Michel: Archäologie des Wissens. Frankfurt/Main 1992 (Original 1970), here pp. 75–82.
11 Ortwin Renn saw a British – or, more to the point, "Anglo-Saxon tradition" of active participation by citizens: Renn, Ortwin, Thomas Webler: A Brief Primer on Participation. Philosophy and Practice, in: Renn, Ortwin, Thomas Webler, Peter Wiedemann (Eds.): Fairness and Competence in Citizen Participation. Evaluating Models for Environmental Discourse. Dordrecht 1995, pp. 17–33, quote p. 17; Ute Hasenöhrl observed a tendency of German institutional actors to push individual citizens aside: Hasenöhrl, Zivilgesellschaft, pp. 208–217; cf. Jasanoff, Sheila: Designs on Nature. Science and Democracy in Europe and the United States. Princeton 2005, p. 31.
12 Welsh, Mobilising; however, this work does not focus on the broader, public rules of debate as a central issue on its own.
13 For the temporally inclined plane: 'Coming to Terms with Conflicts Anchored to the Atom: Methods'.
14 Cf. Welsh, Nuclear Nation, especially p. 7; Herring, Energy, especially p. 9; Oberloskamp, Ambiguities, p. 431; for exemplary contemporary considerations: van Buiren et al., Bürgerinitiativen, pp. 30–31; contributions and debates in: Foley, Gerald, Ariane van Buren (Eds.): Nuclear or Not? Choices for Our Energy Future. A Royal Institution Forum. London 1978; for a contextual background: Jefferys, Politics, p. 16; Seldon, Anthony, Jonathan Meakin, Illias Thoms: The Impossible Office? The History of the British Prime Minister. Cambridge 2021, pp. 125–126, 130–132.
15 The concept of framing allows to understand how things are charged with meaning in social contexts, whereby coexisting or rivalling interpretations are possible, contributing to the dynamics of public debates; cf. Goffman, Erving: Frame Analysis. An Essay on the Organization of Experience. Boston [2]1986 (Original 1974), pp. 27–37; Hoffmann, Jochen, Ulrich Sarcinelli: Politische Wirkungen der Medien, in: Wilke, Jürgen (Ed.): Mediengeschichte der Bundesrepublik Deutschland. Cologne 1999, pp. 720–748, here p. 733.
16 For the information deficit model in science communication: Irwin, Alan, Brian Wynne: Introduction, in: Irwin, Alan, Brian Wynne (Eds.): Misunderstanding Science? The Public Reconstruction of Science and Technology. Cambridge 1998, pp. 1–17, here pp. 1–3; for a more differentiated version: Wynne, Brian: Public Engagement as a Means of Restoring Public Trust in Science – Hitting the Notes, But Missing the Music?, in: Community Genetics 9 (2006), pp. 211–220, here p. 214; for an exemplary contemporary consideration: Goerke, Untersuchung Part I, especially pp. 3–4; for a comparison with concepts of a suggestible 'mass society': Jasanoff, Sheila: A Century of Reason. Experts and Citizens in the Administrative State, in: Skowronek, Stephen, Stephen M. Engel, Bruce Ackerman (Eds.): The Progressives' Century. Political Reform, Constitutional Government, and the Modern American State. New Haven 2016, pp. 382–404, here

p. 384; furthermore: Traube, umschalten, p. 242; Fuller, Steve: The Governance of Science. Ideology and the Future of the Open Society. Buckingham 2000, p. 5; Kaijser et al., Nuclear Energy, p. 4.

17 Cf. Rüdig, Anti-Nuclear, pp. 5–6; Welsh, Mobilising, especially p. 78; for the importance of nuclear energy for the late 1950s' British public perceptions: Hogg, Family, especially p. 536.

18 Cf. Hamblin, Jacob Darwin: Fukushima and the Motifs of Nuclear History, in: Environmental History 17 (2012), 2, pp. 285–299, here p. 290.

19 For which Germany was a vanguard: Radkau, Aufstieg und Krise, pp. 438–440; for studies seeing Wyhl as this development's epicentre: 'The National, the Spectacular, and the Complex Atom: State of Research'.

20 Cf. Mez, Lutz: Nachwort. Der Atomkonflikt, in: Mez, Lutz (Ed.): Der Atomkonflikt. Berichte zur internationalen Atomindustrie, Atompolitik und Anti-Atom-Bewegung. Mit einer Einleitung von Robert Jungk. Reinbek 1981, pp. 339–372, here p. 353; Davies, Sizewell, pp. 22, 29; Cutler, James, Rob Edwards: Britain's Nuclear Nightmare. The Shocking Truth behind the Dangers of Nuclear Power. London 1988, p. 180; O'Riordan et al., Sizewell, pp. 1–3; Rüdig, Wolfgang: Die (gescheiterte) Privatisierung der Kernenergie. Krise des Thatcherismus?, in: Sturm, Roland (Ed.): Thatcherismus. Eine Bilanz nach zehn Jahren. Bochum 1990, pp. 155–178, here p. 158; Rüdig, Low Profile, pp. 82–83; Seldon, Anthony, Daniel Collings: Britain under Thatcher. Harlow 2000, pp. 29–32; Welsh, Mobilising, pp. 180–181; Cooke, Mortal, pp. 351–352; Bösch, Zeitenwende, p. 349; Baker, Keith: Nuclear Power in Britain. A Series of Successful Failures, in: International Review of Public Policy 5 (2023), 1, pp. 26–43, here p. 35.

21 A Programme of Nuclear Power, February 1955 (draft), Paragraphs 25 and 26, p. 6, in: National Archives (NA), CAB 129/73/c/55/31/31; cf. Edgerton, Rise and Fall, pp. 292–293.

22 A Programme of Nuclear Power (cf. note 21), Paragraph 21, p. 5.

23 A Programme of Nuclear Power (cf. note 21), Paragraph 1, p. 1.

24 A Programme of Nuclear Power (cf. note 21), Paragraph 48, p. 11.

25 A Programme of Nuclear Power (cf. note 21), Paragraph 23, p. 5.

26 Cf. There May Be Two Nuclear Stations! Press Conference Announcement, in: *BM*, 5.3.1957.

27 Atomic Power Station Plan. Tests Near Bridgwater Have Proved Satisfactory. Approach Made to Landowners, in: *BM*, 27.11.1956; Nuclear Power Station Project. To Serve County's Growing Demand for Electricity. Start at Bridgwater Bay Site in Near Future?, in: *BM*, 4.12.1956.

28 Collingridge, D.: Lessons of Nuclear Power. US and UK History, in: Energy Policy 12 (1984), 1, pp. 46–67, here pp. 53–54; Taylor, Simon: The Fall and Rise of Nuclear Power in Britain. A History. Cambridge 2016, p. 10; cf. Cooke, Mortal, p. 138.

29 Nuclear Power Station Project. To Serve County's Growing Demand for Electricity. Start at Bridgwater Bay Site in Near Future?, in: *BM*, 4.12.1956.

30 A Programme of Nuclear Power (cf. note 21), Paragraph 37, p. 9.

31 Nuclear Power Station Project. To Serve County's Growing Demand for Electricity. Start at Bridgwater Bay Site in Near Future?, in: *BM*, 4.12.1956.

32 There May Be Two Nuclear Stations! Press Conference Announcement, in: *BM*, 5.3.1957.

33 Central Electricity Authority (CEA), Secretary's Office, to the Secretary, Ministry of Fuel and Power, 7.12.1956, Somerset Heritage Centre (SHC), A\BPR 6/1/1; cf. Report of a Public Inquiry at Town Hall Bridgwater, 14.-15.5.1957, in: NA, HLG120/230

34 Report of the Special Sub-Committee, Somerset County Council, County Planning Committee, on a Proposal by the CEA to Establish a Nuclear Power Station

at Hankley Point, January 1957, p. 5, SHC, A\BPR 6/1/1; CEA, Application under Town and Country Planning Act, 1947 (Sec. 35 (1)), Form B, 7.12.1956, SHC, A\BPR 6/1/1; E. S. Rickards to The Secretary, Ministry of Housing and Local Government, 26.3.1957, SHC, A\BPR 6/1/2; letter to N. P. Bray, 21.3.1957, SHC, A\BPR 6/1/2; Nuclear Power Station Project. To Serve County's Growing Demand for Electricity. Start at Bridgwater Bay Site in Near Future?, in: *BM*, 4.12.1956; cf. Heineberg, Großbritannien, pp. 26, 30–31, 33.

35 Memorandum, County Planning Officer, Somerset County Council, 9.8.1955, in: SHC, A\BPR 3/3/49, as well as further documents here and in NA, HLG 79/1171; Atomic Power Station Plan. Tests Near Bridgwater Have Proved Satisfactory. Approach Made to Landowners, in: *BM*, 27.11.1956; Nuclear Power Station Project. To Serve County's Growing Demand for Electricity. Start at Bridgwater Bay Site in Near Future?, in: *BM*, 4.12.1956; C. J. Hornsby to Clerk, Somerset County Council, 27.11.1956, SHC, A\BPR 6/1/2.

36 Atomic Power Station Plan. Tests Near Bridgwater Have Proved Satisfactory. Approach Made to Landowners, in: *BM*, 27.11.1956; Nuclear Power Station Project. To Serve County's Growing Demand for Electricity. Start at Bridgwater Bay Site in Near Future?, in: *BM*, 4.12.1956.

37 Somerset County Council, County Planning Committee, Special Sub-Committee Appointed to Consider the Proposed Atomic Power Station Near Stogursey, Sub-Committee Meeting 28.1.1957, Report of the County Planning Officer, SHC, A\BPR 6/1/1.

38 Atomic Power Station Plan. Tests Near Bridgwater Have Proved Satisfactory. Approach Made to Landowners, in: *BM*, 27.11.1956.

39 This had been a rather widespread reaction to growing industrialisation in Britain since the 19th century: Sheail, John: An Environmental History of Twentieth-Century Britain. Basingstoke 2002, pp. 103–145.

40 They are documented in SHC, A\BPR 6/1/1.

41 In the public debates about Hinkley Point, it surfaced in the early 1970s.

42 Cf. Linse, Ulrich et al.: Von der Bittschrift zur Platzbesetzung. Ein Vergleich, in: Linse, Ulrich et al. (Eds.): Von der Bittschrift zur Platzbesetzung. Konflikte um technische Großprojekte. Laufenburg, Walchensee, Wyhl, Wackersdorf. Berlin 1988, pp. 219–256; Sheail, Environmental, p. 192; contributions in Brüggemeier, Franz-Josef, Jens Ivo Engels (Eds.): Natur- und Umweltschutz nach 1945. Konzepte, Konflikte, Kompetenzen. Frankfurt/Main 2005; Engels, Naturpolitik, pp. 154–165; Hasenöhrl, Zivilgesellschaft; Brüggemeier, Franz-Josef: Schranken der Natur. Umwelt, Gesellschaft, Experimente. 1750 bis heute. Essen 2014.

43 County Planning Officer to Clerk of the County Council, 9.8.1955, SHC, A\BPR 3/3/49.

44 Atomic Power Station Plan. Tests Near Bridgwater Have Proved Satisfactory. Approach Made to Landowners, in: *BM*, 27.11.1956.

45 For Irving-Bell, cf. Obituary. Mrs Dorothy Irving-Bell, in: Polar Record 23 (1987), 146, p. 612, doi:10.1017/S0032247400008287 (23.1.2023); for her residence next to Hinkley Point, cf. documentation in: NA, HLG 120/230.

46 Cf. letter by Dorothy Irving-Bell, 23.1.1957, SHC, A\BPR 6/1/1; Christopher Rowe & Ash to The Secretary, Ministry of Fuel and Power, 15.1.1957, SHC, A\BPR 6/1/1; S. G. King to Sir Percy Mills, 18.1.1957, SHC, A\BPR 6/1/1; Nuclear Power Station Project. Stormy Protests – and Mild Approvals – at Bridgwater Public Inquiry, in: *BM*, 21.5.1957.

47 Irving-Bell, D. (Bristol): Proposed Nuclear Power Station, in: *BM*, 1.1.1957.

48 Irving-Bell, D.: Whose Choice?, in: *BM*, 30.4.1957.

49 Others were John Prout, Harry Shepherd, and C. T. Govett, who did not oppose the development: Landowner Is Opposed to Power Station Plan. Mr. Colin

Mackenzie Says 'No' to Surveyors, in: *BM*, 5.2.1957; cf. objections listed in SHC, A\BPR 3/3/49.

50 Mackenzie, C. (Stolford): The Atomic Station, in: *BM*, 8.1.1957.

51 Page, Mrs. K. M. (Fiddington): Atomic Power Station, in: *BM*, 22.1.1957; her husband, W. P. Page, Posed Similar Questions in the Bridgwater Rural District Council, K. Margaret Page to the County Council's Chairman A. Whittaker, 16.1.1957, SHC, A\BPR 6/1/1.

52 Richardson, J. ('Archways', Combwich) Points of View (1), in: *BM*, 29.1.1957; Richardson was, in all probability, Joseph Richardson, chairman of Otterhampton Parish Council in the 1970s, cf. *Combwich Clarion*, December 1974, p. 4.

53 Rus in urbe: A Sad Fact, in: *BM*, 5.3.1957.

54 Some of his letters were signed David I. W. Jones, some David G. W. Jones, those to the authorities David I. W. Jones or David J. W. Jones, cf. David I. W. Jones to Chairman, County Planning Committee, December 1956, SHC, A\BPR 6/1/1; David J. W. Jones to Clerk, Somerset County Council, 26.1.1957, SHC, A\BPR 6/1/1.

55 Jones, David J. W. (Bath): Atomic Power Station, in: *BM*, 22.1.1957.

56 At the county council's request (Lord Hylton to E. S. Rickards, 17.9.1955, SHC A\BPR 3/3/49), it had been designated in 1956: Visit Somerset: Natural Beauty Within Somerset, www.visitsomerset.co.uk/discover-somerset/natural-beauty (23.1.2023). Rickards had warned the CEA to expect objections if the Quantock Hills seemed endangered, Report, Clerk of the County Council, Meeting held at the CEA's Offices, 3.10.1955, p. 1, SHC, A\BPR 3/3/49.

57 Jones, David G. W. (Dunkerton, Bath): Power Station Alternative, in: *BM*, 16.4.1957.

58 Munnings, Alfred (Dedham, Colchester): The Atomic Power Station Inquiry, in: *BM*, 14.5.1957; for Munnings, cf. Goodman, Jean: Munnings, Sir Alfred James (1878–1959), Artist, in: Oxford Dictionary of National Biography, www.oxforddnb.com/view/10.1093/ref:odnb/9780198614128.001.0001/odnb-9780198614128-e-35148 (23.1.2023).

59 Nuclear Power Station Project. To Serve County's Growing Demand for Electricity. Start at Bridgwater Bay Site in Near Future?, in: *BM*, 4.12.1956; cf. Report of the Special Sub-Committee, Somerset County Council, County Planning Committee, on a Proposal by the CEA to Establish a Nuclear Power Station at Hankley Point, January 1957, SHC, A\BPR 6/1/1.

60 There May Be Two Nuclear Stations! Press Conference Announcement, in: *BM*, 5.3.1957.

61 Cf. Minutes, County Council Meeting, 26.3.1957, SHC, C/CC/M/2/22, p. 243.

62 There May Be Two Nuclear Stations! Press Conference Announcement, in: *BM*, 5.3.1957; cf. Williams, J. C. (Didcot, Berks): Nuclear Power Station, in: *Somerset County Gazette (SCG)*, 29.12.1956; for Hylton, cf. Hylton (4th Baron, U.K.), William George Hervey Jolliffe, in: Dod's Parliamentary Companion 135 (1967), p. 140. Several Council Officers Shared This View: Elliot to Street, 17.10.1956, NA, HLG 79/1171.

63 County Planning Officer to Clerk of the County Council, 9.8.1955, SHC, A\BPR 3/3/49.

64 There May Be Two Nuclear Stations! Press Conference Announcement, in: *BM*, 5.3.1957; E. S. Rickards to CEA, 5.3.1957, NA, HLG 79/1172.

65 Early Approval Anticipated. Hinkley Point Nuclear Power Station Plan, in: *BM*, 26.3.1957; Main Issue of Controversy. Siting the New Power Station, in: *BM*, 19.3.1957.

66 More Details of Nuclear Power Station. Special Meeting at Stogursey, in: *BM*, 15.1.1957.

67 Power Station: M.P. Questioned. 'Bound to Affect Lives of Local People', in: *BM*, 9.4.1957; for Wills, who at first remained neutral in the debates, to support the plant project after the public inquiry, cf. Pocock, Nuclear, p. 95.
68 'Conservancy Knew about Power Station'. Stogursey and Proposed by-Laws, in: *BM*, 2.4.1957.
69 Hinkley Point Inquiry. Opened This Morning at Bridgwater, in: *BM*, 14.5.1957.
70 Carter, Walter John (Tower Castle, Combwich): Quantocks Will Not Suffer, in: *BM*, 26.2.1957.
71 Hancock, A. (Kendal, Westmorland): The March of Progress, in: *BM*, 12.2.1957.
72 Home Guard (Stogursey): Nuclear Station Controversy, in: *BM*, 12.2.1957; for the electrification process in the 1950s, cf. Edgerton, Rise and Fall, pp. 291–293.
73 Irving-Bell, D. (Bristol): Proposed Nuclear Power Station, in: *BM*, 1.1.1957.
74 Nuclear Power Station Project. To Serve County's Growing Demand for Electricity. Start at Bridgwater Bay Site in Near Future?, in: *BM*, 4.12.1956.
75 More Details of Nuclear Power Station. Special Meeting at Stogursey, in: *BM*, 15.1.1957.
76 Stogursey and the Power Station. Housing Site at Wick Park?, in: *BM*, 19.2.1957.
77 Landowner is Opposed to Power Station Plan. Mr. Colin Mackenzie Says 'No' to Surveyors, in: *BM*, 5.2.1957.
78 There May Be Two Nuclear Stations! Press Conference Announcement, in: *BM*, 5.3.1957.
79 Page, Mrs. K. M. (Fiddington): Atomic Power Station, in: *BM*, 22.1.1957.
80 Page, Mrs. K. M. (Fiddington): Atomic Power Station, in: *BM*, 22.1.1957.
81 Landowner Is Opposed to Power Station Plan. Mr. Colin Mackenzie Says 'No' to Surveyors, in: *BM*, 5.2.1957.
82 Nuclear Power Station Project. To Serve County's Growing Demand for Electricity. Start at Bridgwater Bay Site in Near Future?, in: *BM*, 4.12.1956.
83 There May Be Two Nuclear Stations! Press Conference Announcement, in: *BM*, 5.3.1957.
84 Rowe, Newton A. (Parsonage Side, Otterhampton): Time Will Tell, in: *BM*, 5.3.1957.
85 Stogursey and the Power Station. Housing Site at Wick Park?, in: *BM*, 19.2.1957; There May Be Two Nuclear Stations! Press Conference Announcement, in: *BM*, 5.3.1957.
86 More Details of Nuclear Power Station. Special Meeting at Stogursey, in: *BM*, 15.1.1957.
87 Early Approval Anticipated. Hinkley Point Nuclear Power Station Plan, in: *BM*, 26.3.1957.
88 More Details of Nuclear Power Station. Special Meeting at Stogursey, in: *BM*, 15.1.1957; Minutes, Annual Meeting, Williton Rural District Council, 11.6.1957, SHC, D/R/wil 2/2/15, pp. 349–350, 358–359; Minutes, Meeting, Williton Rural District Council, 9.7.1957, SHC, D/R/wil 2/2/15, pp. 364–365.
89 Hinkley Point Inquiry. Opened This Morning at Bridgwater, in: *BM*, 14.5.1957.
90 County of Somerset, Clerk's Office: Meeting Held at the CEA's Offices, Trafalgar Buildings, London, 3.10.1955, SHC, A\BPR 3/3/49.
91 Bridgwater Broadcast. Experts Answer Questions, in: *BM*, 15.1.1957; for Braithwaite, cf. *The London Gazette*, 40097, 9.2.1954, p. 865.
92 Giles, John S. (Brown's, Stolford, Stogursey): Points of View (2), in: *BM*, 29.1.1957.
93 Local Bricklayer: A Chance for Local Labour, in: *BM*, 2.4.1957.
94 Webb, B. J. (Wembdon, Bridgwater): The Path of Progress, in: *BM*, 5.2.1957.
95 Webb, B. J. (Wembdon, Bridgwater): The Path of Progress, in: *BM*, 5.2.1957.
96 Hutchings, P. (Wood Farm, Stogumber): A-Power Station, in: *SCG*, 2.2.1957.

97 Trades Council and New Power Station. 'Should be Grateful at Its Coming', in: *BM*, 19.3.1957.
98 Atomic Power Station Plan. Tests Near Bridgwater Have Proved Satisfactory. Approach Made to Landowners, in: *BM*, 27.11.1956.
99 There May Be Two Nuclear Stations! Press Conference Announcement, in: *BM*, 5.3.1957; Importance of Nature Reserve. Official at Stogursey Council Meeting, in: *BM*, 21.5.1957.
100 Here and There by Nomad: A New Age of Clean Air?, in: *BM*, 4.12.1956; the controller was probably Arthur Charles Thirtle.
101 Thorsheim, Peter: Inventing Pollution: Coal, Smoke, and Culture in Britain Since 1800. Athens ^{2}2017, pp. 159–192.
102 Here and There by Nomad: A New Age of Clean Air?, in: *BM*, 4.12.1956.
103 Cf. Desmarais, Ralph: Jacob Bronowski. A Humanist Intellectual for an Atomic Age, 1946–1956, in: British Journal for the History of Science 45 (2012), 4, pp. 573–589; I am indebted to Helen Piel for this reference.
104 Bridgwater Broadcast. Experts Answer Questions, in: *BM*, 15.1.1957.
105 A Happy New Year!, in: *BM*, 1.1.1957; cf., for a national British perspective: Pocock, Nuclear, pp. 21, 47; Arnold, Windscale, pp. 21–22; Burton, Bob: Nuclear Power, Pollution and Politics. London 1990, pp. xi–xv; Hall, Nuclear, pp. 38–39; on a side note, the threat of energy shortages was a transnational topic of debate in the late 1950s: Bleidick, Pionier, p. 15. It was also present in Germany: Kitschelt, Kernenergiepolitik, p. 37; here, though, it was less pestering, Radkau et al., Aufstieg und Fall, pp. 82–87.
106 Lawrence, Walter A. (The Central Stores, Combwich): Hankley Power Station, in: *BM*, 5.2.1957.
107 Local Bricklayer: A Chance for Local Labour, in: *BM*, 2.4.1957.
108 Industrialists and Hinkley Project. 'Bridgwater Is Bound to Benefit', in: *BM*, 7.5.1957.
109 Local Products for Power Station? Chamber of Commerce Ask for Lists, in: *BM*, 12.3.1957.
110 Industrialists and Hinkley Project. 'Bridgwater Is Bound to Benefit', in: *BM*, 7.5.1957.
111 Early Approval Anticipated. Hinkley Point Nuclear Power Station Plan, in: *BM*, 26.3.1957.
112 Landowner Is Opposed to Power Station Plan. Mr. Colin Mackenzie Says 'No' to Surveyors, in: *BM*, 5.2.1957.
113 Including British Cellophane Ltd., cf. Maskins, C. W. to the Ministry of Fuel and Power, 22.1.1957, SHC, A\BPR 6/1/1.
114 There May Be Two Nuclear Stations! Press Conference Announcement, in: *BM*, 5.3.1957.
115 Hinkley Point Inquiry. Opened This Morning at Bridgwater, in: *BM*, 14.5.1957; for Mills, cf. Halsbury: Mills, Percy Herbert, First Viscount Mills (1890–1968), Politician and Industrialist, in: Oxford Dictionary of National Biography, www.oxforddnb.com/view/10.1093/ref:odnb/9780198614128.001.0001/odnb-9780198614128-e-35028 (23.1.2023).
116 There May Be Two Nuclear Stations! Press Conference Announcement, in: *BM*, 5.3.1957.
117 Hinkley Point Inquiry. Opened This Morning at Bridgwater, in: *BM*, 14.5.1957; cf. documentation in SHC, A\BPR 6/1/1.
118 Jones, David I. W. (Vernal House, Callingcott, Dunkerton, Bath): Proposed Power Station, in: *BM*, 19.2.1957.
119 Cf. Lord Hylton to E. S. Rickards, 2.1.1957, SHC, A\BPR 6/1/1.
120 Mantle, Lionel P. (Glen Hayes, Holford): No Difficulty, in: *BM*, 4.12.1956; for Mantle, cf. Martin, Ursula: Gladys and Lionel Mantle – Local Heroes, in:

Somerset Genealogy, http://somerset-genealogy.blogspot.com/2018/02/gladys-and-lionel-mantle-local-heroes.html (23.1.2023).

121 Cf. Hogg, Family, pp. 536–539.

122 A Programme of Nuclear Power (cf. note 21), Paragraph 38, p. 9.

123 A Programme of Nuclear Power (cf. note 21), Paragraph 38, p. 9.

124 Welsh, Mobilising; cf. 'The Topics Anchored to the Atom'.

125 Examples include Graul, E. H.: Gefährdung der Zivilbevölkerung durch künstliche Radioaktivität, in: Kühn, Erich, Paul Vogler (Eds.): Medizin und Städtebau. Ein Handbuch für gesundheitlichen Städtebau. Vol. 2. Munich 1957, pp. 421–444; Manstein, Bodo: Im Würgegriff des Fortschritts. Frankfurt/Main 1961; cf. Radkau, Joachim: Angstabwehr. Auch eine Geschichte der Kerntechnik, in: Enzensberger, Hans Magnus, Karl Markus Michel (Eds.): GAU – Die Havarie der Expertenkultur. Tschernobyl – Das größte anzunehmende Experiment. Angstabwehr. Berlin 1986 (Kursbuch 85), pp. 27–53, here p. 36; Piel, Helen: John Maynard Smith and the Fact(s) of Evolution. A Study of Scientific Working Life in Post War Britain. Diss, Leeds 2019.

126 Radkau et al., Aufstieg und Fall, p. 54; for Jordan, cf. Jordan, (Ernst) Pascual, in: Vierhaus, Rudolf et al. (Eds.): Biographisches Handbuch der Mitglieder des Deutschen Bundestages 1949–2002. Vol. 1. A-M. Munich 2002, p. 393.

127 Cf. listing of positions of local authorities in Somerset County Council, County Planning Committee, Special Sub-Committee Appointed to Consider the Proposed Atomic Power Station Near Stogursey, Sub-Committee Meeting 28.1.1957, Report of the County Planning Officer, SHC, A\BPR 6/1/1.

128 More Details of Nuclear Power Station. Special Meeting at Stogursey, in: *BM*, 15.1.1957.

129 Cf. Welsh, Mobilising, pp. 77–90.

130 Page, Mrs. K. M. (Fiddington): Atomic Power Station, in: *BM*, 22.1.1957; in a letter to the local authorities, she had stated her puzzlement at the fact that the station was to be in an area with little population, although the CEA always pointed out the station's safety, letter by K. M. Page, 6.1.1957, SHC, A\BPR 3/3/49. For the growing concern about fallout in British debates during the 1950s: Hogg, British, pp. 75–76.

131 Rowe, Newton A. (Parsonage Side, Otterhampton): Time Will Tell, in: *BM*, 5.3.1957.

132 Jones, David I. W. (Vernal House, Dunkerton, Bath): Could be Better Sited?, in: *BM*, 5.3.1957.

133 Rowe, Newton A. (Parsonage Side, Otterhampton): Warmer for Bathing?, in: *BM*, 19.3.1957; Rowe, Newton A. (Otterhampton): Decadence, Not Progress?, in: *BM*, 26.3.1957.

134 Barnes, R. W. (Bridgwater): Standing the Risk, in: *BM*, 19.3.1957.

135 Bridgwater District Resident: Possible Source of Danger, in: *BM*, 12.3.1957.

136 Rowe, Newton A. (Parsonage Side, Otterhampton): Warmer for Bathing?, in: *BM*, 19.3.1957; Rowe, Newton A. (Otterhampton): Decadence, Not Progress?, in: *BM*, 26.3.1957.

137 Rowe, Newton A. (Otterhampton): Alternative Proposal, in: *BM*, 9.4.1957; for an analysis of the connection between religious beliefs and scientific progress: Noble, David F.: Eiskalte Träume. Die Erlösungsphantasien der Technologen. Freiburg 1998, especially pp. 134–148.

138 Jones, David G. W. (Dunkerton, Bath): Power Station Alternative, in: *BM*, 16.4.1957.

139 Rowe, Newton A. (Otterhampton): Alternative Proposal, in: *BM*, 9.4.1957.

140 Early Approval Anticipated. Hinkley Point Nuclear Power Station Plan, in: *BM*, 26.3.1957.

141 Minutes, County Council Meeting, 26.3.1957, SHC, C/CC/M/2/22, p. 244.
142 Power Station: M.P. Questioned. 'Bound to Affect Lives of Local People', in: *BM*, 9.4.1957.
143 Joint Nature Conservation Committee: History of JNCC. Peterborough 2019, https://jncc.gov.uk/about-jncc/who-we-are/history-of-jncc/ (23.1.2023).
144 'No Danger from Radio Activity'. Conservancy's Reply to Stogursey Council, in: *BM*, 30.4.1957.
145 Local Bricklayer: A Chance for Local Labour, in: *BM*, 2.4.1957.
146 Cf. Flood, Nuclear Dream, pp. 8–10, 28–33, 40–42; Welsh, Mobilising, pp. 31–46, 59–61, 97–113, 147.
147 A Vast Project, in: *BM*, 4.12.1956; indeed, corresponding arguments were published throughout the debate about nuclear power, cf. Hasson, J. A.: The Economics of Nuclear Power. London 1965.
148 Page, Mrs. K. M. (Fiddington): Atomic Power Station, in: *BM*, 22.1.1957; cf. K. Margaret Page to A. Whittaker, 16.1.1957, SHC, A\BPR 6/1/1.
149 W. M. F. Knowles to Lord Hylton, 22.1.1957, SHC, A\BPR 6/1/2.
150 Bridgwater District Resident: Possible Source of Danger, in: *BM*, 12.3.1957.
151 Hogwood, A. M. (Bridgwater): Power from the Moon, in: *BM*, 19.3.1957.
152 Cf. J. F. Hillier to Somerset County Council, 20.3.1957, SHC, A\BPR 6/1/1.
153 Hillier, J. F. (West Buckland): A-Power Station, in: *SCG*, 19.1.1957.
154 The name Hinkley Point was only one of a number of alternatives abound up to early 1957, when it was officially adopted: There May Be Two Nuclear Stations! Press Conference Announcement, in: *BM*, 5.3.1957.
155 The Nature Conservancy: Proposed Nuclear Power Station, Hankley Point, Bridgwater Bay, Somerset, 14.3.1957, SHC, A\BPR 6/1/2.
156 Cf. Report of the Special Sub-Committee, Somerset County Council, County Planning Committee, on a Proposal by the CEA to Establish a Nuclear Power Station at Hankley Point, January 1957, p. 1, SHC, A\BPR 6/1/1.
157 Nuclear Power Station Project. To Serve County's Growing Demand for Electricity. Start at Bridgwater Bay Site in Bear Future?, in: *BM*, 4.12.1956; Power Station: M.P. Questioned. 'Bound to Affect Lives of Local People', in: *BM*, 9.4.1957; cf. Main Issue of Controversy. Siting the New Power Station, in: *BM*, 19.3.1957.
158 Public Notice, Signed by Ernest Long, Secretary, CEA, 13.12.1956, SHC, A\BPR 3/3/49, County Planning Officer to Clerk of the County Council, 9.8.1955, SHC, A\BPR 3/3/49; Note of a Meeting, 12.2.1957, NA, HLG 79/1171.
159 Landowner Is Opposed to Power Station Plan. Mr. Colin Mackenzie Says 'No' to Surveyors, in: *BM*, 5.2.1957.
160 Report of the Planning Committee to the County Council, 4.3.1957, Appendix, para. 24, SHC, D/RA/7/7/8.
161 Cf. letters by D. Irving-Bell, 26.1.1957, C. D. Doubleday and V. J. Doubleday, 26.1.1957, Pauline Adams to the Clerk to the County Council, 26.1.1957, SHC, A\BPR 6/1/1.
162 A. Whittaker to K. M. Page, 18.1.1957, SHC, A\BPR 6/1/1; Minutes, Special Meeting, County Planning Committee, 26.3.1957, SHC, A\BPR 6/1/2.
163 Main Issue of Controversy. Siting the New Power Station, in: *BM*, 19.3.1957.
164 Memorandum, Clerk of the County Council, Proposed Nuclear Power Station, 12.9.1956, p. 3, SHC, A\BPR 3/3/49.
165 Jones, David I. W. (Vernal House, Callingcott, Dunkerton, Bath): Proposed Power Station, in: *BM*, 19.2.1957.
166 Lawrence, Walter A. (The Central Stores, Combwich): The Hankley Miracle, in: *BM*, 26.2.1957.
167 Jones, David I. W. (Vernal House, Dunkerton, Bath): Could Be Better Sited?, in: *BM*, 5.3.1957.

168 Lawrence, Walter A. (Combwich): Hinkley Power Station, in: *BM*, 28.5.1957; cf., for another short and unanswered recourse on the topic, Jones, David I. W. (Dunkerton, Bath): Hinkley, in: *BM*, 12.11.1957.
169 There May Be Two Nuclear Stations! Press Conference Announcement, in: *BM*, 5.3.1957; letters received on the topic are documented in SHC, A\BPR 6/1/1.
170 Main Issue of Controversy. Siting the New Power Station, in: *BM*, 19.3.1957.
171 Main Issue of Controversy. Siting the New Power Station, in: *BM*, 19.3.1957; Early Approval Anticipated. Hinkley Point Nuclear Power Station Plan, in: *BM*, 26.3.1957.
172 Cf. K. Margaret Page to Mr. Whittaker, 20.3.1957, SHC, A\BPR 6/1/1.
173 Mackenzie, Colin (Stogursey): Atomic Power Station, in: *BM*, 22.1.1957.
174 Bridgwater District Resident: Possible Source of Danger, in: *BM*, 12.3.1957.
175 Jones, David I. W. (Dunkerton, Bath): Seaboard Defence, in: *BM*, 14.5.1957.
176 Irving-Bell, D.: Whose Choice?, in: *BM*, 30.4.1957.
177 Cf. Nelkin, Dorothy: Nuclear Power as a Feminist Issue, in: Environment 23 (1981), pp. 14–20, 38–39.
178 Cf. Kaijser et al., Nuclear Energy, p. 11; with a German focus: Davis, Belinda: The Gender of War and Peace: Rhetoric in the West German Peace Movement of the Early 1980s, in: Mitteilungsblatt des Instituts für soziale Bewegungen 32 (2004), pp. 99–130, especially pp. 122–123; more general: Davis, Belinda: What's Left? Popular Political Participation in Postwar Europe, in: The American Historical Review 113 (2008), 2, pp. 363–390, here p. 377; Kirchhof, Astrid Mignon, Jan-Henrik Meyer: Revealing Risks. European Moments in Nuclear Politics and the Anti-Nuclear Movement, in: Wöbse, Anna-Katharina, Patrick Kupper (Eds.): Greening Europe. Environmental Protection in the Long Twentieth Century – A Handbook. Berlin 2022, pp. 331–361, especially p. 333.
179 Cf. Nelkin, Nuclear Power, p. 38; Tennenbaum, Jonathan: Kernenergie. Die weibliche Technik. Marie Curie, Irène Joliot-Curie, Ida Noddack, Maria Goeppert-Mayer, Lise Meitner, Elisabeth Rona, Ellen Gleditsch, Marguerite Perey, Chien-Shiung Wu, Dixy Lee Ray u.a. Wiesbaden 1994; cf. note 20.
180 Even illnesses incurred by women working with radioactive paint in the early 20th century tended to be attributed to psychological factors; cf. Hilgartner, Stephen, Richard C. Bell, Rory O'Connor: Nukespeak. Nuclear Language, Visions, and Mindset. San Francisco 1982, pp. 9–11; Mania, Hubert: Kettenreaktion. Die Geschichte der Atombombe. Reinbek 2010, pp. 82–84; furthermore: Röhr, Ulrike: Vorwort, in: Röhr, Ulrike (Ed.): Frauen aktiv gegen Atomenergie – wenn aus Wut Visionen werden. 20 Jahre Tschernobyl. Norderstedt 2006, pp. 9–10; Biess, Frank: Republik der Angst. Eine andere Geschichte der Bundesrepublik. Reinbek 2019, pp. 381–382.
181 Rowe, Newton A. (Otterhampton): Decadence, Not Progress?, in: *BM*, 26.3.1957.
182 Jones, David I. W. (Dunkerton, Bath): Growing Opposition, in: *BM*, 26.3.1957.
183 Power Station: M.P. Questioned. 'Bound to Affect Lives of Local People', in: *BM*, 9.4.1957.
184 Jones, David G. W. (Dunkerton, Bath): Power Station Alternative, in: *BM*, 16.4.1957.
185 Nuclear Power Station Project. Stormy Protests – and Mild Approvals – at Bridgwater Public Inquiry, in: *BM*, 21.5.1957.
186 E. S. Rickards to County Planning Officer, 18.4.1957, SHC, A\BPR 3/3/49; 25 individuals had raised objections; one list with 600 signatures had been presented; cf. Rough, Elizabeth: Policy Learning through Public Inquiries? The Case of UK Nuclear Energy Policy 1955–61, in: Environment and Planning C: Government and Policy 29 (2011), pp. 24–45, here p. 32.

187 Into the Open, in: *BM*, 30.4.1957.
188 Page, K. M. (Bonson, Fiddington): Rival Claims, in: *BM*, 7.5.1957.
189 Rowe, Newton A. (Otterhampton): Radio-Active Contamination, in: *BM*, 14.5.1957.
190 The county council's planning committee also argued along these lines: Minutes, Special Meeting, County Planning Committee, 26.3.1957, SHC, A\BPR 6/1/2.
191 Letter by D. Irving-Bell, 26.1.1957, SHC, A\BPR 6/1/1.
192 Cf. Farmers Fear a Loss of Labour. 'Fantastic Wages' for A-Power Station Men? County Council to Press for Hinkley Point Inquiry, in: *SCG*, 6.3.1957.
193 Nuclear Power Station Project. Stormy Protests – and Mild Approvals – at Bridgwater Public Inquiry, in: *BM*, 21.5.1957; for a concise description and evaluation of the public inquiry and its development in the context of nuclear debates of the 1980s: O'Riordan et al., Sizewell, pp. 42–71, especially pp. 42–54.; cf. Pocock, Nuclear, p. 95; Chicken, John C.: Nuclear Power Hazard Control Policy. Oxford 1982, pp. 157–158; Welsh, Nuclear Nation, p. 4; Oberloskamp, Ambiguities, p. 431.
194 Nuclear Power Station Project. Stormy Protests – and Mild Approvals – at Bridgwater Public Inquiry, in: *BM*, 21.5.1957; for the inspectors, cf. Pocock, Nuclear, p. 95.
195 The proceedings are documented in: Report of Public Inquiry Regarding Hinkley Point, NA, HLG 79/1172; quotations, if not otherwise mentioned, are taken from: Nuclear Power Station Project. Stormy Protests – and Mild Approvals – at Bridgwater Public Inquiry, in: *BM*, 21.5.1957.
196 Nuclear Power Station Project. Stormy Protests – and Mild Approvals – at Bridgwater Public Inquiry, in: *BM*, 21.5.1957.
197 For the Williton Rural District Council's deliberations, cf. minutes of its meetings, 11.6.1957 and 9.7.1957, SHC, D/R/wil 2/2/15, and 25.8.1959, SHC, D/R/wil 2/2/16.
198 Nuclear Power Station Project. Stormy Protests – and Mild Approvals – at Bridgwater Public Inquiry, in: *BM*, 21.5.1957; for Clark, who later became deputy chairman of the CEGB, cf. Pocock, Nuclear, p. 257.
199 He also opposed the way the CEA proceeded, as some potential objectors never got the chance to be heard, wherein he was backed by Mr. Crouch and Mr. Gripper, both living close to the site; cf. Report of Public Inquiry Regarding Hinkley Point, p. 24, NA, HLG 79/1172.
200 Nuclear Power Station Project. Stormy Protests – and Mild Approvals – at Bridgwater Public Inquiry, in: *BM*, 21.5.1957.
201 Nuclear Power Station Project. Stormy Protests – and Mild Approvals – at Bridgwater Public Inquiry, in: *BM*, 21.5.1957; for Farmer, cf. Report to the Minister of Power, 31.5.1957, p. 12, NA, HLG 79/1172; Laufs, Reaktorsicherheit. Vol. 1, p. 404.
202 Report of Public Inquiry Regarding Hinkley Point, p. 28, NA, HLG 79/1172; cf. E. S. Rickards to Members, County Planning Committee, 15.3.1957, SHC, A\BPR 6/1/2; A. MacC. Armstrong to Street, 13.6.1957, NA, HLG 79/1171.
203 Report of Public Inquiry Regarding Hinkley Point, p. 34, NA, HLG 79/1172.
204 Nuclear Power Station Project. Stormy Protests – and Mild Approvals – at Bridgwater Public Inquiry, in: *BM*, 21.5.1957. One citizen from Stogursey Parish was of a different opinion, cf. Report of Public Inquiry Regarding Hinkley Point, p. 20, NA, HLG 79/1172; for Chidgy, cf. Report of Public Inquiry Regarding Hinkley Point, p. 2.
205 Cf. Pocock, Nuclear, p. 95; from German 'official' eyes, the calming effect British public inquiries had on the debates was rooted in a willingness of critics to adhere to decisions by the authorities: Blickle, D.: Die Öffentlichkeitsbeteiligung im

atomrechtlichen Genehmigungsverfahren aus der Sicht des Verwaltungspraktikers, in: Blümel, Willi (Ed.): Die Öffentlichkeitsbeteiligung bei der Genehmigung von umweltrelevanten Großvorhaben. Zwischenergebnisse eines gemeinsamen Forschungsprojektes. Vol. 2. Speyer 1988, pp. 29–40, here p. 40.

206 Jones, David I. W. (Dunkerton, Bath): Landscaping at Hinkley, in: *BM*, 1.10.1957; Jones, David I. W. (Dunkerton, Bath): Hinkley, in: *BM*, 12.11.1957.

207 Rowe, Newton A. (Otterhampton): Balloons and Atoms, in: *BM*, 26.11.1957.

208 Williams, Enid (Moonfleet, St. Audries): A Farce?, in: *BM*, 21.5.1957.

209 The same holds true for debates by local councils, which, after resolving matters they were responsible for, like roadworks, housing, or water supply, soon and for about two decades almost exclusively considered Hinkley Point in regard of visits to the station or appointments to its Local Liaison Committee (LLC). Cf. Minutes, Williton Rural District Council, SHC, D/R/wil 2/2/15–19, covering the years from 1956 until 1974; Minutes, Somerset County Council, SHC, C/CC/M/2/22–29, covering the years from 1956 until 1989.

210 Rowe, Newton A. (Otterhampton): To the Editor, in: *BM*, 31.12.1957.

211 Rowe, Newton A. (Otterhampton): Plutonium from Hinkley, in: *BM*, 24.6.1958.

212 Notification of Consent, Ministry of Power, 25.6.1957, SHC, A\BPR 3/3/49; September Start on A-Power Plant. 'No Serious Effect on Amenities of the Area', in: *BM*, 2.7.1957; cf. Pocock, Nuclear, p. 95, where 2.7.1957 is given as the date of consent.

213 Yeudole, B. (Bridgwater): A-Power Station, in: *BM*, 24.9.1957.

214 Lawrence, Walter A. (Combwich): The Hinkley Road, in: *BM*, 24.12.1957.

215 Rapid Progress at Hinkley Point. Preparations to Receive 'The Monster', in: *BM*, 5.8.1958.

216 Will Be Biggest in the World. Contracts Placed for £60m. Hinkley Point Project, in: *BM*, 17.9.1957.

217 Cf. The First 500,000 kW Atomic Power Station, Published by English Electric, Babcock & Wilcox, Taylor Woodrow, SHC, A\BPR 6/1/1.

218 £1m Face-Lift for N-Plant, in: *BM*, 19.6.1984; besides, when a nuclear power plant was to be erected near Plymouth, its city planning officer contacted Somerset County Council to ask, not least, about local experiences with concealing such an installation, C. C. Gimingham to R. W. Dale, 5.6.1968, SHC, A\BPR 6/1/4.

219 Hinkley Point. Farmers' Fears Are Unfounded, in: *BM*, 13.8.1957.

220 Lawrence, Walter A. (Combwich): The Hinkley Road, in: *BM*, 24.12.1957.

221 Trade of the Port, in: *BM*, 18.11.1958.

222 Criticisms about Hinkley. 'Cause Considerable Harm', in: *BM*, 24.2.1959.

223 Cf. note 168.

224 A. MacC. Armstrong to E. S. Rickards, 15.7.1960, SHC, A\BPR 3/3/49 (including documentation regarding the rules' formalisation during the following year); Hinkley Point Area. Residential Development Is to Be Severely Restricted, in: *BM*, 3.10.1961; cf. county council files on this debate in SHC, A\BPR 6/1/2.

225 Rowe, Newton A. (Otterhampton): To the Editor, in: *BM*, 31.12.1957.

226 This refers to an energy build-up in the crystalline structure of graphite used in these reactors, which is produced by neutrons hitting the material.

227 For the fire's history: Arnold, Windscale; for the politically motivated censorship around it, Arnold, Windscale, pp. 36–37, 62, 66, 82–86; for the transnational level of the debates, Arnold, Windscale, pp. 94–97; for a contemporary critical perspective on the lacking transparency: Kliefoth, Werner: Atomrundschau. Zusammenfassender Bericht, in: Atomkernenergie 2 (1957), pp. 504–509; for a politically tinted critique: Grümmer, Gerhard: Die Katastrophe von Windscale. Berlin 1989; for the opposite argument: Ritchie-Calder, Lord: The Sociological

Consequences of Nuclear Energy, in: Sherfield, Roger (Ed.): Economic and Social Consequences of Nuclear Energy. Oxford 1972, pp. 76–86; cf. Patterson, Nuclear, pp. 90–92; Hall, Nuclear, pp. 59, 63; Flood, Nuclear Dream, pp. 12–13; Welsh, Mobilising, pp. 94–117; Brown, Invisible, pp. 177–180; Hill, Atomic Empire, pp. 105–137, paints a more friendly picture of the accident and its consequences.

228 Rowe, Newton A. (Otterhampton): Balloons and Atoms, in: *BM*, 26.11.1957.

229 Rowe, Newton A. (Otterhampton): Atomic Power, in: *BM*, 3.12.1957.

230 Jones, David I. W. (Dunkerton, Bath): Not Foul Mud, in: *BM*, 31.12.1957.

231 Bridgwater District resident: Atomic Stations, in: *BM*, 10.12.1957.

232 Onward: Hinkley Project, in: *BM*, 24.12.1957; that is what at least some local councillors and officers did, as the planning officer of West Somerset District Council later admitted that the committee did not understand most technical details, Memorandum, District Planning Officer, 12.1.1978, SHC, A\BPR 6/1/5.

233 Farmers Have Fears. Atomic Fall-Out at Hinkley?, in: *BM*, 29.10.1957.

234 Cf. 'No Interference' Say Farmers. Bridgwater Views on Staghunting, in: *BM*, 21.1.1958.

235 A 'Windscale' at Hinkley? No, Says Resident Engineer, in: *BM*, 5.11.1957. Nonetheless, modifications of the reactor design were made: Hill, Atomic Empire, pp. 234–235.

236 Military Plutonium. To Be Manufactured at Hinkley, in: *BM*, 24.6.1958.

237 Rowe, Newton A. (Otterhampton): Atomic Dangers, in: *BM*, 29.7.1958.

238 Stogursey Protest Meeting, in: *BM*, 29.7.1958.

239 Wittner, Lawrence S.: The Nuclear Threat Ignored. How and Why the Campaign against the Bomb Disintegrated in the Late 1960s, in: Fink, Carole, Philipp Gassert, Detlef Junker (Eds.): 1968. The World Transformed. Cambridge 1998, pp. 439–458, here pp. 440–445; Phythian, Mark: CND's Cold War, in: Contemporary British History 15 (2001), 3, pp. 133–156; Hudson, Kate: CND. Now More Than Ever. The Story of a Peace Movement. London 2005, especially pp. 43–81, 109–111, 167–169; cf. Edgerton, Rise and Fall, pp. 346–347; Kalmbach, Afterlives, pp. 52–53.

240 Stogursey Gets an Assurance, in: *BM*, 5.8.1958.

241 Council Protest at Hinkley Decision. 'Danger in Event of an Accident', in: *BM*, 7.10.1958; cf. Burnham Gets Assurances about Nuclear Station. Council-Representatives Are Told That Atomic Bombs Will Not Be Made There, in: *Weston Mercury and Somersetshire Herald*, 14.12.1956.

242 Haydon, Herbert W. (Weston-super-Mare): Nuclear Power Versus Tidal Barrage, in: *BM*, 21.10.1958.

243 Lafford, A. V. (Chilton Polden): Potential Danger, in: *BM*, 28.10.1958.

244 Stogursey Protest. Military Plutonium from Hinkley, in: *BM*, 11.8.1959.

245 Plutonium Production at Hinkley, in: *BM*, 11.11.1958.

246 Hinkley Hazards. Publicity on Safeguards, in: *BM*, 26.4.1960; cf. minutes and additional papers in SHC, C/FB/50/1 and SHC, C/PHH/201.

247 Grandfield, C. B. (West Somerset District Council Representative on Local Liaison Committee, Holford): Protection of Health and Safety Is Their Main Aim, in: *BM*, 1.4.1987; cf. Minutes, Meeting, Public Health Committee, Thornbury Rural District Council, 4.12.1958, p. 275, Gloucestershire Archives (GlouA), DA38/100/32.

248 Breach, Windscale, pp. 91–92; Chicken, Nuclear, pp. 65–66, 134–135; Dobson, Alan P., Charlie Whitham: Project Lamachus. The Cold War Comes to Scotland. The Holy Loch U.S. Nuclear Submarine Base and Its Impact on Scotland, 1959–1974, in: Engel, Jeffrey A. (Ed.): Local Consequences of the Global Cold War. Washington, DC 2007, pp. 169–192, here p. 174; memories of Marjorie Higham and

Eric Greenslade in: Davies, Sellafield, pp. 41–52, 83–88; cf. Arnold, Windscale, pp. 104–110, 123; for critique regarding the Sellafield LLC as "staged": memories of The Reverend Alan Postlethwaite in: Davies, Sellafield, pp. 125–132, here p. 129.

249 Report to the Minister of Power, 31.5.1957, p. 13, NA, HLG 79/1172.

250 Stogursey Gets an Assurance, in: *BM*, 5.8.1958; Radioactivity Measurements, in: *BM*, 25.11.1975.

251 Quotation from: Progress on District Survey at Hinkley Point, Appendix I to: Minutes, 2nd Meeting, Hinkley Point LLC, 17.4.1962, SHC, C/PHH/201; Bonfield, P. J., Health Physicist, Hinkley Point Power Station: Hinkley Point Local Liaison Committee, Appendix III to: Minutes, Inaugural Meeting, Hinkley Point LLC, 14.4.1961, SHC, C/PHH/201; Environmental Monitoring Report – Hinkley Point. Report No. 1, 1965, p. 1, SHC, C/PHH/201.

252 Hinkley Hazards. Publicity on Safeguards, in: *BM*, 26.4.1960; 1984 brochure 'Hinkley Point Nuclear Power Station', SHC, C/OP/19/18.

253 Press Statement, 10th Meeting, Berkeley and Oldbury LLC, 16.3.1964, NA, HLG 120/230.

254 Minutes, 11th Meeting, Hinkley Point LLC, 8.10.1968, p. 3, SHC, C/FB/50/1.

255 Minutes, Inaugural Meeting, Hinkley Point LLC, 14.4.1961, p. 2, SHC, C/PHH/201.

256 Statement by CEGB Health Physicist B. C. Godbold, Exploratory Meeting, Hinkley Point LLC, 17.2.1960, p. 3, SHC, C/PHH/201.

257 Lederman, H. (Bridgwater): Radio-Active Waste, in: *BM*, 30.4.1963.

258 For example: Blaze at a-Power Station, in: *BM*, 26.7.1966.

259 Hinkley Gets the 'All Clear', in: *BM*, 10.5.1975; Lederman, H. J. (Connington): Hinkley Article 'Well Timed', in: *BM*, 29.8.1978.

260 Viator: Official Opening at Hinkley, in: *BM*, 18.1.1966; for the timing: Nuclear Power Station Officially Opened. Hinkley Point Cost £80 Million. Reception for 500, in: *BM*, 26.4.1966.

261 Exceptions included: Second A-Power Plant Approved. Hinkley Point's No. 2 to have double capacity. Big Job Begins in 1965, in: *SCG*, 28.7.1962; Second Power Station. But Work May Not Begin until 1965, in: *BM*, 17.4.1962; Will Hinkley again Pose Problems?, in: *BM*, 4.1.1966; Radioactive Waste Will Be Dumped in Channel. 'No Risks' Say the Ministry, in: *BM*, 19.10.1971; cf. Somerset County Council to Central Electricity Generating Board (CEGB), 23.3.1966, SHC, A\BPR 6/1/3; for a scarcity of debates in official circles: Outline Application, Williton Rural District, 2.7.1962, SHC, A\BPR 6/1/3; CEGB, Application under Town and Country Planning Act, 1947 (Sec. 35 (1)), Form B, 15.4.1962, SHC, A\BPR 6/1/3; E. S. Rickards to Lord Hylton, 14.5.1962, SHC, A\BPR 6/1/3; Letter of Consent, Ministry of Power, 25.7.1962, SHC, A\BPR 6/1/3; County Planning Officer to Wexford County Council, 13.4.1978, SHC, A\BPR 6/1/5.

262 CEGB, Hinkley Point 'B', Meeting with Members and Officers of Local Authorities and Statutory Bodies, 5.10.1967, Purpose (d), SHC, A\BPR 4/6/2; letter by E. S. Rickards, 20.8.1962, SHC, A\BPR 6/1/3; T. Pratt to R. W. Dale, 15.12.1965, SHC, A\BPR 6/1/3; further documents regarding the process in: SHC, A\BPR 6/1/2, 3&4; cf. Lean, Thomas, Sally Horrocks: Good Nuclear Neighbours. The British Electricity Industry and the Communication of Nuclear Power to the Public, 1950s–1980s, in: Journal of Science Communication 16 (2017), 3, A09, pp. 1–16; it should be mentioned that the 'B' station was planned to become a model for export purposes: Note, Meeting, Board of Trade, 6.1.1967, NA, POWE 14/2014.

263 Nuclear station Misgiving: 'There Should Have Been an Inquiry', in: *BM*, 6.11.1973.

264 There had been some misgivings during the 1970s, especially with the Stogursey Parish Council, which complained that it was not as well informed as it was before, although it seldom received greater public attention; cf. a case of radioactive asphalt buried on site, documented in NA, AT 31/30.
265 Seldon et al., Impossible Office, p. 133; cf. Edgerton, Rise and Fall, p. 444.
266 Flood, Torness, pp. 28–29.
267 Peaceful Protest at Hinkley, in: *BM*, 27.9.1977.
268 Turner, Tom: David Brower. The Making of the Environmental Movement. Oakland 2015, p. 161.
269 Turner, Tom: David Brower. The Making of the Environmental Movement. Oakland 2015, pp. 188–193; initially, climate change had been more prominent: Traube, umschalten, p. 44.
270 Douglas, Mary, Aaron Wildavsky: Risk and Culture. An Essay on the Selection of Technical and Environmental Dangers. Berkeley 1982, pp. 132–137; Kaufman, Scott: The Environment and International History. London 2018, p. 73.
271 Cf. Friends of the Earth International: Our History, www.foei.org/who-are-friends-of-the-earth/friends-of-the-earth-history/ (23.1.2023); Williams, Decisions, p. 262; Radkau, Joachim: Eine kurze Geschichte der deutschen Antiatomkraftbewegung, in: Aus Politik und Zeitgeschichte 61 (2011), 46/47, pp. 7–15, here p. 8; Meyer, Challenging, p. 212; Uekötter, Frank: The Greenest Nation? A New History of German Environmentalism. Cambridge 2014, p. 81; Radkau, Joachim: Der 'Größte Anzunehmende Unfall', in: Uekötter, Frank (Ed.): Ökologische Erinnerungsorte. Göttingen 2014, pp. 50–60, here p. 56; Thomson, Jennifer: Surviving the 1970s. The Case of Friends of the Earth, in: Environmental History 22 (2017), 2, pp. 235–256, here p. 235.
272 Van Buiren et al., Bürgerinitiativen, p. 37.
273 Namely reprocessing, fast breeder and pressurised water reactors, plutonium and waste dumping; cf. Patterson, Going Critical, pp. 25–46, 58, 65–69, 110–128, 158–172.
274 Muton, Garth (Avon and Somerset Ecology Party, Dundon): The Danger Posed by Nuclear Power, in: *BM*, 18.4.1978; Warboys, Nick: Radiation Leak on Hinkley Beach, in: *BM*, 10.4.1979.
275 Nationally, Labour aimed to reduce reliance on nuclear energy in the late 1980s; cf. Butler et al., United Kingdom, p. 7.
276 Cf. Grave Concern over Nuclear Leak, in: *BM*, 17.4.1979; Radiation Fears for Children Dispelled, in: *BM*, 24.7.1979; Hinkley Point Safe, Say Officials, in: *BM*, 15.1.1980; Atomic Waste Move – Group Will Protest, in: *BM*, 17.6.1980; No Change in Nuclear Waste Train Delivery, in: *BM*, 13.10.1981; Phelps, Kathleen (Labour County Councillor, Bridgwater North, Bridgwater): Hinkley: Treat Us Like Adults, in: *BM*, 30.11.1982; Graham, Fiona: Atomic Waste Fears, in: *BM*, 30.11.1982; Graham, Fiona: Campaign for End of Nuclear Waste Trip, in: *BM*, 6.12.1983; Nuke Protesters' Rail Yard Demo, in: *BM*, 2.7.1985; Redfort, Peter: MP Jeered, in: *BM*, 11.12.1985; Ling, Cyril: After Chernobyl . . . the Local Backlash. Why This Secrecy?, in: *BM*, 14.5.1986; Brookes Accused, in: *BM*, 28.5.1986; Beckett, Bill: Yes, We Slipped Up, Say Nuclear Chiefs – But Not Like the Americans Did . . . Nowhere Near a Disaster!, in: *BM*, 23.7.1986; Redford, Peter: 'Confession' Plays on Nuclear Fears, in: *BM*, 4.3.1987; Report Rejects Bad Welds Claim at Hinkley Point, in: *BM*, 22.7.1987; Brookes, Bob (Bridgwater): Do They Really Want to Find These Bad Welds?, in: *BM*, 29.7.1987; for the role played by nuclear waste in mobilising opposition to nuclear power in England: Sheail, Environmental, pp. 224–227.
277 Cf. Puriton Nuclear Waste Dump Fears, in: *BM*, 31.7.1979; Donkin, Brod: Anger over Nuclear Waste Dump Plan, in: *BM*, 22.1.1980; White, P. A. (Highbridge): Nuclear Dumps, Further Points, in: *BM*, 30.9.1980; Donkin, Brod: Test

Bores Move Alerts Critics of Radio-Active Waste Dumping, in: *BM*, 7.10.1980; Peaceful Protest, in: *BM*, 27.1.1981; 'Great Victory' Claimed as Borehole Bid Is Sent Back, in: *BM*, 3.2.1981; Now Surface Storage of Nuclear Waste Is Favoured Option, in: *BM*, 23.6.1981; Beckett, Bill: New Fear over N-Waste Site, in: *BM*, 13.12.1983.

278 Cf. Beckett, Bill: Hinkley Is Cleared in Health Probe, in: *BM*, 27.11.1984; Power Station 'Cancer Link', in: *BM*, 5.2.1985; Hinkley Cleared in Cancer Probe, in: *BM*, 14.5.1986; What If It Happened again . . .?, in: *BM*, 21.5.1986; Redford, Peter: Open Information – on 300 Pages. N-Chiefs Play It By the Book, in: *BM*, 22.10.1986; Rosamond, Jon: Concern over Leukaemia Cases, in: *BM*, 19.11.1986; Link between Hinkley and Leukaemia?, in: *BM*, 20.12.1988; Clusters May Be Caused by a Virus, in: *BM*, 7.3.1989.

279 Cf. 3,000 New Jobs – But Worries Too. Hinkley: Joy and Protests, in: *BM*, 31.8.1982; Hinkley Meetings Planned, in: *BM*, 7.9.1982; Graham, Fiona: 'Community Must Revolt against Nuclear Plans', in: *BM*, 30.11.1982; rumours along this line had already stirred up limited but critical discussions in the middle of the 1970s: Worried Council to Meet Hinkley Superintendent, in: *BM*, 25.12.1973; 'No Plans for Third Power Station', in: *BM*, 15.1.1974; Reactor 'Unsafe' Fears, in: *BM*, 15.1.1974; Fear of the Unknown, in: *BM*, 29.1.1974; Power Stations Probe Welcomed, in: *BM*, 26.3.1974; Happier about Hinkley Security, in: *BM*, 25.11.1975.

280 New Anti-Nuclear Alliance, in: *BM*, 16.10.1979; in Bristol, a Nuclear Free Liaison Group was closely connected to the city council, worried about accidents at nearby nuclear reactors, especially in the wake of Chernobyl: Minutes, Meeting, Nuclear Free Liaison Group, 7.7.1986, Bristol Archives, M/BCC/NFL/1/1.

281 Hinkley C Nuclear Power Station Public Inquiry Opens Today at Cannington College. Battle of Sedgemoor '88, in: *BM*, 4.10.1988; files regarding the inquiry are kept at the SHC, including the over 3,000-pages strong official report: Barnes, Michael: The Hinkley Point Public Inquiries. A Report by Michael Barnes QC to the Secretaries of State for Energy and the Environment. 9 Vols. London 1990, SHC, A\BPR 4/6/202.

282 It was a development quite in contrast with the "more passive citizenship in Britain" observed by Jefferys, Politics, p. 228.

283 Radkau et al., Aufstieg und Fall, pp. 49–50, 62–63.

284 Radkau, Joachim: Mandel, Heinrich, in: Neue Deutsche Biographie 16 (1990), pp. 9–10, www.deutsche-biographie.de/gnd137909020.html#ndbcontent (23.1.2023); Mandel, Heinrich: Die Kernenergie im Spannungsfeld zwischen wirtschaftlicher Nutzung und öffentlicher Beteiligung. Opladen 1978.

285 Radkau et al., Aufstieg und Fall, pp. 138–140; Radkau, RWE zwischen Braunkohle und Atomeuphorie, pp. 189–191; Bm: Kernkraftwerk mit tausend Megawatt Leistung soll bei Frankfurt entstehen. Das RWE Will dafür eine halbe Milliarde DM investieren, in: *Die Welt*, 18.6.1968; dpa: Ein Kernkraftwerk, in: *Wormser Zeitung (WZ)*, 19.6.1968.

286 de: Kommt Riesen-Kernkraftwerk nach Mannheim?, in: *Mannheimer Morgen (MM)*, 20.6.1968; de: Kampf um Brückenkopf auf der Friesenheimer Insel. Das RWE will Süddeutschlands größtes Kernkraftwerk allein bauen/BASF als Hauptabnehmer, in: *MM*, 20.6.1968; de: Zweifel an der Biblis'schen Geschichte des RWE. Der Kampf um das größte Kernkraftwerk nimmt ungewöhnliche Formen an/Wird die BASF unter Druck gesetzt?, in: *MM*, 22.6.1968; cf. Radkau, Aufstieg und Krise, pp. 376–377; Müller, Wolfgang D.: Geschichte der Kernenergie in der Bundesrepublik Deutschland. Vol. 2. Auf der Suche nach dem Erfolg – Die Sechziger Jahre. Stuttgart 1996, p. 402.

287 ff: Kernkraftwerk für 600 Millionen. Bessere Versorgung im Rhein-Main-Neckar-Raum/Anbindung an Umspannanlage Rosengarten/Worms, in: *WZ*, 24.6.1968; cf. Uekötter, Atomare, p. 103.

288 Cf. Laufs, Reaktorsicherheit. Vol. 1, p. 281.
289 Roggenbuck, Klaus J.: RWE hält an Bibliser Kernkraftwerk fest. Nicht als Konkurrenzunternehmen zu Mannheimer Energieversorgung gedacht – Günstiger Standort entschied, in: *Südhessische Post (SP)*, 18.7.1968. During the official hearing, 18 alternative sites were named: hz: Öffentlicher Anhörungstermin zum Bau eines Kernkraftwerks in Biblis. Bedenken jetzt ausgeräumt. Überdurchschnittliche Sicherheitsvorkehrungen – Fachvorträge klärten auf, in: *SP*, 19.1.1970; the *Mannheimer Morgen* later on also claimed there had been 20, -el-: Am Rhein auf Bibliser Gemarkung: Der Welt größtes Kernkraftwerk nimmt seinen Betrieb auf, in: *MM*, 18.7.1974.
290 Radkau, Aufstieg und Krise, pp. 446–448; cf. Tiggemann, Achillesferse, pp. 204–208; for earlier oppositional activities: Radkau, Aufstieg und Krise, pp. 440–445.
291 Noack, Hans-Joachim: Ein Atommeiler kommt ins Dorf. Von Umweltverschmutzung wollen sie nichts hören, in: *Die Zeit*, 8.10.1971; at the official hearing in 1970, an RWE representative claimed that the letter had been published in 15 regional newspapers: hz: Öffentlicher Anhörungstermin zum Bau eines Kernkraftwerks in Biblis. Bedenken jetzt ausgeräumt. Überdurchschnittliche Sicherheitsvorkehrungen – Fachvorträge klärten auf, in: *SP*, 19.1.1970; cf. Radkau, RWE zwischen Kernenergie und Diversifizierung, p. 241.
292 On 18.6.1968: pp: Life-Sendung zum Hessentag. Biblis'che Geschichte auf dem Bildschirm. Hessens Wirtschaftsminister Arndt nahm zum Kerngroßkraftwerk Stellung, in: *MM*, 29.6.1968.
293 RWE: Offener Brief. An die Einwohner von Biblis, Landkreis Bergstraße und Umgebung, in: *MM*, 22.6.1968; the letter was also sent to the people of Biblis on the day before.
294 The *Frankfurter Allgemeine Zeitung (FAZ)*, for example, later on observed a "successful education" of the population by the RWE, hof.: Anders als in Wyhl, in: *FAZ*, 19.4.1975.
295 In that regard, Biblis was typical for power plant projects in Germany at that time: Kitschelt, Kernenergiepolitik, p. 201.
296 Laufs, Reaktorsicherheit. Vol. 1, p. 231.
297 pp: Life-Sendung zum Hessentag. Biblis'che Geschichte auf dem Bildschirm. Hessens Wirtschaftsminister Arndt nahm zum Kerngroßkraftwerk Stellung, in: *MM*, 29.6.1968.
298 lb: Europas größtes Kernkraftwerk. Es soll 1970 in Biblis begonnen werden – Eine aufeinander abgestimmte Atomforschung nötig, in: *SP*, 30.10.1968; for Lenz, cf. Lenz, Carl Otto, in: Vierhaus, Handbuch. Vol. 1, p. 497.
299 Röthlein, Kernenergie, pp. 58–78; Müller, Kernenergie. Vol. 1, pp. 195–199; Deutsches Atomforum e. V. (Ed.): 50 Jahre Deutsches Atomforum e. V. Berlin 2009. One could see, herein, a formalised parallel to the 'Old Boys Network' in Britain (cf. 'The National, the Spectacular, and the Complex Atom: State of Research').
300 Deutsches Atomforum, 50 Jahre; Radkau, Zukunft, p. 150; cf. Bieber, Kernenergienutzung, pp. 37–39, 106; Pringle, Peter, James Spigelman: The Nuclear Barons. The Inside Story of How They Created Our Nuclear Nightmare. London 1983, pp. 168–171, 229–230; Hanel, Tilmann: Die Bombe als Option. Motive für den Aufbau einer atomtechnischen Infrastruktur in der Bundesrepublik bis 1963. Essen 2015, pp. 11, 101–104, 127–132, 191, 202; Bähr, Johannes, Christopher Kopper: Industrie, Politik, Gesellschaft. Der BDI und seine Vorgänger 1919–1990. Göttingen 2019, here pp. 249, 311.
301 MacBarchet [sic, CG]: Ein Atomkraftwerk im Raum Biblis? Erste Fühlungnahme des RWE mit Landrat Dr. Lommel, in: *MM*, 22.6.1968; for Lommel, cf. Lommel, Ekkehard, in: Hessische Biografie, www.lagis-hessen.de/pnd/125629400 (23.1.2023).

302 ff: Kernkraftwerk für 600 Millionen. Bessere Versorgung im Rhein-Main-Neckar-Raum/Anbindung an Umspannanlage Rosengarten/Worms, in: *WZ*, 24.6.1968; the author wishes to thank Sigrid Scheel, who helped to identify Mr Fabrie's by-line, in a mail to the author, 2.10.2019.
303 Große Dinge kommen auf Biblis zu. Bürgermeister Seib berichtete von den Vorhaben der Gemeinde, in: *Bürstädter Zeitung (BZ)*, 18.7.1968; Roggenbuck, Klaus J.: RWE hält an Bibliser Kernkraftwerk fest. Nicht als Konkurrenzunternehmen zu Mannheimer Energieversorgung gedacht – Günstiger Standort entschied, in: *SP*, 18.7.1968; gr: Biblis soll überregionalen Raum versorgen. Kein Kampf um Strom mit Mannheim. Geplantes RWE-Kernkraftwerk gefährdet Projekt auf der Friesenheimer Insel nicht, in: *MM*, 18.7.1968.
304 si: Kernkraftwerk Biblis wird gebaut. Ein Telegramm der RWE aus Essen bestätigt diese Absicht, in: *Darmstädter Echo (DE)*, 19.7.1968.
305 Roggenbuck, Klaus J.: RWE hält an Bibliser Kernkraftwerk fest. Nicht als Konkurrenzunternehmen zu Mannheimer Energieversorgung gedacht – Günstiger Standort entschied, in: *SP*, 18.7.1968.
306 Noack, Hans-Joachim: Ein Atommeiler kommt ins Dorf. Von Umweltverschmutzung wollen sie nichts hören, in: *Die Zeit*, 8.10.1971.
307 Excerpt from the municipal board's protocol, 28.6.1968, Gemeindearchiv Biblis (GAB), KKW Presseartikel 1968 bis 1970.
308 st: RWE wählt Standort. Größtes Atomkraftwerk soll nach Biblis. Die 1000 Megawatt-Station könnte halb Hessen mit Strom versorgen, in: *Frankfurter Neue Presse (FNP)*, 21.6.1968.
309 In der Gemeinde Biblis hat das Atom die Gurke ersetzt, in: *SP*, 18.7.1974; Dörr, Rudi: Bibliser Gurken in alle Welt, www.rhein-neckar-industriekultur.de/objekte/gurkenfabrik-biblis (23.1.2023).
310 Bundesstraße als Wirtschaftsader, in: *MM*, 3.7.1970; cf. Noack, Hans-Joachim: Biblis bekommt den größten Reaktor. Von der Gurke zum Atom. Eine ländliche Gemeinde erwartet frohgemut die Kernspalter, in: *Süddeutsche Zeitung (SZ)*, 10.11.1968.
311 Advertisement, *MM*, 7.9.1968.
312 Pflugmacher, Karl: Atomstrom aus dem Gurkendorf. Biblis sieht sich als Kernenergie-Pionier – Atomei soll Fortschritt bringen, in: *Günzburger Zeitung*, 16.4.1975.
313 Minutes, Meeting, Biblis Municipal Council, 18.9.1969, pp. 8–9, GAB, Gemeindevertretungssitzungen.
314 Vier Mark pro Quadratmeter. Grundstückspreise für das Bibliser Atomreaktorgelände ausgehandelt, in: *SP*, 29.11.1968.
315 ulf: Einzig in der Bundesrepublik. Bibliser Kernkraftwerk kostet 800 Millionen Mark, in: *WZ*, 10.4.1969.
316 Noack, Hans-Joachim: Biblis bekommt den größten Reaktor. Von der Gurke zum Atom. Eine ländliche Gemeinde erwartet frohgemut die Kernspalter, in: *SZ*, 10.11.1968; cf. Radkau, Angstabwehr, p. 32.
317 Leaflet "Im Interesse der Wahrheit", GAB, KKW Presseartikel 1968 bis 1970.
318 Noack, Hans-Joachim: Biblis bekommt den größten Reaktor. Von der Gurke zum Atom. Eine ländliche Gemeinde erwartet frohgemut die Kernspalter, in: *SZ*, 10.11.1968.
319 si: Kernkraftwerk Biblis wird gebaut. Ein Telegramm der RWE aus Essen bestätigt diese Absicht, in: *DE*, 19.7.1968.
320 lb: Europas größtes Kernkraftwerk. Es soll 1970 in Biblis begonnen werden – Eine aufeinander abgestimmte Atomforschung nötig, in: *SP*, 30.10.1968; cf. Barchet, Mac: Bis zum 1. Dezember: Dein Freund das Atom bittet in den Rosengarten. Heute spricht Professor Straßmann über Nutzung von Kern-Energie, in: *MM*, 27.11.1968; Barchet, Mac: Grundsatz für Atomkraftwerke: Erst Sicherheit – dann

Geld verdienen. Professor Straßmann bei der Eröffnung des Atom-Forums im Rosengarten, in: *MM*, 28.11.1968.

321 hz: Öffentlicher Anhörungstermin zum Bau eines Kernkraftwerks in Biblis. Bedenken jetzt ausgeräumt. Überdurchschnittliche Sicherheitsvorkehrungen – Fachvorträge klärten auf, in: *SP*, 19.1.1970.

322 lh: Umweltschützer: Skandal um Bibliser Kernkraftwerk. Sachbearbeiter Mitglied im RWE-Verwaltungsbeirat, in: *WZ*, 2.3.1973.

323 Noack, Hans-Joachim: Biblis bekommt den größten Reaktor. Von der Gurke zum Atom. Eine ländliche Gemeinde erwartet frohgemut die Kernspalter, in: *SZ*, 10.11.1968.

324 hz: Öffentlicher Anhörungstermin zum Bau eines Kernkraftwerks in Biblis. Bedenken jetzt ausgeräumt. Überdurchschnittliche Sicherheitsvorkehrungen – Fachvorträge klärten auf, in: *SP*, 19.1.1970.

325 Noack, Hans-Joachim: Biblis bekommt den größten Reaktor. Von der Gurke zum Atom. Eine ländliche Gemeinde erwartet frohgemut die Kernspalter, in: *SZ*, 10.11.1968.

326 st: RWE wählt Standort. Größtes Atomkraftwerk soll nach Biblis. Die 1000 Megawatt-Station könnte halb Hessen mit Strom versorgen, in: *FNP*, 21.6.1968.

327 Noack, Hans-Joachim: Biblis bekommt den größten Reaktor. Von der Gurke zum Atom. Eine ländliche Gemeinde erwartet frohgemut die Kernspalter, in: *SZ*, 10.11.1968.

328 Cf. rn: Schwierigkeiten mit Schlamm. Auf dem Weg zum Atomkraftwerk Biblis rollen zuerst Bagger und Planierraupen. Vor dem Reaktorbau müssen die Straßen fertig sein – Überflutetes Gelände hemmt, in: *MM*, 1.9.1969; In der Gemeinde Biblis hat das Atom die Gurke ersetzt, in: *SP*, 18.7.1974.

329 Noack, Hans-Joachim: Biblis bekommt den größten Reaktor. Von der Gurke zum Atom. Eine ländliche Gemeinde erwartet frohgemut die Kernspalter, in: *SZ*, 10.11.1968.

330 Wie wird das Kernkraftwerk aussehen? Dem Fortschritt kann man sich nicht mehr entgegenstemmen, in: *BZ*, 19.7.1968.

331 Wie wird das Kernkraftwerk aussehen? Dem Fortschritt kann man sich nicht mehr entgegenstemmen, in: *BZ*, 19.7.1968.

332 June 13, 1969, is the date stated in: Hessisches Ministerium für Umwelt, Klimaschutz, Landwirtschaft und Verbraucherschutz: Genehmigung nach § 7 Abs. 3 des Atomgesetzes zur Stilllegung und zum Abbau von Anlagenteilen des Kernkraftwerkes Biblis, Block A (A 022/12). Wiesbaden 2017, p. 19, https://umwelt.hessen.de/sites/umwelt.hessen.de/files/2021-09/genehmigungsbescheid_block_a_nicht_barrierefrei_.pdf (23.1.2023); regional newspapers pointed to a date in May, cf. bert: Beim 'Tag der offenen Tür'. Bibliser besuchten 'ihr' Kernkraftwerk. Großbaustelle glich einem Ameisenhaufen, in: *MM*, 11.10.1973; hkh: Im Ried entsteht Europas größtes Kernkraftwerk. Das RWE baut bei Biblis/Auftrag an die von AEG und Siemens gegründete Kraftwerk-Union vergeben, in: *MM*, 31.5.1969.

333 rn: Schwierigkeiten mit Schlamm. Auf dem Weg zum Atomkraftwerk Biblis rollen zuerst Bagger und Planierraupen. Vor dem Reaktorbau müssen die Straßen fertig sein – Überflutetes Gelände hemmt, in: *MM*, 1.9.1969.

334 Roggenbuck, Klaus-Joachim: Wird der Strom billiger? Atomfeuer glüht bald im Bibliser Reaktor. Energieversorgung durch die Uran-Pille. Umfangreiche Sicherheitsmaßnahmen vorgesehen – Bahnlinie vom Kreis genehmigt, in: *MM*, 13.1.1970.

335 wn: RWE-Bizeps erlaubt Investitions-Kraftakt. Mehr als 1 Milliarde DM pro Jahr/Warten auf Kapitalerhöhung/Abschreibungsbedarf für Gelsenberg, in: *MM*, 16.1.1970.

336 KR: Glänzende Kugel in der Rheinebene. Atomkraftwerk bei Biblis bereitet erhebliche Sorgen – Gefahr für den Fluß?, in: *FNP*, 28.11.1969.
337 Barchet, Mac: Blick in den Reaktor-Topf. Projekt Kirschgartshausen reift heran. Drei Kernkraftwerk-Typen stehen zur Wahl. Stadträte inspizieren Baustellen an der Elbe und an der Weser, in: *MM*, 21.4.1970; for Ratzel, cf. Ratzel, Ludwig, in: Vierhaus, Rudolf et al. (Eds.): Biographisches Handbuch der Mitglieder des Deutschen Bundestages 1949–2002. Vol. 2. N-Z. Anhang. Munich 2002, pp. 667–668.
338 ff: Tauziehen um Atomkraftwerk. Mannheim sagt Kampf an, in: *WZ*, 27.6.1968; cf. Barchet, Mac: Mit dem Atom auf du. Bürgermeister kletterte auf den Reaktor. Dr. Ratzel, Walter Kübler und Stadträte sammelten englische Erfahrungen, in: *MM*, 7.11.1969.
339 Barchet, Mac: Am Atom spalten sich die Geister. Elektroschock trifft Großkraftwerk. Geplantes Kernkraftwerk bei Oberhausen läßt Dr. Ratzel das Schlimmste befürchten, in: *MM*, 15.3.1969; cf. de: Badenwerk bleibt im Ländle und fürchtet sich nicht. Von Plänen des RWE 'nicht betroffen'/Ab 1970 wieder Ausbaupläne/ Kapital dringen nötig – aber woher, in: *MM*, 25.7.1968 (emphasis by hand); Minutes, Meeting, Bergstraße District Council, 23.6.1969, TOP 9/4, Hessisches Staatsarchiv Darmstadt (HStAD), H2 Bergstraße 7525.
340 Cf. Grawe, Joachim: Probleme des Umweltschutzes im deutsch-französischen Grenzgebiet, in: Zeitschrift für ausländisches öffentliches Recht und Völkerrecht 34 (1974), pp. 299–319, especially pp. 307–311, 316–318.
341 PS: Bescheidene Ergebnisse bei Kernenergieverwertung. Reaktorenentwicklung immer noch in Experimentierphase – Lange Reparatur- und Wartungszeiten, in: *DE*, 28.8.1968.
342 Baum, Karl-Heinz: Atom-Denkmäler allüberall. Fünf Kernkraftwerke sollen im Rhein-Neckar-Raum entstehen, in: *DE*, 7.8.1969.
343 KR: Glänzende Kugel in der Rheinebene. Atomkraftwerk bei Biblis bereitet erhebliche Sorgen – Gefahr für den Fluß?, in: *FNP*, 28.11.1969.
344 ulf: Einzig in der Bundesrepublik. Bibliser Kernkraftwerk kostet 800 Millionen Mark, in: *WZ*, 10.4.1969.
345 Pk: Reaktorleistung läßt sich regeln. Brennelemente im Bibliser Atomkraftwerk werden fernbedient/Wasser ausgewechselt, in: *WZ*, 5.12.1969.
346 Pk: Dampfturbine treibt 1500-MVA-Generator an. Zwei weitere Reaktoren sind geplant. Kapazität des Bibliser Werkes wird dann auf rund 3000 Megawatt ansteigen, in: *WZ*, 19.12.1969.
347 Strom für Hessen kommt bald aus Gemeinde Biblis. Vorbereitungen für Europas größtes Atomkraftwerk, in: *WZ*, 10.1.1970 (newspaper identified based on the page design, CG).
348 A fact later bemoaned: Trinkl, Erika: Südhessens Umweltschützer protestieren: Erweiterung des Kernkraftwerkes bringt zu viele Risiken mit sich. Furcht vor Wiederaufbereitungsanlage/Kritik am Bibliser Bürgermeister Seib, in: *MM*, 10.6.1975.
349 Eirich, George: Erörterungstermin in Biblis. Plakette des TÜV für das Kernkraftwerk. Einwände gegen RWE-Projekt ohne Erfolg, in: *MM*, 16.1.1970; the dates given are January 14 and 15, cf. Roggenbuck, Klaus-Joachim: Wird der Strom billiger? Atomfeuer glüht bald im Bibliser Reaktor. Energieversorgung durch die Uran-Pille. Umfangreiche Sicherheitsmaßnahmen vorgesehen – Bahnlinie vom Kreis genehmigt, in: *MM*, 13.1.1970; sz: 'Wer bezahlt den Umzug bei Atomgefahr?' Öffentlicher Anhörungstermin über Einsprüche gegen Bau eines Kernkraftwerkes bei Biblis, in: *DE*, 16.1.1970.
350 hz: Öffentlicher Anhörungstermin zum Bau eines Kernkraftwerks in Biblis. Bedenken jetzt ausgeräumt. Überdurchschnittliche Sicherheitsvorkehrungen – Fachvorträge klärten auf, in: *SP*, 19.1.1970.

351 sz: 'Wer bezahlt den Umzug bei Atomgefahr?' Öffentlicher Anhörungstermin über Einsprüche gegen Bau eines Kernkraftwerkes bei Biblis, in: *DE*, 16.1.1970.
352 Schutzgemeinschaft Deutscher Wald (Ed.): Schutzgemeinschaft Deutscher Wald. Wald ist unsere Sache. Bonn 2018, www.sdw.de/fileadmin/Bundesverband/PDF_Dokumente/SDW-Imagebroschuere.pdf (23.1.2023).
353 hz: Öffentlicher Anhörungstermin zum Bau eines Kernkraftwerks in Biblis. Bedenken jetzt ausgeräumt. Überdurchschnittliche Sicherheitsvorkehrungen – Fachvorträge klärten auf, in: *SP*, 19.1.1970.
354 hz: Öffentlicher Anhörungstermin zum Bau eines Kernkraftwerks in Biblis. Bedenken jetzt ausgeräumt. Überdurchschnittliche Sicherheitsvorkehrungen – Fachvorträge klärten auf, in: *SP*, 19.1.1970; for Scheuten, cf. Uekötter, Atomare, pp. 223–224.
355 Eirich, George: Erörterungstermin in Biblis. Plakette des TÜV für das Kernkraftwerk. Einwände gegen RWE-Projekt ohne Erfolg, in: *MM*, 16.1.1970.
356 hz: Öffentlicher Anhörungstermin zum Bau eines Kernkraftwerks in Biblis. Bedenken jetzt ausgeräumt. Überdurchschnittliche Sicherheitsvorkehrungen – Fachvorträge klärten auf, in: *SP*, 19.1.1970.
357 Eirich, George: Erörterungstermin in Biblis. Plakette des TÜV für das Kernkraftwerk. Einwände gegen RWE-Projekt ohne Erfolg, in: *MM*, 16.1.1970; Strahlenschutzkommission (Ed.): Verzeichnis der Mitglieder der SSK seit 1974. Bonn 2021, p. 2.
358 sz: 'Wer bezahlt den Umzug bei Atomgefahr?' Öffentlicher Anhörungstermin über Einsprüche gegen Bau eines Kernkraftwerkes bei Biblis, in: *DE*, 16.1.1970; for Meyer, cf. Müller, Kernenergie. Vol. 2, pp. 150, 355–356.
359 hz: Öffentlicher Anhörungstermin zum Bau eines Kernkraftwerks in Biblis. Bedenken jetzt ausgeräumt. Überdurchschnittliche Sicherheitsvorkehrungen – Fachvorträge klärten auf, in: *SP*, 19.1.1970.
360 sz: 'Wer bezahlt den Umzug bei Atomgefahr?' Öffentlicher Anhörungstermin über Einsprüche gegen Bau eines Kernkraftwerkes bei Biblis, in: *DE*, 16.1.1970.
361 Eirich, George: Erörterungstermin in Biblis. Plakette des TÜV für das Kernkraftwerk. Einwände gegen RWE-Projekt ohne Erfolg, in: *MM*, 16.1.1970.
362 hz: Öffentlicher Anhörungstermin zum Bau eines Kernkraftwerks in Biblis. Bedenken jetzt ausgeräumt. Überdurchschnittliche Sicherheitsvorkehrungen – Fachvorträge klärten auf, in: *SP*, 19.1.1970.
363 Carson, Rachel: Silent Spring. Boston 1962.
364 Eirich, George: Erörterungstermin in Biblis. Plakette des TÜV für das Kernkraftwerk. Einwände gegen RWE-Projekt ohne Erfolg, in: *MM*, 16.1.1970; hz: Öffentlicher Anhörungstermin zum Bau eines Kernkraftwerks in Biblis. Bedenken jetzt ausgeräumt. Überdurchschnittliche Sicherheitsvorkehrungen – Fachvorträge klärten auf, in: *SP*, 19.1.1970.
365 Eirich, George: Erörterungstermin in Biblis. Plakette des TÜV für das Kernkraftwerk. Einwände gegen RWE-Projekt ohne Erfolg, in: *MM*, 16.1.1970.
366 hz: Öffentlicher Anhörungstermin zum Bau eines Kernkraftwerks in Biblis. Bedenken jetzt ausgeräumt. Überdurchschnittliche Sicherheitsvorkehrungen – Fachvorträge klärten auf, in: *SP*, 19.1.1970.
367 Eirich, George: Erörterungstermin in Biblis. Plakette des TÜV für das Kernkraftwerk. Einwände gegen RWE-Projekt ohne Erfolg, in: *MM*, 16.1.1970.
368 sz: 'Wer bezahlt den Umzug bei Atomgefahr?' Öffentlicher Anhörungstermin über Einsprüche gegen Bau eines Kernkraftwerkes bei Biblis, in: *DE*, 16.1.1970.
369 hz: Öffentlicher Anhörungstermin zum Bau eines Kernkraftwerks in Biblis. Bedenken jetzt ausgeräumt. Überdurchschnittliche Sicherheitsvorkehrungen – Fachvorträge klärten auf, in: *SP*, 19.1.1970.
370 Cf. Radkau, Aufstieg und Krise, p. 116; Hanel, Bombe, p. 236.
371 Eirich, George: Erörterungstermin in Biblis. Plakette des TÜV für das Kernkraftwerk. Einwände gegen RWE-Projekt ohne Erfolg, in: *MM*, 16.1.1970.

372 hz: Öffentlicher Anhörungstermin zum Bau eines Kernkraftwerks in Biblis. Bedenken jetzt ausgeräumt. Überdurchschnittliche Sicherheitsvorkehrungen – Fachvorträge klärten auf, in: *SP*, 19.1.1970; Eirich, George: Erörterungstermin in Biblis. Plakette des TÜV für das Kernkraftwerk. Einwände gegen RWE-Projekt ohne Erfolg, in: *MM*, 16.1.1970.

373 hz: Öffentlicher Anhörungstermin zum Bau eines Kernkraftwerks in Biblis. Bedenken jetzt ausgeräumt. Überdurchschnittliche Sicherheitsvorkehrungen – Fachvorträge klärten auf, in: *SP*, 19.1.1970.

374 Cf. note 16.

375 sz: 'Wer bezahlt den Umzug bei Atomgefahr?' Öffentlicher Anhörungstermin über Einsprüche gegen Bau eines Kernkraftwerkes bei Biblis, in: *DE*, 16.1.1970; cf. hz: Öffentlicher Anhörungstermin zum Bau eines Kernkraftwerks in Biblis. Bedenken jetzt ausgeräumt. Überdurchschnittliche Sicherheitsvorkehrungen – Fachvorträge klärten auf, in: *SP*, 19.1.1970.

376 Eirich, George: Erörterungstermin in Biblis. Plakette des TÜV für das Kernkraftwerk. Einwände gegen RWE-Projekt ohne Erfolg, in: *MM*, 16.1.1970.

377 hz: Öffentlicher Anhörungstermin zum Bau eines Kernkraftwerks in Biblis. Bedenken jetzt ausgeräumt. Überdurchschnittliche Sicherheitsvorkehrungen – Fachvorträge klärten auf, in: *SP*, 19.1.1970.

378 Eirich, George: Erörterungstermin in Biblis. Plakette des TÜV für das Kernkraftwerk. Einwände gegen RWE-Projekt ohne Erfolg, in: *MM*, 16.1.1970.

379 hz: Öffentlicher Anhörungstermin zum Bau eines Kernkraftwerks in Biblis. Bedenken jetzt ausgeräumt. Überdurchschnittliche Sicherheitsvorkehrungen – Fachvorträge klärten auf, in: *SP*, 19.1.1970.

380 bert: 'Der Brunnen erhält einen Deckel', in: *MM*, 7.10.1970.

381 Marga Müller seems to have continued arguing against nuclear power in Biblis and even been able to activate supporters locally in the long run: et/bert: Akten liegen seit Mitte Dezember im Kreisbauamt. RWE stellt Bauantrag nur für Block C. Vorerst Abstand vom vierten Werk. Gemeinde nimmt Anfang Februar Stellung/Diskussion in Darmstadt um Wasser für Block B, in: *MM*, 15.1.1976.

382 The start of which, in other words, need not be considered 'unexpected', cf. Radkau, Zukunft, p. 161.

383 (kn): Das Kernkraftwerk Biblis und sein Kühlwasser. Schutzgemeinschaft Deutscher Wald schrieb an den Petitionsausschuß des Hessischen Landtags, in: *DE*, 18.9.1970; for Karry, cf. Karry, Heinz-Herbert, in: Hessische Biografie, www.lagis-hessen.de/pnd/12263814X (23.1.2023).

384 Götter, Christian: Accepted to Cool. Conflicts about Cooling Technologies for Riverside Nuclear Power Plants, in: Historical Social Research 49 (2024), 1, pp. 126–147, here p. 134; cf. lb: Schadet Kühlwasser der Rhein-Biologie? Bau von zwölf Atomkraftwerken entlang des Rheins geplant – Ein Wunsch der BASF Ludwigshafen, in: *SP*, 6.5.1971.

385 os: Neuer Pfeiler der Energieversorgung. Anhörungstermin wegen Block B des Kernkraftwerkes in Biblis, in: *DE*, 26.11.1971.

386 Götter, Accepted, p. 134; for Best: Best, Werner, in: Hessische Biografie, www.lagis-hessen.de/pnd/1102942995 (23.1.2023).

387 rl: Lampertheim meldet Bedenken an. Bürgermeister Pfeiffer zu geplantem Atomkraftwerk, in: *DE*, 25.4.1970; MacBarchet: Die kritische Betrachtung. Der Rhein darf nicht zum toten Strom werden, in: *MM*, 11.7.1970; Eirich, George: Kernkraftwerk nimmt Formen an. In Biblis beginnt das Atomzeitalter. Ab 1974 fließt Strom aus dem südhessischen Ried, in: *MM*, 14.12.1970; Der Rhein darf nicht zu Europas Riesenbadewanne werden. Atomkraftwerk Biblis regte Länderkommission an, einen 'Wärmelastplan' zu entwickeln, in: *Hessischer Kommunalanzeiger*, 25.3.1971; cf. Grawe, Probleme, especially pp. 307–311, 316–318; Radkau, Aufstieg und Krise, pp. 397–398.

388 Cf. Rüdig, Anti-Nuclear, pp. 129–131; Löser, Bürgerinitiativen, pp. 19–33; Milder, Fish, pp. 43–57.
389 ff: Umweltschützer nehmen Bibliser Kernkraftwerk aufs Korn, in: *WZ*, 15.1.1973.
390 oh: CDU: Das Ried ist in Gefahr. Lorscher CDU fordert umfassende Hilfsmaßnahmen – Resolution verabschiedet, in: *DE*, 18.10.1972.
391 Laufs, Reaktorsicherheit. Vol. 1, p. 89.
392 Bieber, Kernenergienutzung, pp. 32, 103–104.
393 Laufs, Reaktorsicherheit. Vol. 1, pp. 70–75, 89–92, 477–479; cf. Radkau et al., Aufstieg und Fall, pp. 273–377.
394 Eirich, George: Kernkraftwerk nimmt Formen an. In Biblis beginnt das Atomzeitalter. Ab 1974 fließt Strom aus dem südhessischen Ried, in: *MM*, 14.12.1970; for Eirich, who published under his nickname George, cf. -tin: Gedenken an jene, die nicht mehr sind, in: *MM*, 2.6.2014.
395 art: Ludwigshafener Stadtrat billigt den Bau eines Atomkraftwerks, in: *WZ*, 3.2.1970.
396 Mac: Ist die Sicherheit gewährleistet? Vorsorglich den Fuß in der Atom-Tür. Einspruch der Stadt Mannheim gegen BASF-Kernkraftwerk-Pläne, in: *MM*, 3.3.1970.
397 Lichtenberg, Gerrit: Zusammenballung auf engem Raum nicht angenehm: Beim Bau von Atomkraftwerken in Stadtnähe muß Sicherheit das oberste Gebot sein. Mannheimer Bürger melden Bedenken an/Einsprüche gegen BASF-Projekt, in: *MM*, 3.4.1970; cf. Radkau, Aufstieg und Krise, pp. 376–380.
398 Radkau, Technik in Deutschland, p. 360; Radkau, kurze Geschichte, pp. 9–10; for a more detailed description of the background: Radkau, RWE zwischen Kernenergie und Diversifizierung, pp. 222–224; for Leussink: Behrmann, Günter C.: Leussink, Hans, in: Kempf, Udo, Hans-Georg Merz (Eds.): Kanzler und Minister 1949–1998. Biografisches Lexikon der deutschen Bundesregierungen. Wiesbaden 2001, pp. 432–437.
399 dpa: Minister Leussink: Keine absolute Sicherheit von Kernkraftwerken. Projekt BASF und Bau eines Schnellen Brüters zurückgestellt, in: *DE*, 18.8.1970; cf. Laufs, Reaktorsicherheit. Vol. 1, p. 479.
400 Wer soll was noch erforschen? Die BASF möchte keine Zeit verlieren. Der Bau des Kernkraftwerks ist für die Ludwigshafener ein 'dringliches Vorhaben', in: *MM*, 20.8.1970.
401 RN – Eigener Bericht: Kritische Reaktor-Reaktionen in Mainz. Neubauer: Aufschub des BASF-Kernkraftwerkes schwer verständlich, in: *MM*, 29.8.1970; another critic was Karl Winnacker, cf. Winnacker, Karl: Schicksalsfrage Kernenergie. Düsseldorf 1978, p. 240; for Neubauer: Neubauer, Dr. rer. pol. Hanns. CDU, in: Der Präsident des Landtags Rheinland-Pfalz (Ed.): Die Stellvertreter des freien Volkes. Die Abgeordneten der Beratenden Landesversammlung und des Landtags Rheinland-Pfalz von 1946 bis 2015. Wiesbaden 2016, p. 490.
402 Cf. Radkau, Aufstieg und Krise, pp. 380–382; Jung, Matthias: Öffentlichkeit und Sprachwandel. Zur Geschichte des Diskurses über die Atomenergie. Opladen 1994, pp. 74–75; Müller, Kernenergie. Vol. 2, pp. 405–407; Radkau, RWE zwischen Kernenergie und Diversifizierung, pp. 229–230; for the parallel development of the Würgassen protests: Radkau, RWE zwischen Kernenergie und Diversifizierung, pp. 445–449; Radkau, kurze Geschichte; for the final end of the BASF-project: Laufs, Reaktorsicherheit. Vol. 1, pp. 480, 486–487.
403 Cf. Kirchgässer, Wilhelm (Mannheim): Das große Risiko, in: *MM*, 16.4.1971; Böttger, Peter (Mannheim-Feudenheim): Sicherheit von Kernkraftwerken, *MM*, 27.2.1971; Josso, Jakob (Speyer): Die Ohnmacht der Wahrheit, in: *MM*, 16.4.1971.
404 Cf. Beck, Prof. Dr. Hanno: Kernreaktor der BASF, in: *MM*, 16.12.1971.

405 spd: Kernkraftwerke und Umweltschutz. Anfragen und Anträge der SPD-Landtagsfraktion, in: *SP*, 14.9.1971; for Bugert, cf. Bugert, Erwin, in: Hessische Biografie, www.lagis-hessen.de/pnd/1023786583 (23.1.2023).
406 kh: 'Bau einer Mischanlage verhindern'. IG Umweltschutz für den Südkreis legte Einsprüche ein, in: *DE*, 14.3.1973.
407 Sternglass, Ernest J.: Low-Level Radiation. New York 1972; for his role, cf. 'Rational Atoms'.
408 k: Schäden durch Atomkraftwerke? Umweltschützer am Weißen Turm zum Kernkraftwerk, in: *DE*, 12.7.1972.
409 K: Umweltschutz darf keine Partei- und Ländergrenzen kennen. Gespräch mit M. Schlappner (MdL) über Wert von Bürgerinitiativen, über Umweltschutz und Kernkraftwerk in Biblis, in: *DE*, 2.2.1973.
410 Loff, Birgit-Ingeborg: Der Tag, der niemals kommen darf. Deutschlands größtes Kernkraftwerk und seine Sicherheit, in: *Frankfurter Rundschau*, 28.9.1974.
411 ff: Umweltschützer nehmen Bibliser Kernkraftwerk aufs Korn, in: *WZ*, 15.1.1973.
412 tom: Atomkraftwerk Biblis beunruhigt die Nachbarn. Protestkundgebung der Aktionsgemeinschaft für Umweltschutz Rheinhessen-Süd in Worms, in: *MM*, 9.5.1975; the *Darmstädter Echo* later on claimed that the initiative had been founded to fight the Biblis power plant in 1972, mig: Unter 'Polizeischutz'. Landesverband Bürgerinitiativen gegen Kernkraftwerk, in: *DE*, 15.4.1975; for Staab, cf. Franke, Nils M.: 40 Jahre BUND. Die Geschichte des Bund für Umwelt und Naturschutz Deutschland e. V., 1975–2015. Cologne 2015, www.bund.net/fileadmin/user_upload_bund/_migrated/publications/150721_bund_ueber_uns_40_jahre_bund_geschichte_des_bund.pdf (23.1.2023), pp. 22, 24, 36.
413 Ff: Umweltschützer verlangen Baustopp in Biblis. Resolution in der Hammer Kundgebung verfaßt. Auf Anhieb etwa 300 Unterschriften/Wissenschaftliche Beiträge gaben Einblick, in: *WZ*, 22.1.1973 (date and newspaper deduced from context), GAB, Presseartikel KKW 1977; k: Umweltschäden durch Atomenergie? Aktionsgemeinschaft Umweltschutz diskutierte mit RWE-Vertretern über Kernkraftwerk, in: *DE*, 31.1.1973.
414 ff: Umweltschützer nehmen Bibliser Kernkraftwerk aufs Korn, in: *WZ*, 15.1.1973.
415 Rieder, Barbara: Der Bundesverband Bürgerinitiativen Umweltschutz. Geschichte, Struktur und Aktionsformen einer Dachorganisation der Ökologiebewegung. State Examination Thesis, Berlin 1980, especially pp. 14–15, 28–32; Nelkin, Dorothy, Michael Pollak: The Atom Besieged. Extraparliamentary Dissent in France and Germany. Cambridge 1981, p. 126; Kempf, Udo: Der Bundesverband Bürgerinitiativen Umweltschutz (BBU), in: Guggenberger, Bernd, Udo Kempf (Eds.): Bürgerinitiativen und repräsentatives System. Opladen 21984, pp. 404–423, here pp. 405–407, 412; Paul, Anti-AKW-Bewegung, pp. 20–21; Brüggemeier, Schranken, pp. 256–257; Metzger, Birgit: 'Erst stirbt der Wald, dann Du!' Das Waldsterben als westdeutsches Politikum (1978–1986). Frankfurt/Main 2015, pp. 76–80; Uekötter, Frank: Ökologische Verflechtungen. Umrisse einer grünen Zeitgeschichte, in: Bösch, Frank (Ed.): Geteilte Geschichte. Ost- und Westdeutschland 1970–2000. Göttingen 2015, pp. 117–152, here pp. 126–127; Becker-Schaum, Christoph: Bürgerinitiativen und die Geburt der Grünen, in: Lakeberg, Beata Dorota, Hans-Christian Pust (Eds.), Atom. Strom. Protest. 50 Jahre Wyhl und anderswo. Ostfildern 2023, pp. 142–151, here pp. 146–149; cf. Brüggemeier, Tschernobyl, pp. 210–211; for Wüstenhagen: Becker-Schaum, Bürgerinitiativen, p. 148; Jensen, Annette: Ein Bürger im Kampf gegen AKW. Hans-Helmut Wüstenhagen, in: *taz*, 4.12.1996.
416 Mez, Lutz: Bundesrepublik Deutschland. Der unaufhaltsame Aufstieg zur Atommacht, in: Mez, Lutz (Ed.): Der Atomkonflikt. Berichte zur internationalen

Atomindustrie, Atompolitik und Anti-Atom-Bewegung. Mit einer Einleitung von Robert Jungk. Reinbek 1981, pp. 29–53, here pp. 45–53.

417 Renner, Bauplatz, p. 95; Renner, Jens: Die Bundesrepublik im Ausnahmezustand. 24. September 1977. Die Demonstration gegen den Schnellen Brüter in Kalkar, in: Borchers, Uli (Ed.): Die AKW-Protestbewegung von Wyhl bis Brokdorf. Hamburg 2011, pp. 131–149, here p. 136; Paul, Reimar: Unberechenbar auch für sich selbst. Eine Bilanz der Anti-AKW-Bewegung von Wyhl bis Tschernobyl, in: Borchers, Uli (Ed.): Die AKW-Protestbewegung von Wyhl bis Brokdorf. Hamburg 2011, pp. 197–217, here pp. 207–208.

418 Ff: Umweltschützer verlangen Baustopp in Biblis. Resolution in der Hammer Kundgebung verfaßt. Auf Anhieb etwa 300 Unterschriften/Wissenschaftliche Beiträge gaben Einblick, in: *WZ*, 22.1.1973 (date and newspaper deduced from context), GAB, Presseartikel KKW 1977; k: Umweltschäden durch Atomenergie? Aktionsgemeinschaft Umweltschutz diskutierte mit RWE-Vertretern über Kernkraftwerk, in: *DE*, 31.1.1973. The DE identified Wüstenhagen's deceased father, Karl Wüstenhagen, as the discussion's chairman.

419 k: Umweltschäden durch Atomenergie? Aktionsgemeinschaft Umweltschutz diskutierte mit RWE-Vertretern über Kernkraftwerk, in: *DE*, 31.1.1973; ff: Weniger Sonne durch Dampffahne des Bibliser Kernkraftwerkes?, in: *WZ*, 22.1.1973; for Muth: Fuchß, Peter: Muth, Reinhard (1931–2015), in: Gesellschaft für Geschichte des Weines e. V. (Ed.): Persönlichkeiten der Weinkultur, www.geschichte-des-weines.de/index.php?option=com_content&view=article&id=727:muth-reinhard-1931-2015&catid=45:persoenlichkeiten-a-z&Itemid=83 (23.1.2023).

420 Ff: Umweltschützer verlangen Baustopp in Biblis. Resolution in der Hammer Kundgebung verfaßt. Auf Anhieb etwa 300 Unterschriften/Wissenschaftliche Beiträge gaben Einblick, in: *WZ*, 22.1.1973 (date and newspaper deduced from context), GAB, Presseartikel KKW 1977.

421 ff: Wachsame Umweltschützer entdeckten die Flugschneise über Bibliser Atomwerk, in: *WZ*, 17.5.1973.

422 ff: Umweltschützer nehmen Bibliser Kernkraftwerk aufs Korn, in: *WZ*, 15.1.1973.

423 Probably the geologist László Trunkó, from the Karlsruhe Museum of Natural History.

424 k: Umweltschäden durch Atomenergie? Aktionsgemeinschaft Umweltschutz diskutierte mit RWE-Vertretern über Kernkraftwerk, in: *DE*, 31.1.1973.

425 Probably Hans Klumb, from the University of Mainz; cf. Hans Klumb, in: Verzeichnis der Professorinnen und Professoren der Universität Mainz, http://gutenberg-biographics.ub.uni-mainz.de/id/d373a232-bb00-4a9c-bd37-8eb6196a8612 (23.1.2023).

426 k: Umweltschäden durch Atomenergie? Aktionsgemeinschaft Umweltschutz diskutierte mit RWE-Vertretern über Kernkraftwerk, in: *DE*, 31.1.1973.

427 ff: Umweltschützer nehmen Bibliser Kernkraftwerk aufs Korn, in: *WZ*, 15.1.1973; for Schmitt-Vockenhausen: Albrecht, Willy: Schmitt-Vockenhausen, Hermann, bis 1960 Schmitt, in: Vierhaus, Handbuch. Vol. 2, pp. 765–766.

428 ff: Weniger Sonne durch Dampffahne des Bibliser Kernkraftwerkes?, in: *WZ*, 22.1.1973.

429 k: Umweltschäden durch Atomenergie? Aktionsgemeinschaft Umweltschutz diskutierte mit RWE-Vertretern über Kernkraftwerk, in: *DE*, 31.1.1973.

430 ff: Umweltschützer nehmen Bibliser Kernkraftwerk aufs Korn, in: *WZ*, 15.1.1973.

431 Probably Dietmar Kuhnt, then legal advisor at the RWE's headquarters in Essen, who later on became its CEO; cf. Baentsch, Wolfram: Gemeinsame Wertvorstellungen – das ist unser Imperativ, in: Schweer, Dieter, Wolf Thieme (Eds.): RWE – 'Der gläserne Riese'. Ein Konzern wird transparent. Wiesbaden 1998, pp. 97–104, here p. 104.

432 k: Umweltschäden durch Atomenergie? Aktionsgemeinschaft Umweltschutz diskutierte mit RWE-Vertretern über Kernkraftwerk, in: *DE*, 31.1.1973.
433 ff: Umweltschützer verlangen Baustopp in Biblis. Resolution in der Hammer Kundgebung verfaßt. Auf Anhieb etwa 300 Unterschriften/Wissenschaftliche Beiträge gaben Einblick, in: *WZ*, 22.1.1973 (date and newspaper deduced from context), GAB, Presseartikel KKW 1977.
434 ff: Umweltschützer nehmen Bibliser Kernkraftwerk aufs Korn, in: *WZ*, 15.1.1973.
435 k: Umweltschäden durch Atomenergie? Aktionsgemeinschaft Umweltschutz diskutierte mit RWE-Vertretern über Kernkraftwerk, in: *DE*, 31.1.1973.
436 ff: Umweltschützer nehmen Bibliser Kernkraftwerk aufs Korn, in: *WZ*, 15.1.1973.
437 Friedel Fabrie counted 300 signatures, Ff: Umweltschützer verlangen Baustopp in Biblis. Resolution in der Hammer Kundgebung verfaßt. Auf Anhieb etwa 300 Unterschriften/Wissenschaftliche Beiträge gaben Einblick, in: *WZ*, 22.1.1973 (date and newspaper deduced from context), GAB, Presseartikel KKW 1977.
438 dpa/lrs: Kernkraftwerk-Baustopp gefordert, in: *MM*, 29.1.1973.
439 ff: Die Umweltschützer alarmieren die Minister. Hessen erteilte Genehmigung für Atomblock B, in: *WZ*, 3.2.1973; for Meyer: Meyer, Otto. CDU, in: Der Präsident des Landtags Rheinland-Pfalz (Ed.): Die Stellvertreter des freien Volkes. Die Abgeordneten der Beratenden Landesversammlung und des Landtags Rheinland-Pfalz von 1946 bis 2015. Wiesbaden 2016, pp. 461–463.
440 hs: Keine Angst vor dem Atomkraftwerk. Ministerien und RWE: Eine ungünstige Beeinflussung des Mikroklimas in Biblis ausgeschlossen, in: *SP*, 6.2.1973.
441 ff: Wirklich Sturm im Wasserglas, in: *WZ*, 21.2.1973.
442 hs: Keine Angst vor dem Atomkraftwerk. Ministerien und RWE: Eine ungünstige Beeinflussung des Mikroklimas in Biblis ausgeschlossen, in: *SP*, 6.2.1973.
443 For Buch, cf. Buch, Georg, in: Hessische Biografie, www.lagis-hessen.de/pnd/119342200 (23.1.2023).
444 This was another issue that was also prominent in the Wyhl debates.
445 lh: Der Hessische Landtag soll Biblis überprüfen, in: *SP*, 2.3.1973.
446 lh: Umweltschützer: Skandal um Bibliser Kernkraftwerk. Sachbearbeiter Mitglied im RWE-Verwaltungsbeirat, in: *WZ*, 2.3.1973.
447 dpa/lrs: Genscher soll Farbe bekennen. Aktionsgemeinschaft verlangt Auskunft über Atomkraftwerke, in: *MM*, 7.4.1973.
448 kh: 'Bau einer Mischanlage verhindern'. IG Umweltschutz für den Südkreis legte Einsprüche ein, in: *DE*, 14.3.1973.
449 Cf. Noack, Hans-Joachim: Ein Atommeiler kommt ins Dorf. Von Umweltverschmutzung wollen sie nichts hören, in: *Die Zeit*, 8.10.1971.
450 Lang, Walter (Rohrhof): Leserbrief, in: *WZ*, 9.3.1973.
451 R.: Viel Unzufriedenheit und harte Diskussionen um das Kernkraftwerk vor der Wormser Tür, in: *WZ*, 9.4.1975.
452 os: Umweltverschmutzer? SPD Biblis-Wattenheim mit erfreulicher Bilanz, in: *DE*, 26.1.1973.
453 -gz-: Arbeit soll intensiviert werden. Bibliser Umweltschützer entdeckten Flugschneise, in: *WZ*, 9.5.1974.
454 -gz-: Umweltschützer fordern Antwort, in: *WZ*, 7.8.1974.
455 Block A des Kernkraftwerkes Biblis offiziell übergeben, in: *Unidentified Newspaper*, 15.4.1975, in: GAB, KKW.
456 ff: Brandversicherung klammert Risiken bei Schäden durch Kernenergie aus, in: *WZ*, 28.4.1973; cf. Wehner, Christoph: Die Versicherung der Atomgefahr. Risikopolitik, Sicherheitsproduktion und Expertise in der Bundesrepublik Deutschland und den USA 1945–1986. Göttingen 2017.

457 Gesetz über die friedliche Verwendung der Kernenergie und den Schutz gegen ihre Gefahren (Atomgesetz), 23.12.1959, in: Bundesgesetzblatt 1959 I, Nr. 56, 31.12.1959, pp. 814–828; bert: Ministerialrat Wagner gab Auskunft. Bedenken gegen den Bau des Kraftwerkes werden von Verantwortlichen nicht geteilt. Bibliser Gemeindevertretung befaßte sich mit befürchteten Gefahren, in: *MM*, 15.5.1973; cf. Bleidick, Pionier, p. 116; Göppner, Nadine: Vorgeschichte und Entstehung des Atomgesetzes vom 23.12.1959. Frankfurt/Main 2013; Wehner, Versicherung.

458 os: Rheinerwärmung, Kühltürme, Versicherung. Information über Genehmigungsverfahren für Kernkraftwerk, in: *DE*, 16.5.1973.

459 ff: Wachsame Umweltschützer entdeckten die Flugschneise über Bibliser Atomwerk, in: *WZ*, 17.5.1973.

460 Podiumsdiskussion über das Kraftwerk. Auch nach dem Pro und Contra blieben die Meinungen verschieden. Umweltschutz-Initiatoren bestritten die Feststellungen der Verantwortlichen, in: *MM*, 7.6.1973; ff: Junge Union verhalf den Umweltschützern, ihre ablehnende Haltung zu demonstrieren, in: *WZ*, 7.6.1973.

461 os: Wie sicher ist nun ein Kernkraftwerk? Erstes Bibliser Kernkraftwerk-Gespräch – Fragerunde mit RWE-Verantwortlichen, in: *DE*, 26.5.1973.

462 Cf. the report from the second instalment in September 1973: -bert-: Die ersten Brenn-Elemente werden im Oktober erwartet, in: *SP*, 25.9.1973; for the fifth: -et-: Probeläufe an Block A abgeschlossen. Im Kernkraftwerk Biblis nähert sich der Tag X. Noch im Juni soll Reaktor zum ersten Male kritisch werden, in: *MM*, 1.6.1974.

463 os: Mehr als 15 000 Besucher im Kraftwerk. Zwei Tage der offenen Tür beim KKW Biblis, in: *DE*, 10.10.1973.

464 Cf. Redlich, Maria (Mannheim): Wettlauf in den Untergang zum Problem der Kernreaktoren, in: *MM*, 5.10.1973.

465 ff: Exemplar über meteorologisches Gutachten nicht herausgegeben, in: *WZ*, 18.3.1974; hei: Mahnung an Minister Karry. Untersuchung über Kraftwerk Biblis steht noch aus, in: *DE*, 20.6.1974.

466 ni: Bürgerinitiative warnt vor Reaktor: Im Katastrophenfall 100000 Tote, in: *WZ*, 22.6.1974.

467 -el-: Am Rhein auf Bibliser Gemarkung: Der Welt größtes Kernkraftwerk nimmt seinen Betrieb auf, in: *MM*, 18.7.1974.

468 ni: Bürgerinitiative warnt vor Reaktor: Im Katastrophenfall 100000 Tote, in: *WZ*, 22.6.1974.

469 Cf. opd: Gibt es eine Alternative für die Kernenergie? Tagung mit dem Thema 'Für und wider die Atomkraftwerke am Oberrhein'/Sonnenenergie fördern, in: *MM*, 28.1.1975; Leicher, Günther: Verquere Fronten in Wyhl. Umweltschützer und linke Demonstranten vereint gegen Kernkraftwerk, in: *WZ*, 25.2.1975.

470 Block A des Kernkraftwerkes Biblis offiziell übergeben, in: *Unidentified Newspaper*, 15.4.1975, GAB, KKW.

471 Trinkl, Erika: Insgesamt vier Kernkraftwerks-Blöcke geplant: Erstes Atom-Ei wird heute übergeben. RWE bekennt sich zum Standort Biblis. Block A kostet 850 Millionen/Umweltschützer gegen Erweiterung, in: *MM*, 15.4.1975.

472 Minutes, Meeting, Biblis Municipal Council, 1.9.1970, p. 9, GAB, Gemeindevertretungssitzungen; os: Baubeginn für Block C im nächsten Jahr? Pressegespräch in Biblis – Schärfere Kontrollen seit Panzerfaust-Gag, in: *DE*, 19.6.1975.

473 et: Kernkraftwerk in Biblis wird weiter ausgebaut, in: *MM*, 8.3.1974.

474 os: Katastrophenabwehrpläne bleiben in der Schublade. Sechstes Bibliser Kernkraftwerksgespräch – Probebetrieb verläuft planmäßig, in: *DE*, 10.9.1974; cf. Minutes, Meeting, Bergstraße District Council, 25.8.1975, TOP 4/4, HStAD, H2 Bergstraße 7527.

475 Siegelmann, Bert: Probebetrieb von Reaktor und Turbine. Kernkraftwerk läuft ohne Pannen. Thema Katastrophenabwehr unbeliebt. Notfallpläne sollen nicht veröffentlicht werden/Fischsterben gebremst, in: *MM*, September 1974 (no specific date given).
476 Siegelmann, Bert: 160 Meter hohe Kühltürme am Bibliser Rheinufer geplant. Baubeginn für Block C Frühjahr '76? RWE rechnet nicht mit hartem Protest. Kernkraftgegner werden 'verdammt ernst' genommen/Grundwasser bleibt unberührt, in: *MM*, 28.5.1975.
477 mig: Unter 'Polizeischutz'. Landesverband Bürgerinitiativen gegen Kernkraftwerk, in: *DE*, 15.4.1975.
478 Protest gegen Nuklearpark Biblis, in: *WZ*, 3.5.1975; R.: Ausbaustopp gefordert. Gegen Erweiterung des Atomkraftwerks Biblis. Kundgebung und Unterschriftenaktion, in: *WZ*, 9.5.1975; tom: Atomkraftwerk Biblis beunruhigt die Nachbarn. Protestkundgebung der Aktionsgemeinschaft für Umweltschutz Rheinhessen-Süd in Worms, in: *MM*, 9.5.1975.
479 mg: Gegen Reaktoren in Ballungs-Gebieten. IG Umweltschutz fordert Baustopp von Kernkraftwerken, in: *DE*, 9.6.1975.
480 Cf. Aus Briefen an die WZ. Zum Bibliser Kernkraftwerk: Gegen jede Verharmlosung, in: *WZ*, 24.5.1975; Jung, Hartmut and Lieselotte (Biblis): Nein zu 'brüderlicher Umarmungstaktik', in: *MM*, 27.5.1975.
481 There were others, however, cf. gr: RPS: Keine Auswirkungen auf das Klima. Zwischenbericht über Auflagenerfüllung des Kernkraftwerks, in: *DE*, 22.1.1975; Kitschelt, Kernenergiepolitik, p. 201.
482 Urban, Christel: MM-Gespräch mit SPD-Fraktionsführer Karl Lutzi. Zwei Atommeiler in Biblis sind Groß-Rohrheim genug. Gemeindevertreter legen gegen weiteren Ausbau Einspruch ein, in: *MM*, 24.4.1975.
483 Worms droht mit Klage gegen Kernkraftwerk, in: *SP*, 14.6.1975.
484 Karbiner, Hans: Stadtverordnetensitzung griff aktuelles Thema auf: Lampertheimer 'Nein' für weitere Reaktoren gründet sich auf Furcht vor Umweltschäden. Künstliche Panikmache abgelehnt/Kernkraftwerke dezentralisieren, in: *MM*, 20.5.1975.
485 gr: Gegen Kernkraftwerk-Ausbau, in: *DE*, 19.6.1975.
486 In der Gemeinde Biblis hat das Atom die Gurke ersetzt, in: *SP*, 18.7.1974.
487 BS: SPD läßt sich nicht in linksextreme Ecke schieben. Opposition in Biblis verurteilt Äußerungen von Bürgermeister Seib. Warnung vor 'gedankenlosen, enthusiastischen Befürwortern' weiterer Kernreaktoren, in: *MM*, 22.5.1975.
488 os: 'Wir wollen kein zweites Ruhrgebiet'. Auch Bibliser Gemeindevertretung fordert Aussetzung der Genehmigung für KKW, in: *DE*, 21.6.1975.
489 Minutes, Meeting, Biblis Municipal Council, 18.6.1975, p. 7 f., in: GAB, Gemeindevertretungssitzungen.
490 Maubacq, Francoise, Claudette Tailland: Elecnuc. Les Centrales Nucleaires Dans le Monde. Nuclear Power Plants in the World. 1997, 31.12.1996, p. 64, in: IAEA-Collection, CEA-DSE-SEE – 1997-k, FR9803738, https://inis.iaea.org/collection/NCLCollectionStore/_Public/30/020/30020260.pdf (23.1.2023).
491 Cf. Kitschelt, Kernenergiepolitik, pp. 201–205; Lieb, Aktivismus, p. 233, observing the dissolution of the local Green Party group.
492 One could argue that this kind of openness was facilitated by the strict secrecy around the military core of the country's early nuclear activities.
493 Cf. note 20.
494 Cf. Welsh, Nuclear Nation, pp. 1–2, 11.
495 Cf. Welsh, Nuclear Nation, pp. 21–22.

2 Emotional Atoms[1]

The need to maintain public support for the nuclear arms race at the heart of the Cold War stood behind the invention of one of the most consequential techno-political campaigns of the 20th century: When US President Dwight D. Eisenhower presented the concept of "Atoms for Peace" to the world in 1953, his explicit goal was to make the "fear of the atom . . . disappear from the minds of the people and the governments" by turning "this greatest of destructive forces . . . into a great boon, for the benefit of all mankind." He pitted hope against fear: "The United States knows that peaceful power from atomic energy is no dream of the future. That capability, already proved, is here – now – today."[2] As David E. Nye aptly put it in his seminal study of the "American Technological Sublime", the sublime character of nuclear explosions was to be "domesticated" by the campaign for the "Peaceful Atom", highlighting humankind's ability to employ what Thomas P. Hughes described as large technological systems to control the vast powers of nature, slumbering at the core of the atom.[3]

Obviously, hopes and fears were interwoven with nuclear energy from the start[4] and during its early days, its proponents seemed considerably more emotional about it than its critics.[5] Therefore, it is no surprise that hopes and fears have commanded the attention of historians of atomic energy for quite some time. Most prominently, they were the focus of Spencer R. Weart's study on "Nuclear Fear". He presented the history of radiation as interwoven with emotive images of apocalypse, feared by the atom's opponents, and salvation, hoped for by its advocates.[6] To be sure, it was not that clear-cut a split. As can already be gleaned from the emotional strategy behind the "Peaceful Atom", hopes and fears were present on both sides in the debates about the atom, and even at the same time, as was emphasised, not least, by Joachim Radkau.[7]

Nonetheless, typical phases of early enthusiasm and later fears have often been described, even by Radkau, with the scales tipping around the 1960s.[8] A number of studies have been undertaken into both of these phases, and in the transition from one to the other – from, one could say, the era dominated by the hopeful writings of Ernst Bloch to that epitomised by Hans Jonas' "heuristics of fear".[9] When it comes to the euphoric early years, the

DOI: 10.4324/9781003583677-3

practical implementation of the "Atoms for Peace" campaign tended to be in focus, with highlights such as the 1955 nuclear exhibition in Geneva. Some attention is also paid to the combination of the technology with concepts of societal planning and the "Peaceful Atom's" special appreciation on the political Left, as well as the role the campaign as such played in overcoming fears of nuclear weapons.[10] Indeed, fear can be seen as the emotion that has received most of the attention from historians of technology.[11] It is also the one that has been most widely researched regarding the atom's emotional history. In particular, its relationship to nuclear weapons has gained much attention.[12] But fears around the "Peaceful Atom" were also studied,[13] as were the attempts to manage the emotion, not only in the early days of the "Atoms for Peace" campaign but throughout the period under consideration here.[14] It is worth noting that conceptions of a specific and irrational "German Angst" around the atom were contested, not least by Radkau and Andrew S. Tompkins.[15] Finally, the transition from one phase to the other has often been examined. Regarding the emotional dimension of this development, Radkau argued that, during the 1970s, people realised that the early euphoria for atomic energy had been a mere "empty phrase . . . no reality". Based upon this recognition, the "greatest anger" grew "out of disappointed love".[16] Somewhat more common is the observation of a return of the fears, originally sidelined by Eisenhower's campaign, caused by a growing knowledge about the dangers of nuclear power spreading in the public, diminishing the trust placed in pro-nuclear experts.[17] Focusing on the emotional level of these changing relationships of trust, Albrecht Weisker highlighted the emotional dimension of the rising mistrust from the 1970s.[18]

Overall, then, historians did recognise that understanding the history of nuclear energy without considering the role played by various emotions would be nigh impossible.[19] This holds true for both nations under consideration here, even though there is a common conviction that British anti-nuclear activists were less prone to emotionality than their continental counterparts.[20] Yet, in practice, most historians did, like President Eisenhower, focus on hopes and fears while ignoring other emotions connected to and shaping the history of the "Peaceful Atom".

To broaden the conception of nuclear emotions for this analysis, an approach to emotions derived from the works of Carroll Ellis Izard is used here. According to Izard, a number of "basic emotions" can be described which are shared by all human beings, namely "interest, joy/happiness, sadness, anger, disgust, and fear".[21] They appear in varying intensities and combinations involving cognitive components, which are culturally dependent, longer lasting, and very potent motivators for action.[22] The latter, called "emotion schemas" by Izard, may include what James R. Averill described as "intellectual emotions", for example, hope and pride – or variations of basic emotions that are not restricted to short-term experiences, but drawn out over longer periods of time.[23] Coming together under the more general term of "emotions", they can be grouped in positive ones, which tend to broaden the outlook of those experiencing them, including joy as well as interest, and negative ones, tending to focus attention and including anger or fear.[24] To identify these emotions within the source documents, I lean upon a model developed by Deborah D. Danner, David A. Snowdon, and Wallace V. Friesen. They postulated that the presence of emotions could be deduced from explicit mentions as well as descriptions of them or of actions resulting from the emotions.[25] It is a model that fits in rather well with the assumption of historians of emotion that the expression of emotions was an integral part of experiencing them, thereby making them available to historical analysis. Adding to this approach, which was devised with autobiographical texts in mind, descriptions by third persons, namely journalists, and contextual information – for example, clusters of articles on a certain topic – are also used to identify emotions within the debates. Beyond the emotions themselves, the often closely entwined expression of emotions is an important part of the analysis, especially when it comes to the diagnosis of emotions as arguments and the construction of "emotional regimes" (see later).[26]

In pursuit of this book's main goals, in this chapter I focus on three central arguments: *First*, I demonstrate that the debates about nuclear energy were indeed, to a considerable degree, shaped by emotions. These served as motives or arguments, spontaneous reactions, or parts of strategies.[27] The local perspective applied in this study reveals that this holds true for both sides of the debates for their whole course – and for a scope of emotions clearly exceeding the hopes and fears examined most often in the emotional history of nuclear energy.[28] In consequence, it becomes clear that this was more complex than the oft observed two consecutive phases would imply. Indeed, the latter's ups and downs were shaped by an interplay of national as well as local factors.

Second, leaning upon a concept used by historians of emotion, most prominently Ute Frevert and William M. Reddy, I argue that the expression of these emotions within the debates, as well as the effects they had on them, were, to a large extent, governed by "emotional regimes". These unwritten rules regarding the expression (and experience) of emotions were specific to Britain on the one hand and Germany on the other. They turned the emotional influence on the debate about the atom into a motor for the development of these debates in the two countries.[29] In Germany, not least because of the spectre of Weimar, the emotional regime in force at the beginning of the time under consideration here was quite rigid when it came to the expression of emotions, construing them as irrational.[30] This held true especially if they ran counter to standing political decisions (cf. "Nuclear Democracy").[31] "Positive" emotions on the other hand, backing up those decisions, were more readily accepted in the public debates.[32] After all, as Holger Nehring emphasised, nuclear energy had initially been hailed as a way into a rational, plannable future and, thereby, a way to leave behind the highly emotional memories of National Socialism.[33] Emotions backing up such a position, in a way, just had to be rational. Up to the 1970s, even the critics of nuclear energy mostly adhered to this regime and presented themselves and their critique of nuclear power as unemotional, to demonstrate their rationality.[34] In the end, this very rigidity fuelled a conflict about the emotional regime itself. It was questioned, not least, by Robert Jungk in 1977, who argued that the possible consequences of nuclear energy were so drastic that they simply had to be debated in an emotional way.[35] The conflict escalated whenever nuclear catastrophes injected high levels of emotionality into the debates. After the nuclear catastrophe of Chernobyl, this led to a revolt by, not least, mothers. They demanded not only the right to express their emotions but also the necessity of attaching political value to them.

In Britain, the situation was different. Admittedly, as in other western countries during the age of (high) modernity, the interpretation of emotions as an irrational phenomenon in a world best served by rationality was not unknown.[36] This was especially true because rationality was deemed a core component of the democratic system and, thereby, of the country's triumph over National Socialism in the Second World War.[37] However, the analysis of the local debates about nuclear power reveals a rather high incidence of emotional expressions in everyday statements, exemplified in the quite common foundation of one's view in a "feeling". In short, the predominant emotional regime was open to a large variety of emotional expressions and treated them as acceptable arguments in political dialogue – if no claims were made that they had a higher intrinsic value than other types of arguments. Those making emotional arguments must adhere to the principle that these and "factual" arguments or measures could mutually influence one another, be it in a supportive or contradictory way. By leaving much more leeway for emotions within the public debates, the British emotional regime turned out to be quite resilient and much better able to withstand the challenges it encountered during the debates about atomic power, especially so in the wake of nuclear accidents and an intensified (trans-)national debate from the late 1970s onwards. The relationship between the two countries' emotional regimes, then, was not all that dissimilar from that of their conditions of debate (cf. "Nuclear Debates") and clearly pointed to their different democratic cultures. Overall, the two emotional regimes helped to shape the course of the nuclear debates they had become anchored to in both countries, contributing to an escalation of the societal split in Germany and a quieter course in Britain. The emotional regimes, in turn, were influenced by the debates about nuclear power, resulting in a (limited) revolution in the Federal Republic of Germany (FRG) and an affirmation in the United Kingdom of Great Britain and Northern Ireland (UK).

Third, with a view to some striking similarities between the two countries despite their differing emotional regimes, I argue that some of the emotions observed here were less closely intertwined with these regimes, and more with the specifics of nuclear power and its artefacts. These "tech-emotions" in the purest sense, as one could say with Martina Heßler,[38] seemed to appear relative to the temporal phases of the respective local nuclear power plant projects, rather than the specific state of the national emotional regimes. They were, mostly, varieties of interest and joy or happiness, appearing when new nuclear power plants were under construction. As emotional reactions to the dimensions and complexity of nascent reactors, they show that the technological sublime argued for by Nye was not limited to America but could be found attached to a specific form of technology at specific points in its life cycle.[39]

To unpick the complex history of emotions throughout the dynamic debates about nuclear power and uncover the emotional regimes and their challenges in Britain and Germany, this chapter focuses on Biblis in Hesse, and Oldbury-on-Severn in Gloucestershire. In the latter case, the sub-chapter is divided into three sections that loosely follow the diachronic unfolding of the debate, outlining the different sets of emotions prevalent in different phases of the history of the local nuclear power plant and the gradual appearance and final unsuccessfulness of challenges to the dominant emotional regime against the background of the sharpening (trans-)national conflicts about nuclear power. In the case of Biblis, the same number of sections exists. Here, however, to render more comprehensible the workings of an emotional regime that idealised the containment of emotional expressions while, at the same time, a great variety of them were present, two passages through the course of the debates are made: The first section demonstrates the established and upheld emotional regime with its ideal of being factual, from the beginning of the debate to the end of the period under investigation here. It then serves as a foil against which the second and third sections highlight the way emotions and emotional statements were nonetheless made and utilised in the debates relative to this regime, first up to and second after the nuclear catastrophe of Chernobyl, in the wake of which the emotional regime was openly challenged by a group of local mothers – and all but faltered.

Biblis – The Rise of Public Emotionality

In Germany, the construction and operation of commercial nuclear power plants was governed by the 1959 Atomic Energy Act. It was designed to advance the development of nuclear energy on an industrial scale while at the same time being shaped by concerns about the technology's potential hazards. In its first paragraph, the protection of "life, health and goods against the dangers of nuclear energy" was named among its main goals.[40] Explicitly, it demanded that new nuclear installations had to conform to the "state-of-the-art of science and technology" as a "provision against damages".[41] Clearly, in the formulation of the law, the atom's dangers had been recognised and taken into account. The law could even be understood as one component of a rational system designed to rein in worries about nuclear safety, felt by those hoping to advance the technology.[42] No explicit mention was made, however, of hopes or fears regarding the atom within the law. Nonetheless, these abounded during the decade, exemplary expressed in the Social

Democratic Party of Germany's (SPD) "Godesberger Programm" from the same year, which made both emotions explicit in its first paragraphs.[43] This difference is obviously owed to the different types of documents those two texts represent. The latter was a general vision put out for public consumption, while the former was a law, setting down rules to be applied, directly influencing the technology's history.

This differentiation points to the emotional regime in place in Germany when the public debate about nuclear energy started in and around the small town of Biblis in the Hessian reed landscape, the "Ried". Emotions, although a great variety was present and influential, were not to be openly mentioned when it came to political decisions with practical consequences. They were only deemed acceptable in laid-back statements – or when they backed the decisions taken in a demonstrably unemotional, factual way. This ideal, which favoured atomic power's proponents, was initially accepted by Biblis' critics of nuclear energy coming forward during the 1970s. Up to the catastrophe of Chernobyl, they tended to be more interested in pointing out infringements of the regime by their opponents than attempting to dismantle the construct of nuclear-averse emotions as irrational. They shared the ideal of being factual. Therefore, emotions were commonly treated with distancing strategies, their expression watered down, their existence only accounted for in others, to be analysed, admonished, or even denied point-blank.

The Ideal of Being Factual

In the beginning, the public debate about nuclear power in Biblis became ostentatiously set in a framework of thoughtful, argumentative, factual deliberations, of dispassionate objectiveness, encapsulated by the terms "Sachlichkeit" or, less prominently, "Nüchternheit". It was an emotional regime that prioritised controlling emotions. As such, it was well-suited to the scientific and technological arguments favoured by the proponents of nuclear power plants, who were able to employ it in a way that impeded their opponents' access to the debate. The earliest champion of this ideal of being factual was the Rheinisch-Westfälisches Elektrizitätswerk AG (RWE), the company planning the Biblis power plant on the banks of the river Rhine. In the summer of 1968, when it jump-started the public debate about nuclear power in Biblis with an open letter, published in a number of regional newspapers (cf. "Nuclear Debates"), it explicitly encouraged people to "think" about its plans to form an opinion: "Please consider the opportunities arising out of this power station, for your community and its population, today as well as tomorrow, think about it."[44] The company anticipated that people would have "questions" and promised "that we will answer".[45] Emotions did not come to the fore in their proposal, although they lurked in the background: The RWE toyed with implicit promises for an economically better future in a community which had lost much of its gherkin-based industry.[46] Thereby, in all probability, it was deliberately trying to incite hopes in the small town in the Ried. On the explicit level, however, the letter was firmly rooted in the factual, rational sphere typical for Western Europe's culture of high modernity, where "the rational manipulation of the environment", as David S. Landes put it, was the strived-for ideal.[47] The public debate that ensued seemed to take the course charted by the RWE. It focused on the technology's safety and, backed by experts from the industry and the authorities, positioned steel, concrete and calculations against doubts or worries, albeit leaving some room for hope for the municipality's and its citizens' financial prospects.[48]

Indeed, fears of the atom were only expressed in laid-back forms during these early years. Mostly, they were confined to "worries" or "concerns", sometimes even "considerable concerns", namely by the river authorities regarding the effects of the planned power plant on the Rhine.[49] Taking part in the licensing process, the authorities very likely were among those actors the RWE lauded as "willing to argue in a factual manner" – code for their willingness to subordinate any fears to the factual arguments of the pro-nuclear experts. The company was willing to always talk to this kind of participant in the public debate, while "[f]ighting back" against "every polemic and unobjective statement".[50] Of the latter, there were few. When looking back on the early debate in the middle of 1971, with the power plant's construction well underway, Biblis' mayor Josef Seib from the Christian Democratic Union of Germany (CDU) even praised the people of Biblis for "having behaved in such a reasonable way". The small number of fearful people, the mayor pointed out, had melted away because of the widespread "education" about nuclear power plants and the regular contact with the RWE's experts, who mingled with the population in local pubs, spreading their factual knowledge and, thereby, overcame any residual emotions. Accordingly, when asked about their feelings regarding the power station by a journalist in 1971, some of Biblis' inhabitants pointed out that the power station would "be there without our feelings".[51]

This kind of disregard for emotions was typical for the early debate in Biblis. The reverence for the factual and rational was woven into its very fabric, and the emotional regime focused on the containment of emotions. This became particularly clear in the region's newspaper reports about the official hearing in January 1970 (cf. "Nuclear Debates"). Although Marga Müller, a Biblis housewife who was one of the earliest local opponents of nuclear power, had used the opportunity to describe, in the words of the *Südhessische Post*, "worries of a general character", not least regarding dangers for "people's health and lives", the event was dominated by facts and rational explanations by experts. Müller was even reprimanded by the meeting's chairman, the Ministry for Economic Affairs and Transport's permanent secretary Heinz Wagner, "that the majority of those fears had been allayed by the talks and explanations given earlier on".[52] In other words, he argued that the "factual information" given out should have reined in her emotions. Quite in line with this mindset, most newspapers pointed out the factual character of the debate as a quality feature. The *Darmstädter Echo* described: "The debate was very factual",[53] the *Mannheimer Morgen* lauded "the expertise" of the power plant's opponents,[54] and the *Südhessische Post* judged that "the very factual discussion" had made it clear that all parties involved would do everything possible to prevent any dangers from materialising out of the project.[55] Clearly, factual arguments were the strived-for ideal, and emotions – especially those expressed by critics of nuclear power – had to be subordinated to it if one wanted to take part in the early public debates, which were dominated by the technology's advocates.[56]

This emotional regime was upheld, even when the debate intensified after regional critics had begun to join together at a public discussion in Hamm, opposite the Biblis power plant in the state of Rhineland-Palatinate, on 18 January 1973 (cf. "Nuclear Debates"). Friedel Fabrie, who reported on the meeting in the *Wormser Zeitung*, pointed out that, while regional winegrowers were concerned about the various potential dangers of nuclear energy, these had been talked about in a very factual way.[57] In the wake of this discussion, further events were organised in the region around Biblis and in the Hessian town itself, often including critics as well as proponents of nuclear energy. Concerning one of them, the *Darmstädter Echo* explicitly lauded the "endeavour to hold a factual debate", even though critics and proponents of nuclear energy had been unable to reconcile their views.[58] The latter in particular consistently continued to emphasise the importance of being factual: In May 1973, the RWE initiated the so-called power plant talks (Kraftwerksgespräche) with a group of local politicians, "to talk about questions raised in a factual manner".[59] Here, in the view of the *Darmstädter Echo*, every critical point raised was "refuted in a factual manner, not completely convincing in every instance, but skilfully".[60] A few days later, an event organised by the Biblis CDU youth organisation brought together representatives from the Hessian government as well as various authorities and the RWE on the one hand, and those from regional citizens' initiatives and their national umbrella organisation, the Bundesverband Bürgerinitiativen Umweltschutz on the other. In the words of the newspaper, the event's moderator, Member of the State Parliament Erna-Maria Geier, "asked for a factual debate, free from emotions, for the sake of the topic".[61] This wording put the publicly construed role of emotions within the debates and their relationship to the factual and rational in a nutshell: Emotions were seen as opposed to factual and rational arguments, which were held to be the only basis for valid opinions and (political) decisions about an issue as complex as nuclear energy.

This interpretation was not restricted to Biblis and its surrounding region.[62] The most prominent example of its presence on the national plane of debate was the Bürgerdialog Kernenergie (Citizens' Dialogue on Nuclear Energy). Initialised by Federal Minister for Research and Technology Hans Matthöfer in early 1975,[63] the Citizens' Dialogue aimed at reining in the technology's critics and their surging expressions of fears regarding nuclear reactors.[64] To that end, it provided informative texts and talks about nuclear power, in particular "Kernenergie. Eine Bürgerinformation", a book that saw numerous editions through the following years.[65] It was designed to answer the (assumed) questions behind expressions of emotion, instead of considering them as relevant in their own right. It was also, similar to the RWE's original approach to the Biblis debates about nuclear power, more of a monologue by official experts in the know, directed at citizens considered as uninformed, than a dialogue.[66] As the Federal Minister for Economic Affairs Hans Friderichs put it, citizens had to be "factually informed" and "educated" about the technology, while the decisions had to rest with their representatives, not with a "plebiscite".[67] Among the technology's opponents, consequently, the Bürgerdialog tended to be considered rather sceptically.[68] In April 1975, however, Biblis was, to Friderichs, something of a national model for the rational, unemotional debate desired by the technology's proponents all over the Federal Republic. To him, the fact that the people of Biblis stood behind the now officially opened power plant proved "that citizens can be convinced of the security and necessity of nuclear power plants by factual and comprehensive information".[69]

This ideal, however, came under pressure a few weeks later, when the debate about nuclear energy in and around Biblis became much more heated and critical after the RWE's announcement of their application for the licence to construct two additional reactors, in the context of the festivities celebrating the start of operations at the first (cf. "Nuclear Debates"). Once again, the company aimed for a quiet, factual debate, focusing on the safety of the technology and already established suitability of the location.[70] Their proposal, however, immediately led to an escalation of the number of regional and even local critics, with the volume of critique rising, too, emotional statements included.[71] Reacting to this escalation, proponents of nuclear power immediately demanded "to return to a factual debate from a partially emotionally laden discussion".[72] In addition, accusing one's opponents of trying to stir up emotions, stepping outside the accepted emotional regime's boundaries and thereby devaluing one's own argument became a popular strategy within the debate. It was applied by advocates as well as critics of the technology.

In principle, this kind of mutual accusation was not new to either side. As early as April 1973, the Hessian Minister for Economic Affairs, Heinz-Herbert Karry from the liberal Free Democratic Party, had warned the public against giving in to "exaggerated fears" regarding nuclear power, as he believed the technology to be necessary to fulfil the rising demand for electricity.[73] This statement, in turn, had provoked one of the earliest regional citizens' initiatives, the Aktionsgemeinschaft für Umweltschutz Rheinhessen-Süd e. V., to state that it was warnings about future electricity shortages that formed the true "tendentious panic mongering", which would "complicate a factual debate of environmental problems", not their concerns.[74] In other words, the atom's critics adhered to the ideal of the factual. At least they were willing to adopt it as a tool in the debate. At any rate, the ideal was actively upheld by the regional newspapers: In July 1973, the *Darmstädter Echo* advised against talks of a looming "energy crisis", as this term at the time seemed too dramatic. It demanded "a debate less burdened by emotions" instead.[75] In the following February, journalist Gerd-Peter Schulze, writing in the *Südhessische Post*, accused both sides of having strayed from the ideal of the factual, "[w]ith unobjective panic mongering the critics, whitewashing and trivialisation of risks the proponents".[76]

In the heated atmosphere after the RWE's announcement about the planned expansion of the power plant, such accusations of deviating from the emotional regime of the factual surged. At a discussion in Biblis, organised by the recently founded local Aktionsgemeinschaft für Umweltschutz Südliches Ried (AGUSR), the environmental activist Wolfgang Staab declared claims about nuclear energy's necessity for securing the supply of electricity or jobs to be "lies and panic mongering".[77] Walter Pfeiffer, an SPD member of Lampertheim City Council, argued that the RWE was trying "to spread fear . . . with the aim of generating profits for the RWE's shareholders", as the *Mannheimer Morgen* summed up his position.[78] Werner Twardzik, from Biblis' SPD, implied that the RWE used "methods of advertisements to manipulate the population" instead of a factual debate, which his party would prefer.[79] From the other side of the nuclear divide, the CDU in the Bergstraße District accused the SPD of emotionalising the debate,[80] while Seib criticised the Bergstraße's newly sceptical district administrator Dr Ekkehard Lommel for expressing his position in a way that was "suited to stirring emotions against the RWE's plans for Biblis",[81] later on adding in more general terms that "[t]he population has been unsettled by, as I see it, irresponsible panic mongers."[82]

Besides accusing one another and, for good measure, the media,[83] of emotionalising the debate, critics as well as proponents of the expansion plans strived to show that their own position was the one that best fitted the ideal of being factual. In that vein, Seib argued that, "[g]iven an objective educational campaign, the majority of Biblis' population will positively back the further expansion."[84] In unison, his party group in the municipal council declared that "objective information" was an effective counter to "panic and fear".[85] Contrary to that assumption, Günter Mössinger, a local critic of nuclear power and member of the AGUSR, believed that growing concerns about nuclear power were not irrational emotions, but the result of "increased education and information", enabling people to critically reevaluate early "assertions of security".[86] Similarly, the local and district SPD jointly stated that their rejection of the expansion was "[t]he result of a factual debate".[87]

The ideal of being factual was, in other words, very much intact. After a discussion in Biblis at the beginning of 1976, the *Mannheimer Morgen*'s Bert Siegelmann explicitly pointed out that it had been "[f]actual and without incidents",[88] a characteristic which the *Bibliser Blatt*, the local municipal newspaper, explicitly called "[p]leasant" and considered a precondition for it having "delivered real information".[89] To keep it that way, the Informationskreis Kernkraftwerk (Nuclear Power Plant Information Circle) was founded in Biblis in May 1976.[90] It was to serve as a platform for regular communication between representatives of the municipality and the RWE, with additional participants from neighbouring communities, state authorities and others. Herein, representatives from the Hessian government saw a chance for the "factualisation of the debate",[91] and Hans-Peter Seubert from the *Darmstädter Echo* pointed out the body's notable value:

> Factual, yet critical education about the nuclear power plants is necessary. In many places, polemics dominate the debate. Uncertainty, fear, lack of technical understanding and, as a result, indifference among the population are traces of a vociferous discussion, primarily focusing on the dangers of nuclear energy in power plant construction. Even experts . . . have a hard time filtering out objective arguments. It is therefore gratifying that Biblis' local politicians, who enjoy the trust of the citizens and who also have access to the authorities and operators of the nuclear power plant, are now embarking on explaining to the population the problems with which they must live . . . It would suit both sides' interests well if it was possible to relax and factualise the debate about the Biblis nuclear power plant.[92]

While the ideal of being factual was upheld in the heated debate, it was now also openly employed as a discursive strategy by both sides specifically to devalue their opponents' perspectives. It continued to be used in this way throughout the following years, particularly so whenever discussions peaked in the context of unexpected occurrences at the local power plant or when external events drew attention to it. For example, Christian Rettweiler from the workers' council at the power plant, who was to become a regular participant in the local and regional debates, wrote to the *Bibliser Blatt* after problems had been revealed during the first regular inspection of "Block A". He demanded "more objectivity in the debate" about them, especially from critics belonging to the local SPD.[93] Critics of nuclear power, on the other hand, demonstrated their adherence to the dominant emotional regime by accusing members of the industry or the administration of violating the ideal of a debate based on facts by lacking transparency.[94] After the accident at the Three Mile Island (TMI) nuclear power plant in 1979, proponents of nuclear power like the *Darmstädter Echo*'s Herbert Lilge confirmed the factual ideal,[95] as did open critics of the Biblis power plant, like the district administrator of neighbouring Groß-Gerau, Willi Blodt, who spoke of the "commandment of a factual and fair debate".[96]

The emotional regime as such came under serious pressure, however, when anti-nuclear protests surged in the wake of the catastrophe of Chernobyl in 1986. Now, fears visibly came to the fore in all aspects of the German nuclear debates.[97] As the *Wormser Zeitung* put it regarding Biblis, "the debate about nuclear energy has reached an emotional dimension that hardly allows for factual arguments to be heard."[98]

This development is taken up in detail in this sub-chapter's final section, "Legitimised Emotions". Here, it is important that, nonetheless, the ideal of being factual continued to be upheld by many actors on both sides of the debate. They even increased its employment as a discursive strategy to diminish their respective other's arguments. Rudi Kilgus, writing in the *Wormser Zeitung*, blamed the SPD and Green Party for using "[t]he Chernobyl shock as a lever to turn fear into party-political updraft".[99] Similarly, Biblis' conservative Hans Bitsch accused the SPD of "building their business upon fear",[100] and Heppenheim's CDU was convinced that "[t]he population's fear . . . has been stoked up in an irresponsible way: through panic-mongering by self-acclaimed green and red 'experts' . . ., through demagogic performances and paroles . . . and, most of all, through the sensation-media."[101] Similar accusations of fearmongering were raised by nuclear power's opponents. When, for example, the Bergstraße District Council debated about a motion by the Greens to shut down the Biblis nuclear power plant in June,[102] against which Biblis' new mayor Siegmund Piegsa had brought into play the suggestion that it was the party's goal to achieve a "phase-out of the industrial society",[103] Klaus Kübler, the SPD's Member of Parliament for Bergstraße, accused Piegsa and the CDU of "fear-mongering" with a view to job-losses.[104]

Besides its aggressive utilisation as an offensive weapon within the debate, the ideal of being factual continued to be positively highlighted. Only weeks after Chernobyl, the workers' council at Biblis power plant had published an article to "thank [the local media, CG] for the mostly factual reports, free from emotion, in the daily papers. This factuality was soothingly present within the leading articles . . . We . . . appeal to all citizens to discuss the future of the energy supply with factual arguments only, and not in form of a war of beliefs."[105] With them, Biblis' priest Winfried Penk was convinced that it was important "to factualise the debate",[106] and the district's SPD, though sceptical of nuclear energy at that point in time, vowed to "discuss the problem quietly and carefully".[107] Writers of letters to the editors of regional newspapers advanced the opinion that fear was a burden to be overcome. "What we need, with all due understanding for emotional insecurities that occurred, is [sic, CG] a dispassionate, factual analysis of the situation and prudence."[108] To support this goal, Rettweiler, who in a later statement explicitly pointed out that "he was only for an emotion-free discussion with factual and reasonable arguments",[109] became very active. He wrote articles for the *Bibliser Blatt*[110] and produced flyers with "factual information", arguing for the continued use of nuclear power in Germany.[111]

Even at the end of the 1980s, therefore, after the catastrophe of Chernobyl and in view of additional scandals upsetting the German nuclear industry (cf. "Atomic Trust(s)"), the emotional regime remained visible, with its ideal of being factual and its construction of emotions as irrational, the expression of which was held to be a sign of an argumentative weakness. Nevertheless, the growing number of accusations against those seen as being in breach of the ideal in phases of intensified debate had already indirectly shown that the debate could not have been free from emotions. In fact, even the RWE's letter from June 1968 had indicated their presence at its very beginning. Most of the time, however, up until the catastrophe of Chernobyl, they had been expressed in laid-back form, with distancing formulations accompanying their statements, or they had been only hinted at. In a way, quite in concordance with the reigning emotional regime, they had been presented as renounced emotions.

Renounced Emotions

Starting the public debates about nuclear power in Biblis, the RWE, in its open letter, had not only laid out the potential for hopes for a better future for a nuclear Biblis. The company had also freely used superlatives to describe itself and its plans with expressions like "most modern machines" or the wish to build a "super-large nuclear power station".[112] While these were not explicitly expressed emotions, the descriptions were fittingly chosen to generate awe for the company's technological prowess. Soon, this way of describing the planned power station was adopted by the region's newspapers. In July 1968, the *Bürstädter Zeitung* predicted that the installation would be "gigantic", followed by an overview of its expected dimensions, the thickness of its walls and the amounts of material used for its construction.[113] The *Wormser Zeitung* pointed out that it would be the biggest – and most expensive – power plant in Germany,[114] while the *Mannheimer Morgen* widened that claim to Europe and emphasised that there would be no bigger installation even in the USA, calling it a "milestone",[115] "a super project", and "gigantic factory".[116] Once the construction had started, this pattern continued with reports about the "colossal building site",[117] where hundreds of workers were busy at the reactor building, weighing 140,000 tons, resting upon a foundation of ferro concrete 2.6 meters thick, which consisted of 7,500 cubic metres of concrete and 700 tons of steel, with similar details given for various aspects of the ongoing work time and again.[118] In March 1971, the *Mannheimer Morgen* found the "precision" with which "1000 workers" laboured on site to be "admirable".[119] Obviously, the project was met with interest and forms of joy, namely admiration and awe. Beyond that, Hans-Joachim Noack, writing for the national newspaper *Die Zeit*, observed in August that "at the pub, men were debating the wondrous world of superlatives with pride in their throats".[120] This mix of emotions, even if it was rarely explicitly pointed out, was a characteristic reception of the technological feat of containing and controlling the power of the atom, taking shape in a setting of high modernity. It was what Nye described as the technological sublime, especially "the geometrical sublime", concretised in buildings that "appeared to dominate nature through elegant design and sheer bulk" and "expressed the triumph of reason in concrete form".[121] It was an emotional cocktail, which was typical for the early years of a debate about a new power plant, before the reactors-to-be could become an established element of the region's everyday life,[122] and before groups of vocal critics could organise themselves and pressure the proponents of nuclear power to adhere to the emotional regime in the same way they themselves were told to.

Clearly, the positive emotions felt by many of nuclear power's proponents in those early days constituted a strong motivation to back the technology's development. First and foremost, this holds true for Biblis' mayor Seib, who came to be known for being "fire and flame" for the power plant.[123] In fact, he was one of the few well-known proponents of nuclear energy who openly expressed his emotions. Explicitly, for example, the mayor presented the first reactor's construction as the "proudest project" to be realised within the municipality's borders when he informed the public about upcoming activities in Biblis during the summer of 1968.[124] Later on, he looked back at the beginnings of Biblis' nuclear history with the words: "I still rejoice that I could get the nuclear power plant for Biblis. It was a big success for me".[125] With a similar outspokenness, in 1980, Lommel's successor as a district administrator, Lothar Bergmann called for "the courage of those in the know to stand to the principle of hope" in view of a decision regarding compact storage facilities for used fuel elements at the Biblis power plant.[126] Other advocates of nuclear energy tended to be less outspoken regarding their own feelings. Nonetheless, in all probability, they did not feel much differently when they made statements about emotions observed in others. One example is a protocol from Biblis' authorities in the summer of 1968, according to which the administration's heads were satisfied that the local people's "mood" regarding the RWE's plans was "entirely positive".[127] In retrospect in 1976, the *Darmstädter Echo*'s Michael Grimm went so far as to state that during these early years, "[o]verjoyed about safe jobs and growing tax income, no one, in the context of a general euphoria for progress, thought about the reverse side of the medal."[128] This may have been exaggerated, although it seems reasonable to assume that most local people were, as the local CDU put it in a 1972 brochure, "open-minded and interested in the project".[129] In all probability, this was also true for the Bergstraße district at large.[130] In any case, employing the sexually laden word "erregend", the Echo called the concept of nuclear energy "arousing",[131] and the *Süddeutsche Zeitung* spoke of a divine "promise", using the religiously connoted term of "Verheißung" when it reported about the municipality's expected profits.[132]

Obviously, in the debate's early years, it was quite normal for those welcoming the atom to Biblis to express the emotions motivating them before vociferous critics joined the debate, holding the atom's proponents to the same standards they expected of the critics when it came to the emotional regime. These emotions were, nonetheless, mostly stated in an indirect form, as an observation in others, or in retrospect, with Seib being the vocal exception to this rule. He continued to be outspoken about his positive emotions regarding the nuclear power plant, even after the debate turned into a public argument in 1973. In 1975, for example, he stated: "We people of Biblis are proud of this building . . ., which carried our municipality's name out into the world",[133] adding that the municipality was awaiting the future full of hope.[134] By then, most other proponents of nuclear power were concentrating first and foremost on emphasising the absence of fear in Biblis. Indeed, Seib did so, too. He was adamant that his fellow citizens were not worried about the reactor, because they "can trust" the responsible experts.[135] This was also the picture the *Bild am Sonntag* painted in an article about Biblis in 1976. According to the paper, Maria Höger, the innkeeper of Biblis' "Darmstädter Hof", reacted to the question whether or not she was afraid of nuclear power by exclaiming: "AFRAID? Don't make me laugh!"[136] Most of the people interviewed shared her outlook.[137] Beyond that, the paper reported that some guests in the inn were happy, some farmers even joking about "atomic gherkins", and on the menu, the reporter spotted a cocktail named "reactor water".[138] Seib's successor, Piegsa, who assumed office in 1984, when visiting the power plant for the first time, also stated that he had been "impressed" while not developing "[a] feeling of fear or something similar". When describing the emotional stance prevailing in Biblis, according to the *Bibliser Blatt*, he found analogous words: "Piegsa explained that none of the municipality's citizens had, so far, reported worries or fears about the nuclear power plant's operations to him."[139] This situation seemingly remained unchanged even in the first few days after the catastrophe of Chernobyl. When, soon after the accident had become known, the *Bergsträßer Echo* conducted a telephone survey in Biblis, the newspaper observed that there was no "panic" because of the accident. The newspaper's sources were a local pharmacy, a priest, a teacher, a doctor, an insurance agent, representatives of the power plant and the city hall. "Siegmund Piegsa is satisfied that the people of Biblis feel secure in the shadow of their nuclear power plant."[140] The next day, however, the newspaper observed that there had been some worried callers in pharmacies outside Biblis,[141] foreshadowing the more prominent role of "critical" emotions regarding nuclear power after the nuclear catastrophe in the Soviet Union (see later).

That is not to say that emotions such as concerns, worries, or fears regarding the power plant entered the debate for the first time in 1986. On the contrary, just like variations of joy and interest, pride, and hope, they had been present as potent motivations and as expressions in the debate for quite some time. This holds true even within the group of nuclear power's proponents. On the one hand, as argued in this sub-chapter's introduction, there were some signs that fears of the atom were present and considered by them at the very beginning of the "Peaceful Atom's" career in Germany. On the other hand, they were concerned, or even afraid, *for* atomic energy, whenever there seemed to be obstacles for its expansion, or when those drawing its use into question seemed to gain political tailwind.[142] Once more, Seib was one of those outspoken about the emotional side of the atom – although he was less explicit on this count than he was with the positive side of his feelings. After the number of critics of the Biblis power plant had swelled in 1975, he occasionally focused on the expected loss of income the municipality would incur without the plant and its further expansion.[143] At least once he also hinted at the negative consequences abandoning nuclear energy would entail for the job situation and the environment.[144] It seems probable that he deliberately worked with the implication of concerns – even if he was not consciously trying to evoke these emotions in the wider public. The latter aim can more readily be presumed when it came to statements by leading representatives of the nuclear power industry. One example is the RWE's Heinrich Mandel. He was very present in the (trans-)national debates about nuclear power, not least by shaping the differentiation between risks that were so high they had to be considered when designing nuclear power plants, and those that were not. He also injected his arguments into Biblis' debates: Cheap electricity, which could only be delivered by atomic power plants, was necessary "for the continued existence and the further development of our economy, our civilization and our social order".[145] Therefore, nuclear technology could not be hampered by the "objections of small groups opposing nuclear power mostly on emotional grounds and without sufficient factual knowledge".[146] This combination of accusing nuclear power's opponents of acting outside of the accepted emotional regime, while at the same time presenting a picture of a bleak future that at least condoned the evocation of emotions, was a prime example of the discursive strategy widespread with proponents of nuclear power after the technology's expansion had been met with strong resistance. In a somewhat less refined way, similar statements were issued by local and regional proponents of nuclear power, too. In the *Darmstädter Echo*, journalist Thilo Koch argued in November 1976 that, when it came to atomic energy, "an irrational primal fear resists the rational requirements of our civilization with its unavoidable risks". Opposing nuclear power, to him, was a sign of lacking "sense of proportion, reason and . . . responsibility", leading to the question: "[D]o the protesters seriously want to accept the long-term effects of the energy crisis hitting us head on? Do they know what that means, in practice? It means: freezing, starving, being unemployed, perishing."[147] Certainly, intense expressions like these were the exception; even in phases where the critique against the local power plants peaked. Mostly, nuclear power's proponents remained restrained in their emotional expressions, for example, when it came to worries about job losses expressed by the workers' council at the Biblis power plant in 1984, in view of a cooperation between the SPD and the Greens in the Hessian state government.[148] Nevertheless, the strategy to accuse the technology's opponents of "emotionalising" the debate while doing exactly that was quite common. Early in 1985, Piegsa saw "[w]ith great worry" that the SPD of southern Hesse was considering a nuclear phase-out. To him, it was unacceptable "that a large political party . . . allows itself to be blackmailed by a small political group which practices politics with the emotion of fear". At the same time, he complained that it was out of line to toy with jobs in that way and to have "about 600 people directly concerned living in permanent fear about their secured existence".[149] This line of argument remained prominent after the catastrophe of

Chernobyl, not only with the local CDU[150] but also with the SPD. Debating a proposed resolution by the Greens in Biblis' Municipal Council to shut down the power plant, Twardzik emphasised that "[t]housands of job [sic, CG] would be lost, misery would befall many families." Above and beyond that, coal- and oil-fired plants would have to be used more intensively, damaging the air and the environment, turning the planet into "a greenhouse".[151] In view of the change of position his party-group undertook a few years later, it is questionable whether he personally was afraid of the developments he described. That his descriptions were well suited to instigating the emotion, that they were designed as a discursive tool to be used against the opponents of nuclear power, to put fear against fear, is all too likely.

When it comes to the critics of nuclear power, variations of fear, were, of course, not felt *for*, but because *of* the atom. As most of the critics who took part in the widening public debate of the middle 1970s adhered to the established emotional regime, these fears were in general expressed in milder terms of "worries" or "concerns" or attributed to others. Early instances of fears of nuclear power, expressed in this way, but nevertheless refuted or admonished during the licensing process for the first reactor in Biblis, have already been mentioned. That notwithstanding, the citizens' initiatives joining the debates from 1973 onwards also adhered to the emotional regime in place. They were keen to point out the factual basis of their position and kept the expression of the emotional motivation behind their engagement low-key, for example, regarding the effects of the power plant's cooling towers on the local climate.[152] It is very likely that many of the critics expected to be taken more seriously when formulating their adverse emotions in restrained forms. After all, proponents of nuclear power like Mandel differentiated between critics of nuclear power plants motivated "by genuine concern" and "open to a factual debate" on the one hand, and others adhering to a "sectarian faith".[153] Certainly, this kind of demarcation, which also stood behind the Bürgerdialog Kernenergie and was shared by political institutions up to the European Parliament,[154] carried the implicit expectation that the critics would "overcome" their fears with the help of "education" and let go of their opposition. True enough, most critics of nuclear power at least sought to express their nuclear-critical emotions in a way that adhered to the emotional regime of being factual. Twardzik, for example, explained the SPD's opposition to the immediate expansion of the power plant in June 1975 by referring to worries felt by large parts of the population in view of the RWE's plans,[155] and included himself in this group.[156] Five months later, when a delegation of the party visited Hesse's Prime Minister Albert Osswald, Twardzik pointed out "that the population of Biblis, by accepting the first two reactors, had demonstrated a high degree of objectivity". Having thus firmly placed the local population within the realm of the rational, he emphasised that now there were "concerns . . . about a further concentration of reactors".[157] So contextualised, he obviously held them to be legitimate within the emotional regime governing the debate. Fittingly, when critics of nuclear power expressed fear more explicitly, they tended to distance themselves from the emotion by ascribing it to others or with generalisations. Staab, in a letter to the editor in April 1974, claimed that "[a]ll critics of this project . . . wish, full of fear, that not a single one of their fears may ever come true."[158] In August 1976, Ludwig Bauer from the AGUSR, argued that a meeting of the group earlier in the month had shown "that the fear of the atomic power plant is on the rise in the population of Biblis".[159] These attributions of strong emotions to groups that one then claimed to represent offered the chance to use the strong motivational force of the emotion, while personally remaining inside the accepted boundaries of the emotional regime. It was a rather typical strategy for leading figures in citizens' initiatives, for newspapers reporting on minor incidents that would have been quite uninteresting without a group of fearful onlookers[160] and for proponents of nuclear energy attempting to devalue their opponents' position as emotional.[161]

The widespread adherence to the emotional regime should not lead to the assumption that none of the critics of nuclear energy were afraid or willing to say so in the public debate. Mostly, however, when one of the technology's opponents openly resorted to frightening descriptions of the possible consequences of nuclear power, it was a single voice, speaking out once but not establishing itself in the public debates. One example is Walter Rettig from Alsbach, about 10 kilometres from Biblis, who wrote: "A maximum credible accident would, when the security installations did not react, wipe out all life in a 15km radius and condemn millions of humans to infirmity." Even without such a catastrophe, nuclear power plants would cause "the radioactive poisoning of all agricultural products, including gherkins and cabbage!"[162] When similar statements were made by regular participants in the public debate, they tended to be reactions to specific events, like the chemical catastrophe of Seveso,[163] after which Bauer from the AGUSR pointed out that it had "demonstrated what innocent humans have to suffer when living next to an inhuman installation" and that a similar event in Biblis would have far worse consequences.[164] They were also often reactions to decisions directly concerning the situation locally, such as one to keep the emergency plans for the Biblis power plant secret. This, Bauer alleged, could only mean that the plan made provisions for "hermetically sealing up the surrounding area with its population", "as a radioactive contamination of humans and the countryside poses a deadly hazard for the rest of mankind".[165] In other words: Cases where fear of the atom motivated nuclear power's opponents and was only thinly veiled in their public statements did exist. Sometimes, they even spoke about their emotions explicitly – notably regarding anger towards those who did not engage against nuclear power or even advocated it.[166] Most of the time, however, there were no explicit or strong emotional statements whenever participants in the debate referred to themselves. This pattern could be found over the course of the whole period under investigation, especially in phases of intensified debate, in consequence of events or decisions concerning the local power plant, or in the wake of industrial catastrophes globally – and even more so after spectacular accidents in nuclear power plants. The catastrophe of Chernobyl, however, turned out to be an opening for an attempt to change the way emotions could be expressed, to modify the emotional regime as such by way of its connection to nuclear power. The main driving force behind this attempt in Biblis was a group of mothers, unwilling to continue submitting to the emotional regime and suppress their fears for the sake of a technology, but demanding that they be taken seriously in the political arena, for the sake of their children.

Legitimised Emotions

Sure enough, there had been moments during the Biblis debates whence the dominant emotional regime, with its ideal of being factual, had been challenged and, to a degree, emotions accepted as valid arguments. In 1974, for example, during a heated conflict regarding whether or not to publish the emergency plans for the Biblis nuclear power plant, Rudolf Wolter from the Biblis Municipal Council had argued that publishing a rough outline of the plans could well help to "calm down the population a little", which he thought worthwhile.[167] His stance, which had accepted the emotions stirred by the secrecy around the plans as something that should be taken seriously enough to justify a change of policy, had been shared by some,[168] while others, not the least district administrator Lommel, had argued that, on the contrary, publishing the plans would "worry" the region's population, which was to be avoided.[169] Notwithstanding this line of argumentation, widespread among the authorities, after three years of disputes the Hessian Minister of the Interior Ekkehard Gries was willing to allow the publication of the foundational principles behind the emergency plans, "to counteract the population's feeling of insecurity".[170] Small retractions from the established emotional regime like these, aimed at calming down the conflict, were a widespread reaction to the escalation of the nuclear debate. On the national plane of debate, in November 1976, Helmut Schmidt, the Federal Chancellor, asked industry and authorities to show more "compassion" for the "comprehensible fear" the citizens felt.[171] This gesture did not mean, of course, that the use of nuclear energy or even a single nuclear power plant project should be given up because of the emotions of the technology's critics. It did not change the dominant conviction that decisions had to be based solely upon rational and factual arguments to be legitimate and that rational arguments would, in the end, convince all those who were worried, while those who would not be convinced continued to be seen as having deeper-lying, sinister motives.[172] It was, in other words, in line with the emotional regime governing the debate.

An attempt to change that very regime in the Biblis debates did occur, however, in the wake of Chernobyl. It formed against the background of a wave of fears more distinct and widespread than any before.[173] As early as

7 May 1986, an article by the AGUSR and Biblis' Greens was published in the municipal newspaper, asking its readers:

> What do you feel when your children return from school or kindergarten and tell you they were not allowed to play in the sandpit, although the sun is shining? . . . We do not feel safe next to the atomic power plants in Biblis, we do not feel safe as long as atomic power plants are operated at any place in the world . . . Let us not forget that a catastrophe is possible at any time in Biblis, too. We demand . . .: Stop using atomic energy![174]

Here, emotions were openly stated as a motivation for the demands of nuclear power's critics.[175] One week later, in another article, the AGUSR even ridiculed those of their fellow citizens who adhered to the emotional regime as having been "educated by the RWE, living happily and satisfied in the atomic power plants' shadow", by pointing out that, the weekend before, "the chest freezers in the supermarkets were cleaned out. Did you prefer frozen vegetables, too? Surely, you were not afraid? After all, nothing can happen, because of our safety regulations." They asked their neighbours to stop accepting "the so-called 'inherent risk'" and to start demonstrating, to "enforce the only rational option".[176] In other words: The AGUSR not only implied that others felt the way they did but were unwilling to say so in the open. It even asked them to change that behaviour and join them, as, in the words of Christa Hübel, chairwoman of the AGUSR, it was pointless "to stay put in your room with your fear".[177] More importantly, the group argued that admitting fear of the nuclear power plant and acting accordingly was "the only rational option". It was a notion that had, since the 1970s, begun to be established on the (trans-)national plane of the debates about nuclear power,[178] not least in the context of the technology's military side,[179] and especially so regarding the North Atlantic Treaty Organization's (NATO's) Double-Track decision.[180] In Biblis, at this point in time, the AGUSR's statement remained ambivalent: It could have been an attempt to classify fears as rational, and thereby acceptable, within the ideal of being factual. It could also be interpreted as an open challenge to the emotional regime by claiming that acting in accordance with one's emotions was rational while controlling them was not. No ambivalence remained, however, when a group of mothers formed a new, outspoken participant in Biblis' debates about nuclear power.

On 15 May 1986, the IG Frauen und Mütter aus Biblis und Umgebung (IGFM) (Interest Group Women and Mothers from Biblis and its Surroundings), was founded by eight women and mothers, worried about Chernobyl's consequences and implications.[181] In that regard, Biblis was moving in step with most of the German national and even the transnational debate. It was largely, although not entirely, because of nuclear accidents from the late 1970s onwards, most prominently the accident of TMI,[182] that critical groups were formed consisting of women, specifically mothers, but also coming to include fathers. This indicated a critique of nuclear power less rooted in ostensibly female emotions, but those of parents, objecting to nuclear power out of worry about the well-being of future generations.[183] The IGFM soon designed a flyer, explicitly stating that they were motivated by emotions, that they derived their programme and their actions from these emotions, and asking others to join them: "We are afraid, we think about our children and children's children. We turn against nuclear power. Women and mothers, help us!"[184] On 24 May 1986, they staged their first "action" by talking to people in front of a store "to demonstrate that one can and must publicly own one's fear of the reactor, even in Biblis". Interestingly, the report about this in the municipal newspaper was put under the headline "Contemplate – A Rethink is Worth it!"[185] The wording indicated that, in its early days, the IGFM paid some reverence to the established emotional regime. Nonetheless, it challenged it by establishing emotions as relevant causes for political decisions and actions.

In the following weeks, this idea spread. When the Bergstraße District Council debated a motion by the Green Party to shut down the Biblis nuclear power plant,[186] the party's councillor from Lampertheim-Hofheim, Frank Rothenhäuser, argued that it was the job of politicians to act upon the population's fears: "Who, if not elected representatives, should pick up the citizens' fears."[187] Soon after, in the Groß-Rohrheim District Council, the SPD's Jürgen Kehl demanded a worldwide nuclear phase-out because "the fear that attacks you is a reason to act". Even Walter Öhlenschläger from the CDU admitted that "no one is allowed to put this fear aside" – although he still demanded an "unbiased" debate.[188] Besides the regional parliaments, the argument that fear as an emotion was a legitimate motivator for actions and a sound basis for political decisions spread on the streets: In an advertisement in the *Wormser Zeitung* on 14 June 1986, a petition was published, designed to collect signatures, listing various quotes without clear authors, including the statements: "I am afraid that we humans are forcing upon our children a future of illness and hopelessness", "My assistance was silence" and "I will not be silent anymore."[189] Nuclear power and its dangers, illustrated by the Chernobyl catastrophe, had clearly led to open challenges to the ideal of being factual.

The most active champions of this challenge in the Biblis area remained the IGFM. After a talk with Mayor Piegsa and Dean Gerhard Muth, the group even admitted being "very disappointed, because none of their fears or worries was shared by Mayor Piegsa or Dean Muth".[190] Three months later, when the majority of Biblis' inhabitants had still not come over to the side of the opponents of nuclear power, the group was convinced that the reason behind this was that "many people's worries about their children's health and lives are played off against the fear of losing jobs in the atomic power plant", assuming that they felt "fear of a possible accident", but also "of losing their societal standing".[191] To them, the factual ideal seemed to have lost any meaning. It seemed to have been dissolved into the emotional dimension, which was deemed the only relevant one upon which to base one's opinion on nuclear power. Those who were not opposed to the technology, in their view, had to be more afraid of letting it go than they were of its potential hazards. Other reasons not to show fear of nuclear power seemed almost unthinkable. Consequently, the members of the IGFM demanded that their emotions be taken seriously by others. In that vein, they collected about 2,000 signatures in and around Biblis, calling for the local reactors' shutdown, which they presented to members of the federal government, the political parties, and other societal actors. "This campaign is supposed to be an appeal to the politicians, church and energy suppliers to take serious our fears and stop relying on a technology that can endanger and kill human beings and will render the country uninhabitable for decades."[192] With equal consequence, they were unwilling to "overcome" their own emotions. Accordingly, when Rettweiler contacted them to point out that the risks of phasing out nuclear power could well be worse than those of its continued use, especially criticising the idea of switching to a coal-fired plant in Biblis, and demanding that they "learn to live with risks in a rational way", his offer to "help dismantle your fears"[193] was met with silence.

Even when the debate about nuclear power quieted down some months after Chernobyl, the IGFM continued its activities, convinced that "[p]regnant women, breast-feeding mothers and mothers whose children up to then [Chernobyl, CG] were able to play in the sandpit in the garden undisturbed, were afraid and still are."[194] In a brochure published in late spring, 1987, the group even attempted to read the whole history of nuclear power in Biblis through the lens of emotions, arguing that, in the beginning, when nuclear energy had been described as positive, "[c]oncerns and fears in the population were brushed away with phrases, as they still are today. Opponents of atomic electricity generation were turned into undesirable people, weirdos, troublemakers and radicals by defamatory statements."[195] Because of this, "many people of Biblis suppressed their fears of this inhumane technology; the fear of societal and human isolation was stronger in that moment.[196] Secrecy regarding the emergency plans and false alarms had brought those fears back to the surface.[197] When, after Chernobyl, the women and mothers had come together, they "knew: We have to seize the fear so that it does not seize us!"[198] Therefore, they went public with their emotions, started collecting signatures and talking to people.[199] The results of the elections in Hesse on 5 April 1987, then, which saw the CDU come to power for the first time, ousting the nuclear-critical coalition between the SPD and the Greens, had them "shocked", as talks about Biblis C consequently resumed.[200] The group held out against attempts by the RWE to bring them back into the realm of the primacy of the factual, publishing an advertisement in Biblis' municipal newspaper, stating: "Many – especially mothers – worry. This is understandable, but unnecessary. Data proves that the additional radiation is low within the range of fluctuation of natural background radiation."[201] Ignoring such efforts, the IGFM pursued the opposite goal: On the occasion of their first anniversary, the group explained that its aim was "to point out to the population the dangers of nuclear electricity generation" and "to encourage more people to publicly state their worries. A retreat into the idyll is no longer possible!"[202]

By now, other actors had adopted the more determined approach of the IGFM – although not every group was willing to step quite as far outside the ideal of being factual as they were. After it had become public knowledge, in November 1988, that an incident regarding the irregular treatment of a wrongly positioned valve that had occurred in the Biblis A power plant in December 1987 could have resulted in a larger release of radioactivity, the AGUSR emphasised that, besides being "horrified about what could have happened", they were "outraged by the wording of the attempts to calm us down once more! . . . We do not want to continue living with those accidents that happened and will happen . . . We feel put under tutelage by the information policy of operators and the government."[203] The affair even led to a change in the local Social Democrats' position regarding the existing power plants,[204] although three of their councillors stepped down in protest.[205] Twardzik explained that the party "had to recognise that, in the population in its broadest form as well as inside the party, unrest and fears took hold", identifying the industry's secrecy regarding incidents as the most important reason for this.[206] They were motivated by "understandable concern for mankind, its future and the coming generations".[207] In detail, the party called for Biblis A to be shut down until the incident was cleared up and both reactors to be investigated by experts, including some critical of nuclear power, and shut down permanently if any irreparable problems came up.[208] In other words: The local SPD used the debate about an incident in the local power plant to bring its position regarding these reactors' future in line with the party's majority outside Biblis – and it explicitly argued for this shift with a reference to emotions, to "fears" which had to be "recognised" and "concerns" which could be "understood". When, however, members of the CDU criticised the SPD's new position,[209] it became clear that, while accepting the legitimacy of emotions as a motivation, the party did not mean to give up the ideal of being factual. Twardzik argued that the party's position was well worked out and had been explained in the council "factually and without emotion", so it was uncalled for to call the SPD "an agitator" – the party "merely articulated the population's fears".[210] With this interpretation of the role of emotions within the debate, they had aligned with the municipality's Green List, which continued to demand the local power plant's shutdown, pointing out: "We are no fear-mongers, we just express the worries and concerns we and many other people harbour."[211]

In a way, then, the continuous efforts by the IGFM to get politicians "to take serious our worries and fears" and act accordingly,[212] had made some progress. Various party groups in the region and even within Biblis modified their positions regarding the role of emotions in the debate – as something to be taken seriously, to be acted upon by politicians. And even if the resulting action was presented as being based on rational deliberations, this was something quite different from the rationalisation of emotions that had been dominant in the years before. Nevertheless, this balancing act implies that the ideal of being factual, albeit challenged, was still present. This holds true especially for the local power plant's staunchest supporters in the RWE and the local CDU. Even they, however, had become less rigorous. In the years after Chernobyl, the number of references to emotions made by those groups when it came to local jobs being endangered – although still overwhelmingly in an implicit form, mainly attributed to the power plant's workers – rose considerably.[213] This did not change the position of the Ried town regarding nuclear power as such, however. After the municipal elections of March 1989, Hübel, as a member of the Green List, stated that she was "disappointed that the CDU continues to have a majority in Biblis",[214] seeking an explanation for that result in the RWE employees' "fear" of losing their jobs. An open letter to Piegsa added: "We understand that fear. But: After a beyond-design accident in Biblis, no one will ask for . . . any kind of job, including yours." And: "We are afraid of the steadily growing probability of the last big incident. From many personal conversations we know that the disquiet in the population is on the rise. We therefore ask you to take our worries and fears seriously and . . . commit yourself to the shutdown of blocks A and B in Biblis."[215] The change in the role of emotions within the debate had also gained a solid foothold beyond Biblis: In July 1989, the Bergstraße District Council talked about a motion by Greens and SPD to do all in its power to effect a shutdown of the Biblis power plant. The SPD's Norbert Bauer from Bensheim argued that this was the only way to guarantee the population's security, admitting that his party "had fallen for the irrational belief in progress of the 60s and 70s".[216] "Fears", "resulting from the use of nuclear energy, had been dismissed as irrational, although they had presented themselves, little by little, through various incidents and accidents . . . as absolutely well-founded fear." Therefore, he argued, his party wanted to phase out the use of nuclear power, to get rid of the unavoidable risks to be confronted with unbearable consequences, even if that meant, for a time, using more fossil fuels – which, he thought, could be largely avoided by efforts to conserve energy.[217] In the end, the motion was carried against the CDU.[218]

In the years after Chernobyl, then, the emotional regime in Biblis and its surrounding region had undergone a significant change via its anchoring to the topic of nuclear power, resulting in a marked split. While emotions had played an important role from the beginning, not least as a motivation for action, they had long been cast as something opposed to the factual, objective, and rational ideal. They had been portrayed as something to be controlled in order to have a legitimate say within the debate and to be able to formulate a valuable argument. As such, signs of emotionality had long served as a tool within the debate, usable against those who showed them, initially restricted to emotions critical of nuclear power, but also applied to emotions that backed it once the critique rose in the region. In the wake of

Chernobyl, for some, emotions were transformed into something that had to be taken seriously, something one could openly act upon. The dichotomy that had been construed time and again between rational thought and (strong) emotions was, to a considerable degree, overcome, and in part even turned on its head. It was replaced by the notion that rational decisions could be legitimately founded upon emotions, that emotions could well be rational, while the factual argument could, when it ignored emotions, become irrational. As the IGFM put it at the end of 1989: Using nuclear power was "madness", while "[w]ithdrawing from the atomic industry . . . is a sign of rationality and intelligence".[219]

This change had been no discursive coup d'état by the IGFM. There had been precursors, phases within the debate on its various levels, during which the role of emotions as legitimate arguments for political actions had already been accepted. Examples include the publication of the emergency plans for the Biblis power plant, calls by Helmut Schmidt to consider the emotional side of one's actions, and the very existence of the Bürgerdialog Kernenergie as a reaction to publicly expressed forms of fear. And when, after Chernobyl, the IGFM came together and relentlessly began to refer to emotions as a motivation to act and as something that had to be taken seriously as an argument within the debate and in the political sphere, they were not alone in doing so. Other citizens' initiatives and the Green Party were with them; some had even taken this path before the IGFM's foundation. In their wake, however, more and more participants followed this course. And while a similar shift could be observed on the higher planes of debate, quite independent of their actions in the Biblis area,[220] the IGFM was unquestionably the staunchest champion of the idea that emotions were a legitimate part of the debate – and maybe even the only true foundation of a given position.

Despite this rather revolutionary and consequential attack on the established emotional regime, the ideal of being factual remained an important pillar of the debates in and around Biblis. This held true, of course, with the ingrained proponents of nuclear energy, continuously trying to use factual arguments to educate their counterparts with a view to overcoming their fears.[221] It also held true within the group of nuclear power's opponents, who advocated that emotions be taken seriously while making a point of emphasising the intense thinking behind arriving at this demand. It even held true for those who tried to frame[222] acting upon emotions as the rational thing to do while deeming the suppression of emotions as irrational. Undeniably, however, the time when emotions had to be renounced, watered down, or attributed to others to be able to take part in the debate had come to an end. In Britain, meanwhile, a similar dilemma had never quite existed.

Oldbury – The Resilience of Measured Emotions

In February 1955, the history of British civilian nuclear energy was, so to speak, officially launched with the presentation of the White Paper

"A Programme of Nuclear Power" to the Houses of Parliament.[223] Although its text was rather technical, elaborating on reactor types and production costs, various emotions became visible as motivations behind its measures.[224] To the "energy of the future", hopes were pinned to cope with "the country's great and growing demand for energy" and to secure Britain's "future as an industrial country".[225] At the same time, it was set against fears that without nuclear power, "our position as a leading industrial nation" could be endangered,[226] while fears of the atom, with its "known dangers", namely its "destructive possibilities" and the "hazards . . . caused by the concentration of highly radioactive materials",[227] were to be met with "vigour, imagination and courage".[228] This simultaneous, visible presence of factual, rational statements on the one hand and various, even contrarian emotions and feelings on the other was characteristic for the British emotional regime dominating the debates about nuclear power about to unfold in and around the places where the new atomic reactors were to be built – even though the "factual reasons" for the selection of these very places were hardly seen as open to challenge.[229]

The example of the debates in and around Oldbury-on-Severn demonstrates not only the wide spectrum of emotions connected to atomic energy in Britain, but also the important role they were allowed to openly play in the debates. Explicitly expressed by opponents and advocates of the technology alike, emotions became central for the power plant's integration into the regional self-image and important for the way the invisible radioactivity came to be considered controllable and, in turn, acceptable. That emotional arguments did function in this way was due to the prevalent emotional regime, according to which emotions were not seen as dichotomous to high modernity's signature rationality, but understood as essentially on par with factual statements: They were accepted as reasons for decisions and actions, as long as facts and information were accepted as being able to influence emotions in turn. Besides, expressions of feelings were widely present in frequently used figures of speech utilised to justify one's point of view in the British debates. It was, therefore, "normal" to make explicit statements about emotions.

This balanced emotional regime came under pressure, and a dichotomy between emotions and rationality began to be constructed when the debate in and around Oldbury went beyond issues limited to, and "treatable" in, the region itself, such as the power plant's design, its consequences for the water balance or regional radioactive pollution. Then, the debate began focusing on more fundamental and harder-to-resolve questions – including the legacies of the atomic industry, namely nuclear waste or the fate of disused reactors – and started to be more and more codified along party lines closed to negotiation locally. This process was foreshadowed by singular episodes during the 1970s, but developed fully from the end of that decade through the 1980s, under the government of Margaret Thatcher, with political representatives from the national and European plane as an important driving force. Nevertheless, the acceptance of emotional statements and arguments in the debates did not end. The dominant emotional regime, open to the statement of emotions on par with factual ones, proved to be resilient in the face of its challenges.

To carve out the role played by emotions in Oldbury's nuclear history, governed by the British emotional regime, as well as the challenges this was confronted with, this sub-chapter's three sections follow the chronological course of the debate in and around Oldbury-on-Severn. The parish with a few hundred inhabitants on the southern bank of the River Severn was part of the Thornbury Rural District in the County of Gloucestershire until 1974, when it became part of the district of Northavon in the County of Avon. A peculiarity of the debate about nuclear power in and around Oldbury is that it did not start in March 1959, with the Central Electricity Generating Board (CEGB) announcing its intention to build and operate a power station with gas-cooled Magnox reactors on the rocky grounds at the Severn's southern shore. It had begun in 1956 with the construction of a nuclear power plant at Berkeley, about eight kilometres upriver, which was also part of the Thornbury Rural District.

Everyday Emotionality

"THE FIRST HEAT EXCHANGER". The words under a picture on the *Thornbury Gazette's* front page on 19 April 1958 invoked the impression of being screamed at. This, for sure, would have been the only way to communicate on the construction site of the Berkeley nuclear power plant depicted here. The black-and-white photography was dominated by said component part being hauled up into position. "The exchanger weighs 125 tons", the caption explained, "and the figures at the base show its size".[230] Indeed, a small group of people becomes visible on a closer look at the picture, construction workers, three of whom, standing on each other's shoulders, would have been necessary to match the heat exchanger's diameter – which, in turn, amounted to just about a quarter of its height. The picture, with its caption, the explicitly pointed-out data and the by-the-by mention of the group of human beings, almost vanishing next to the metal cylinder dangling above them, did not explicitly point out any emotion. Nonetheless, it clearly belonged to a group of articles that was prevalent during the construction phases of nuclear power plants, at the same time expressing and provoking awe for the technological feats they were covering. Together with the new power plants, the technological sublime was under construction.[231]

For the debates about atomic energy in and around Oldbury, the construction of the Berkeley nuclear power plant meant that, years before talks about Oldbury's reactors began, nuclear power had already been established in the area as a representation of the technological sublime, documented in the *Thornbury Gazette*, the weekly newspaper widely read in the Thornbury Rural District.[232] Typical articles pointed out technical achievements like "new welding techniques" and the use of "X-ray equipment", the personnel's proficiency and the buildings' dimensions,[233] emphasised "the very immensity of the permanent works" – that "look very attractive indeed", clad with "Portland stone cement slabs"[234] – or dwelt upon the positive influence the growing power plant had on "British prestige",[235] with visitors from about 40 countries coming to the construction site "of one of the world's first stations from which power is to be derived from nuclear fission", able to supply "millions of homes" with electricity.[236] Descriptions like these, spread over numerous articles during the power plant's construction not only conveyed – and boosted – interest in the huge artefact being built near Berkeley, turning its concretisation into a regional media event.[237] They also oozed with awe and enthusiasm for the technological feat that was the rising nuclear power plant, taking shape on the Severn's banks as a signifier of British industrial prowess, which thereby seemed to rub off on the surrounding area, presenting it as one of the country's most progressive places.[238]

As with most nuclear projects, joy, interest, and pride were mixed up with feelings of fear, expressed at least in milder variations. In 1958, local farmers from the Berkeley area were reported as being "anxious",[239] and farmers from Coleford, about 10 kilometres away on the other side of the Severn, stated their concerns about "the possible effects of radioactivity on farm stock and crops", which were then publicised by Gloucestershire's branch of the National Farmers' Union (NFU).[240] Besides the farmers, the Thornbury Rural District Council's medical officer of health Dr W. J. Douglas Cooper[241] "was very concerned about" "the subject of atomic radiation".[242] Worries like these neither came unexpectedly nor was their expression met with critique, as the "Programme of Nuclear Power" had openly documented potentially worrisome aspects of atomic power, resulting in the provision that "the first stations, even though they will be of inherently safe design, will not be built in heavily built-up areas."[243] Consequently, the expression of adverse emotions was rather expected and accepted under the prevailing emotional regime – as long as security measures were considered as a way to overcome them.

Accordingly, the authorities promised to prove the power plants' safety to all who were as worried as they were themselves, especially so in the wake of the Windscale Fire of 1957 (cf. "Nuclear Debates").[244] Namely, the CEGB announced regular measurements of radioactivity in a 20-kilometre radius.[245] A Local Liaison Committee (LLC) with "all Local Authorities concerned . . . to be represented on" was also convened (cf. "Nuclear Debates").[246] Worries expressed locally and in the region, in other words, were to all intents and purposes taken seriously by the CEGB as the authority commissioned to spread the generation of electricity via nuclear power in Britain. Although the measures promised were neither specific for Berkeley nor designed as a direct reaction to the stated concerns, they were presented as an answer to them, promising transparency, preparation for all eventualities, and certainties and knowledge about the impact of the extra-sensory radiation that was to be utilised in the new power plant. In that way, they demonstrated that the fears expressed regarding the reactor were taken seriously and acted upon, with a view to allaying them, to render the local concretisation of the "Peaceful Atom" acceptable to all.[247]

Indeed, this strategy seemed to work: When, in March 1959, the CEGB announced its plan to build and operate another nuclear power plant near Oldbury-on-Severn,[248] Dr Cooper, who had, by then, received more information about radiation,[249] welcomed the board's proposition. Estimating that the level of radioactivity would not rise, he pointed out: "We are lucky to have the stations here",[250] thereby implicitly referring to the expected economical and reputational premiums for the region.

As had been the case with Berkeley, Dr Cooper's joy was not unequivocally shared in the region. At the county level, the health committee stated its "concern" regarding the lack of consistent information about "[r]adiation hazards" for the public.[251] District councillor Warren Fowler cautioned against the looming vision of a possible concentration of even more reactors,[252] and an anonymous reader of the *Thornbury Gazette* opposed even the one, calling it "an adulteration of the Severnside farmlands".[253] Elizabeth Barrand and Albert R. Child, two of five farmers who were directly "affected", meaning that they would lose at least part of their land, explicitly pointed to their emotions in an interview with the *Gazette*, with Child stating: "It is quite a blow. It is very worrying . . . If you have to sell, it is the finish. I have got this place up to carry 50 cattle. All that work is to be chucked down the drain."[254] Others would lose their fishing grounds and, like the farmers, were "obviously worried to a greater or lesser extent".[255] When, on 21 April 1959, Oldbury's Parish Council under its chairman W. G. Ford took up the topic, "it was unanimously agreed the council protest to the proposed building"[256] because of the loss of valuable stretches of land and water, because of potential flooding and concerns about radiation[257]: "The council also feels there will be continual risk to persons & animals through radioactive fall out [sic, CG]."[258] In other words: Local people, from farmers to councillors, publicised their adverse feelings, their sadness and fear, in combination with factual arguments. In the council's case, the emotions were even part of an official statement.

This open combination of factual issues and emotions, virtually interchangeable on the same level of validity, was not perceived as improper – quite the contrary: When the Thornbury Rural District Council took up the topic in view of the position of Oldbury, its members volunteered to stand by the local council and back its decision. While this offer was not taken up, and the power station's development accepted,[259] this decision by the district council was clearly no rebuke for the parish council having based its decision on emotions. Instead, the rural district council took the locals' position seriously and was, in effect, following a divergent emotional reading of the plans. It had been proposed by Mrs H. E. G. Wells, who had stated: "I cannot see what we are all so frightened about. It is a thing to be proud of and this is the best place to have them."[260] In fact, incorporating emotional aspects in the decision-making process was not only accepted. The *Thornbury Gazette* explicitly lauded it: "It is impossible to do aught but admire the members of the Oldbury-on-Severn Parish Council who have resolved to place on record a decision that they are not in favour of building a nuclear power station . . . the members are to be congratulated on saying what they feel about it."[261] One of the newspaper's readers was even convinced that a broad enough – in other words, democratically measurable – emotional refusal would be effectual on the highest political level, wherefore he called on "all who feel concern" regarding the plans "to make their opinions known to those who represent them", as "if sufficient individual concern is expressed" the project could be halted.[262]

Clearly, concern as an adverse emotion was valued and considered as a legitimate foundation for political positions. In fact, six individual objections and a petition signed by about 300 people and handed to the Minister of Power in July 1959, were one of the reasons for holding a public inquiry on three days in April and May 1960, under A. H. F. Linton, Deputy Chief Engineering Inspector from the Ministry of Power, and C. D. Buchanan, Principal Inspector from the Ministry of Housing and Local Government.[263] During the inquiry's course, too, emotions were stated quite openly.[264] The NFU's Gloucestershire branch, for example, pointed out that "it was felt undesirable to offer such a concentrated strategic target in the event of international hostilities . . . Local people felt thoroughly uneasy about the way this area was being built up into one of the world's plum strategic targets."[265] It also argued with "a natural lay concern at the ever rapid stepping up in the production of a potential Frankenstein monster, full control of which did not appear to be guaranteed, and could not be guaranteed until considerably greater experience had been acquired."[266] Besides the NFU, the Council for the Preservation of Rural England, founded in 1926 as part of a broader movement aiming to preserve parts of England's historical landscape against industrialisation and urbanisation, which made important contributions to the creation of national parks and similarly designated areas from the 1950s onwards,[267] "felt that of the four sites which the Board had been investigating, Oldbury was the worst from their point of view".[268] In addition to these groups, "Mr. J. N. Padbury, a resident of Alveston, was concerned about the danger to the local population."[269] Finally, the inspectors in their recommendations to the Minister of Power, Richard Frederick Wood, freely used expressions of emotions or feelings.[270] What is more, Linton, in his personal commentary to the minister, did not devalue emotions in any way but took them seriously, for example, referring concerns about the station's safety to the judgement of the Chief Inspector of Nuclear Installations.[271] Nevertheless, five months after the public inquiry, in October 1960, the Minister of Power approved the construction of Oldbury nuclear power station.[272] The *Thornbury Gazette's* commentator "Double Glo'ster" found the prospect "exciting and interesting",[273] foreshadowing a renewal of articles dwelling in the technological sublime, with statistics and numbers proving the project's – and, by extension, the nation's – greatness, which would then reflect back upon the local area and people.

Before this could become manifest, however, the start of operations at the Berkeley power plant provoked a wave of concern about radioactive emissions in the region, even prompting the local conservative Member of Parliament, Frederick Vernon Corfield, to make some explicit emotional statements in letters to Wood. On 16 January 1961, Corfield informed him that he was

> very disturbed indeed to learn from the local press and individual constituents that radio-active waste is to be discharged from Berkeley nuclear power station both into the Severn and into the air . . . One of my constituents who came to see me privately on this matter happens to be something of an expert himself, and is particularly concerned with regard to the discharge into the Severn – as indeed am I – not because it is necessarily any more dangerous than discharge into the air, but because it is something of which more seems to be known or at least can be deduced.[274]

He emphasised that his acceptance of the station had rested on assurances by the CEGB's experts "that there would be no question of such discharges".[275] Here, a Member of Parliament not only openly stated his own emotions, in line with those of his constituents, he explicitly identified one of them as an "expert" and pointed out this individual's emotions as something to consider in the matter of discharges. Based upon these concerns, he added a list of further questions regarding the discharges and the way they were communicated, which he thought "outrageous . . . I shall expect an apology . . . delivered in person at my convenience, and by appointment, through my Secretary", adding, by hand: "Sorry about this but I really am cross!"[276] Clearly, Corfield wanted the minister to know his emotions regarding the nuclear discharges themselves and the way the CEGB had handled their communication. This was not meant to imply that he would neglect the board's arguments. In the wake of the minister's answer, he made that explicit:

> I of course accept the scientific advice . . . I am afraid I still feel very aggrieved by the attitude of the C.E.G.B. I am perfectly prepared to accept that they made these statements in good faith, but as soon as they realised that as a result of their misapprensions [sic, CG], I had been misinformed, it seems to me that as a matter of the most elementary courtesy they should have informed me . . . Since it is clear from your letter that they now know how I feel on the subject, I am absolutely astonished that, having been away for a fortnight, there is no letter in my files from the Chairman or anyone concerned, expressing their apologies personally, and I shall certainly take any opportunity that arises to make some fairly acid remarks in the House on the subject.[277]

While the focus of Corfield's emotions had changed, from the fact that there would be radioactive emissions to the way the CEGB had communicated it, they were taken seriously by Wood. The minister had a member of his staff write to the CEGB to ask them to apologise to Corfield or to allow him to do so in their name.[278] In fact, the CEGB's chairman Christopher Hinton wrote to Corfield himself, informing him "how sorry I am that you feel aggrieved" and explaining why the misunderstanding had happened – namely, because, early on, "atomic energy people didn't realise how carefully they had to avoid saying 'no radioactivity' when in fact what they meant was 'no harmful radioactivity'."[279] Clearly, Corfield's expression of emotions had not reduced the importance attributed to his letters. On the contrary, he had received an apology and an explanation from the very top of the industry, while at the same time allowing himself to be reassured regarding the question of the radioactive emissions to be expected. With that, he gave a prime example of the workings of the emotional regime at play.

He was not alone: The *Thornbury Gazette* had initially pointed out that the emissions had not been communicated well by the CEGB, so that the realisation that there would be radioactive discharges came as a "shock".[280] Later on, it had emphasised that, "we are all fearful of the effect which radioactivity may have on the human species", not least because so many things about it were unknown to most people.[281] That notwithstanding, it let go of the topic after the CEGB had explained that releasing a small amount of radioactivity into the environment was considered normal. The Thornbury Rural District Council's public health committee, too, was convinced that herewith "a true picture had been given to the council and that there should be no cause for concern".[282] Its chairmen during the 1960s, J. H. Cooke[283] and John Spratt,[284] observed that, in general, people were reassured. Their view was shared by Minister of Power Frederick Lee, who visited Berkeley in 1965 and stated that "[p]ublic fear of them [nuclear power stations, CG] was quickly dispersing and the public was becoming more educated in its attitude to atom stations."[285] The *Gazette* concurred, writing in 1968 that "people in the Severn Vale have accepted with remarkable aplomb the safety assurances which have been reiterated time and time again by the experts – assurances which have so far been justified up to the hilt."[286]

The episode about Berkeley's radioactive emissions is quite telling of the emotional regime governing the early British debates about nuclear power: Emotions and their expression were accepted as an important part of the debate that had to be considered. They were not seen as being at odds with rational arguments, but as normal reactions to facts, as part of a dialogue. That included the expectation that they could be invoked, strengthened, or allayed by further facts and rational arguments.

During the construction of Oldbury nuclear power plant, fears, worries, or concerns about radiation retreated from the public debate. Instead, the technological sublime returned, with interest and joy, fascination, and awe related to the new building rising that was to contain and control the atom's power. There were descriptions of the "tremendous, controlled activity",[287] on the construction site of a nuclear power plant that would command, as a world first, "a pre-stressed concrete pressure vessel", which – as was emphasised for the benefit of those worried about radioactive emissions – "means that there will be no measurable radiation around the reactor vessels during operation".[288] Therefore, in theory, as the *Thornbury Gazette* put it, the plant could "be built in the centre of a city with no fear of any danger. Should there be an explosion in a concrete pressure vessel – the odds against are 671 million to one – the walls would not explode – the cables would hold them in position".[289] Besides wording that spoke of men controlling nature, typical for the framing of nuclear power,[290] special elation was triggered by modern computers,[291] namely in the control room. "This automatic brain will be manned by only three or four people. It is the first computer of its kind."[292] Regarding the various novelties, world-firsts and gigantisms, a reporter of the *Thornbury Gazette* described his visit to the site in 1964 as "stimulating and exciting" and named an "inspection of the interior of the pressure vessel for Reactor No. 1" probably "the most fascinating experience".[293] One year later, the paper pointed to "an ever-increasing sense of awe and wonder at the ingenuity of the conception of this complex project – one of the technological wonders of this age" and "a scene of light and shade which Rembrandt would have delighted to have recaptured with his brush".[294] When the power station was officially opened by Minister of Technology Anthony Wedgwood Benn in 1969, the Minister saw it as "a matter for pride that in the peaceful exploitation of this tremendous source of energy Britain took a decisive lead",[295] pointing to the grandeur that a newly constructed power plant could encompass. The only downside, regarding emotions in the public debate, was caused by the fact that Oldbury Parish Council had not been invited to the festivities, causing the councillors to write to the CEGB, "stating that they . . . feel very slighted and consider that it was a matter of courtesy to have them invited."[296]

More than one decade after nuclear energy had come to the area around Oldbury with the Berkeley power plant, emotions, including variations of interest, fear, sadness, pride, and joy, were a fixture of the public debates. None of them had been continually present. Nonetheless, they had been expressed: sadness when the power plant's price in terms of lost homes and livelihood was realised, and interest, joy, and pride – the technological sublime – during the years of construction. Fear came to the fore whenever potential consequences of the power plant for its environment, including the surrounding population, moved to the centre of attention. This implied that during the years the power plant was in operation, sadness, joy, interest, and pride played a diminished role. On the other hand, the potential for fears being triggered grew, with more information about the possible consequences of the atom spreading, or incidents suggesting an increased local likelihood of dangers.

Undertones of Anti-Emotionality

While there was no renewal of the emotional debate about radioactive emissions that had occurred when Berkeley had become operational, soon after the Oldbury nuclear power plant was officially opened, plans by the CEGB to expand the power plant,[297] upheld until 1978,[298] provoked another critical public debate.[299] It included a wide number of factual questions and arguments, ranging from effects on the region's water to radioactive emissions. Beyond that, questions of radiation-induced sicknesses became an important topic. Once more, the debate included open emotional statements, too. Now, however, an undertone of critique of using emotions as a discursive strategy came to the fore, contouring the borders of the emotional regime.[300]

Unquestionably, critical positions regarding nuclear energy had increased in the region since the first power plant projects, especially in view of the rising number of atomic reactors around the Severn Estuary. This occurred against a background of open questions without convincing answers being debated (trans-)nationally, including the handling of the nuclear industry's legacies, from atomic waste to disused reactor-buildings. In the public works and planning committee of Thornbury Rural District Council, councillor "Joan James expressed fears about the heating up of water returned to the river". Also in that vein, Douglas Pearce from the river authorities was "not happy that we should allow a further station to be built". Moreover, he abhorred future "nuclear tombs on the edge of the Severn".[301] While James' statement about her fears appears to have been regarded as a regular contribution to the debate, albeit one not shared by the majority, the "tombs"-metaphor was criticised. It was perceived not as an emotional reaction to a factual development, which in turn could be reacted to with safety measures or information, but as a "scare story", an attempt to "emotionalise" the debate, to end the dialogue between emotions and factual arguments. Councillor Leslie Walter Bosisto from Thornbury's Rural District Council was "quite satisfied the safety regulations are such as to be non-worrying to the people. We have, in fact, had the ban on development lifted because we are satisfied there is no real danger". While not sharing them, he took up the fears referring to specific, measurable consequences of the reactors. He repudiated, however, "scare stories of nuclear tombs", which "can only upset the public and have no relevance whatever". Other members assented, seeking an explanation for current "public worry" in a recently televised interview,[302] thereby pointing to a connection between different planes of the debate, with "unreasonable" emotionality being imported from the higher strata. It was a phenomenon that would rise to prominence at the end of the decade.

For the time being, besides this one occurrence, most participants in the Oldbury public debates adhered to the established emotional regime, accepting emotional statements as valuable contributions to the discussion, while abstaining from overly pointed, emotionalising ones.[303] The Oldbury Parish

Council, for example, was "expressing concern", while the "Gloucester Harbour Trustees were worried" about the effects an expansion might have on the water. In South Gloucestershire's County Planning Committee, Dr Dennis Fox stated, regarding planned chimneys emitting radioactivity: "I am very worried about putting radio-active material up into the atmosphere at this particular point . . . I want considerable reassurances about radio-activity which is going to be put out through these chimneys." His colleague Pat "Wells was concerned about the dangers on the road leading to the Oldbury station . . . Residents faced traffic danger and they worry about the transportation of atomic and radioactive material past their homes."[304] When the committee, notwithstanding the critique, approved the CEGB's proposal with a slight majority,[305] Dr Fox protested that he felt "very disturbed" that a slim majority was accepted in "such an important decision . . . when we know the people in Oldbury are very much against it",[306] thereby hinting at the questions about the functionality of established representative institutions, which were often tied to decisions about nuclear energy (cf. "Nuclear Democracy"). Reacting to the emotionally based critique, the CEGB promised that "fullest opportunities would be afforded to the planners and other bodies in the district to show that their fears about radioactivity were unfounded." There would be "no significant increase in the background level of radioactivity", which would be "constantly monitored" and "reported to the local Liaison Committee".[307]

Expressions of emotions as part of or reason for critique, in other words, were still not construed as a reason not to take the critics seriously. On the contrary, fears continued to be seen as a reason to act, to prove, or, more to the point, to disprove or explain. Emotions remained an accepted part of the public debates: "People are naturally worried to know the answers to their questions and we shall make the fallout information available", the board promised.[308] Indeed, the CEGB's willingness to talk to members of the local and regional authorities regarding their fears was taken up after "local representatives" in the county planning committee "felt there was not sufficient evidence to give a final decision": The Thornbury Rural District Council invited representatives of the board "to answer questions and deal with fears aroused by the proposal".[309] Councillor Fowler "felt it was up to the CEGB to . . . assure people they were not living in a potential danger area either today or tomorrow. There were fears of cancers" and other dangers. Dr Fox added that "some of the fears aroused . . . could be answered at a meeting and if assurances could not be given, people could state that they didn't want any more atomic power plants in this part of the country." He supported Fowler, who suggested "a meeting with plain speaking and plenty of time so that they could get good answers to their questions", and "said he had been amazed at the number of people who were extremely concerned about this".[310] In the view of the *Thornbury Gazette*, "Thornbury Rural District Council is well advised to echo the concern being felt along Severnside".[311] On a side note, the newspaper asked for more transparency at the LLC, which "could do much to allay the growing fears curently [sic, CG] troubling the area",[312] highlighting connections between nuclear emotions and questions of trust.

The tried and tested strategy of taking fears seriously as a reason to act, then, was once more applied by the CEGB. The demanded meeting was scheduled for 19 June 1970.[313] At its beginning, Thornbury Rural District Council's chairman F. W. Perry stated that "they were very concerned for the safety of the area in which they lived", not least in view of "any excess radiation". He explained that the council "hoped as an outcome of this meeting to find there was no excess". In the end, from his point of view, this goal was reached.[314] Dr Fox, too, thought that at least "some of our fears have been allayed" and, for the time being, concentrated more upon other types of industrial environmental pollution – especially a large installation of Imperial Chemical Industries.[315] Of course, not all fears had been eradicated. From Bristol, H. S. Davies wrote to South Gloucestershire's Member of Parliament Corfield, calling the expansion plans a "criminal folly" and asking him "to prevent the disaster that threatens a large part of your constituency",[316] and Oldbury's Parish Council continued to oppose the power station's expansion.[317] Here, "feelings ran high" because of the way the county had handled the CEGB's proposal "without full consultation with the people most concerned, namely the local inhabitants",[318] and the council "unanimously decided to continue to fight against the proposals to build a second nuclear power station in the parish. The meeting felt that recent events at Windscale[319] have proved the validity of their criticism and a further letter of protest is to be sent to the Ministry stressing the dangers of siting so many stations in one locality."[320] Besides being another pointer to the impact wider frames of debate could have on the local or regional ones, the CEGB's reaction to the documented emotional reception of their plans by the parish council was a further example of the continued dominance of the British emotional regime: In November 1970, it promised that the objection would "be carefully considered before the C.E.G.B. decides to make an application for consent to its construction".[321] In the end, the expansion was cancelled in the middle of 1978. And while the local debates were not held to have been the decisive reason,[322] it is unlikely that the emotional rejections played no role at all.

In the meantime, while the debates about nuclear power in and around Oldbury during most of the 1970s were not very lively, they flared up on occasion, time and again displaying the impact of emotions on them. On the one hand, there were open days at the local installations, for example, in the autumn of 1971, drawing "a tremendous crowd of interested local people" and allowing the *Thornbury Gazette* to revive the enthusiasm for the power plant's technological feats,[323] besides indicating the technology's widespread acceptance. Explicitly, regarding an open day in 1973, when about 3,000 people visited, the *Gazette* wrote that "[w]ith nuclear power stations within their midst at Oldbury-on-Severn and Berkeley the people of the Thornbury Rural District who live in their vicinity have lost what fears they may have once had of any possible dangers from them. That there is still an intense interest in nuclear power stations was amply demonstrated on Saturday."[324] On the other hand, there were more controversial phases, including another test of the emotional regime's openness: In spring 1971, Dr Fox had triggered a renewed debate about the preparations for nuclear emergencies and their implications for the risks borne by the people living nearby, which was settled by the CEGB pointing to the LLC,[325] and Fox admitting that "a small risk" was acceptable.[326] Later that year, a letter-writer from Alveston, just south of Thornbury, objected plans to dump low-activity waste from Oldbury on a regular waste dump nearby.[327] The position was shared by Fox, who emphasised "his fears about the disposal of slight radioactive waste on the refuse tip", saying: " 'I live near to this tip and I am extremely unhappy at any radio-activity emanating from this tip".[328] He could not convince the Thornbury Rural District Council of his point, however.[329] On the contrary, G. Strange even claimed that "it is utterly wrong for people to arouse fears of this nature – because people do have a certain fear of radio activity".[330] He accused, in other words, Dr Fox of trying to stir up emotions, which was frowned upon as an infringement of the emotional regime. As the *Thornbury Gazette* put it, it was "perfectly right and proper that any person having doubts should express them. Complacency is the last thing that is wanted in such matters. Yet the doubts should not be magnified out of all proportion of their true worth."[331] And, directed at Fox:

> His attitude was no doubt prompted by the fact that the very word "radioactive" has an emotive effect. Because radioactivity cannot be seen, and its measurement is a matter for experts, its reported presence automatically gives many folk cause for concern. Yet their fears are not necessarily justified. The nuclear power industry is a new industry. From the very start its participants have been hedged around with official safeguards that have always erred on the side of extreme caution.[332]

At the end of the 1970s, similar accusations of attempting to induce emotions in others, or to construct a dichotomy between emotionality and rationality, rose sharply, amounting to a serious test of the established emotional regime. It was caused by the rising prominence of fundamental questions regarding nuclear power, which could not be answered or taken care of locally. This was a consequence of connections between the local, the regional, and the (trans-)national planes of debate coming to the fore more frequently while the issue was aligned with party-political conflicts, especially under the Thatcher government.

The Emotional Regime's Challenge and Resilience

On 22 October 1977, the *Gazette* pointed out that "[a] correspondent in last week's Gazette joined the chorus of international protest against nuclear power. It is an emotive subject, and protests invariably gain ready support. It must be admitted, however, that this support mainly comes in ignorance rather than a full understanding of what few would deny is a very complex and technical new industry."[333] While the original letter to the editor by Mrs J. E. Wheeler could not be found, in all probability it was criticising the concept of a closed fuel cycle with its reliance on breeder reactors and plutonium against the background of the debate about new reprocessing facilities at Sellafield making national headlines.[334] The newspaper's reaction implies that her statements lay, in its estimation, outside the accepted range of emotional expressions. The paper thereby demonstrated that emotions, evoked against a background of lacking "factual" information, were not appreciated but bemoaned as endangering the established, balanced emotional regime. There was more to the case, however, than there had been during similar situations earlier in the 1970s. By ascribing the source of the rejected emotionality to a higher level of debate – in this case, the "international" one – the letter was construed as part of a larger development coming to the Oldbury debates from the outside, where emotions seemed to be utilised in an unfounded, irrational way. By considering it in this way, the paper took up a strategy to counter anti-nuclear protests applied by pro-nuclear political and scientific actors on those higher planes of debate, strengthening the connection to them.[335]

It was not only the *Gazette* that assessed Wheeler's letter in this way. David Buckley, a reader working at the nuclear power plant, accused her of "cheap and ignorant scaremongering", while claiming that opponents of nuclear power often used "gross exaggerations and misleading statements to frighten public opinion against the true expertise of many who have devoted their professional lives to the safe and efficient exploitation of nuclear power. Comment and criticism is [sic, CG] truly stimulating but only when it is properly informed."[336] With the legitimacy of emotions bluntly questioned in this way, with them being construed as irrational and inadequate responses to a debate based on facts, the conflict in the newspaper's columns escalated quickly. It brought to the fore more hardened positions and a narrower emotional regime, wherein emotional statements on both sides were more readily repudiated. At first, Wheeler insisted that her worries about breeders and the widespread usage of plutonium were not emotional in an irrational way, but based upon arguments made by scientists, upon "facts which cause one great concern".[337] While this statement surely fit into the established emotional regime, the tone of the exchange between readers sharpened when Mary Brown from Dursley supported Wheeler by pointing to more fundamental problems connected to the use of nuclear energy, such as the rising number of people who fell under rules of secrecy or the necessity of deciding moral questions on a broad basis and not leaving them to a small number of experts.[338] This, to Buckley, was "melodramatic". He recommended reading "sensible and properly researched information" instead of "the more lurid anti-nuclear 'penny-dreadfuls'", to realise that nuclear power generation was saving lives when compared with coal.[339] Later on, he accused opponents of nuclear power of being "primarily motivated by antipathy to the basic nature of industrial society . . . With friends like the Friends of the Earth and their ilk, the earth has no need of enemies."[340] To this, Peter Brown from Dursley reciprocated that, although important problems with nuclear energy were unresolved,

> people in the industry . . . are pushing on in the hope that "something will turn up" while radioactive materials accumulate at Windscale needing constant monitoring and maintenance. This is only part of the case against nuclear power put by Friends of the Earth. David Buckley's contemptuous remarks about this organisation are scarcely worth comment – many highly qualified scientists have doubts about nuclear power.[341]

Buckley retorted by calling these statements "good scaremongering stuff", adding: "The problem with many of the anti nuclear pseudo theorists is that at first sight many of their statements seem entirely reasonable . . . I myself was caught out a few times until I learned to look deeper into these statements, which almost always turned out to be facile and baseless."[342] Mary Brown answered that, indeed, some of nuclear power's critics wanted to change society as such and had "hope that enough people will come to realise that there are things more important in life than colour television, frozen vegetables and two cars in a centrally heated garage. And let them say this loudly enough, before the scientists plunge us into the plutonium economy, which . . . might well be a near-Police State."[343]

Clearly, when it came to more fundamental questions connected to nuclear power that went beyond matters that could be reacted to locally with specific measures, like practices of democracy (cf. "Nuclear Democracy") – topics that were hotly debated (trans-)nationally in the late 1970s – quite a few of the participants in the local and regional debates were increasingly unwilling to accept emotional statements. Instead, they rejected the "blind hopes" of atomic power's proponents' on the one hand, and "scaremongering" by the technology's critics on the other. The long-established emotional regime had obviously come under pressure.

During the 1980s, similar general aspects of nuclear energy came to be appropriated into the local and regional debates time and again. One example for this process was the aspect of nuclear waste, which permeated the separate planes of debate in a very physical form, as it was regularly transported from the power plants to reconditioning facilities at Windscale. In Stonehouse, a parish in Gloucestershire about 30 kilometres from Oldbury, where trains transporting nuclear waste from Hinkley Point stopped regularly, "John Marjoram, leader of the local anti-nuclear lobby", explained: "We are not taking this matter lightly . . . the people of Stonehouse have every right to be very alarmed about this potential for nuclear disaster'". In addition, "John West, vice-chairman of Stonehouse Parish Council, called on his colleagues to write to British Nuclear Fuels to express anger at not being informed, and to ask for more details." Remarkably, the reason given by West for the importance quite suddenly attached to the transports was that "[e]veryone thought the transportation of nuclear waste was safe until they had a scare in France last autumn when a cylinder tipped over and there was a terrible problem righting it."[344] Obviously, information from the transnational plane of the debate that highlighted potential problems then influenced the regional one, causing emotional responses there.

Now, however, the CEGB – and nuclear power's proponents at large – were not reacting to these emotional statements as they had when similar situations developed locally earlier on, when they gave assurances they would take local voices seriously, held special debates and established an ongoing, reasonably transparent monitoring programme. Instead, the critique was met with resistance from members of the ruling Conservative and Unionist Party, which bet on nuclear power not least as a way to weaken the unions with which it was locked in a bitter conflict, further sharpening the tone in the debate.[345] On the national level, Secretary of State for the Environment Nicholas Ridley later even denigrated the technology's opponents as "selfish and irrational".[346] In the region surrounding Oldbury, South Gloucestershire's Conservative Member of the European Parliament, Richard J. Cottrell, was convinced that emotions expressed by critics of nuclear power – even if they were understandable – had to be resisted as the spearhead of an irrational, transnational political movement against nuclear energy.[347] The UK, he pointed out "must not make the mistake now being made in many European countries of giving in to 'a tidal wave of so-called Green Power opposition to nuclear power'", which

> was about to invade Britain. The Three Mile Island accident in the United States had naturally frightened people – but the plain fact was that no-one had ever closed down all the coal mines despite a long history of appalling accidents and loss of life . . . We live in a real world where our very survival depends on access to energy and we must make ourselves increasingly less dependent on oil. The Government's energy minister warned only this week that the world scramble for oil could lead to war. It would be utterly senseless for Britain, for Europe, to opt out of the practical development of alternative sources of energy such as nuclear power.[348]

Cottrell, who repudiated fear of the atom on the one hand while openly describing scenarios that lent themselves to inducing fear for the atom – a tactic that was obvious to the technology's contemporaneous critics[349] – stuck to this kind of argument during the following years. Regarding protests that were statements of adverse "feelings" against the dumping of nuclear waste at sea, which became an important topic for critics of nuclear power in the region because of transports starting from Sharpness, about five kilometres from Berkeley,[350] he argued that "the anti-nuclear movement peddling myths and misunderstandings" was doing "rational debate" a disservice. He thought it "hypocritical . . . to condemn nuclear power . . . while ignoring totally the risks of the underground mining of coal" and fossil fuels' "possibly harmful affects [sic, CG] in the upper atmosphere". To him, "the continued expansion of nuclear power was essential for the country's future prosperity. 'We are talking about power . . . to run hospitals, homes and factories which create jobs . . . Those who seek to frighten and dismay the nation . . . belong in the same category as those in previous generations who have resisted any technological advance.'"[351] Cottrell's emotive presentation of nuclear power, with all its risks and problems, as the rational choice when compared with other sources of energy, or the scenario of a lack of energy, while condemning critics for trying to use emotions to mobilise support for their irrational purposes, was soon adopted by other proponents of nuclear power in the region – and, in fact, widespread in Britain around 1980.[352] In March 1981, the *Gazette* had already pointed out that the nuclear industry was an important employer in the region, so that, if there was "any mishap of the type feared by pressure groups opposed to the nuclear industry", they themselves would be affected. This, the newspaper implied, meant that they would take care to avoid such a "mishap". To the *Gazette*, "[u]nfortunately" the "pressure groups are very vociferous and skilled in the use of publicity media to put over their viewpoint to the population at large",[353] while, in fact, the continuous monitoring of the region's radioactivity proved that there was nothing "to cause concern",[354] and that "some of the critics of nuclear power are worrying somewhat too much".[355]

Clearly, the increasingly prominent connections between the local and regional debates and their (trans-)national pendants, established in the wake of more fundamental questions regarding nuclear power coming into focus, went along with a way to regard emotions and emotional expressions within the debate that was not covered by the established emotional regime in Britain. Instead, it resembled the one dominant in the case of Biblis up to 1986. The declaration of emotions tended to be seen as irrational, and a fault line regarding the acceptance of corresponding statements run between proponents and opponents of nuclear power. This fell in line with main party-political affiliations, additionally heating up the conflict. To some proponents of nuclear power, most often representatives of the CEGB or Conservatives like Cottrell, it was now irrational to be afraid of nuclear power – and especially condemnable to express any kind of fear in this regard while saying nothing about the dangers of the coal industry, with which the party was at loggerheads. Doing so was deemed "black magic hysteria",[356] and synonymous with fearmongering for political reasons. To critics of nuclear power, on the other hand, concerns about the technology were rational and a reason to act. These were represented, not least, by Labour,[357] by the centrist SDP-Liberal Alliance between the UK's Social Democratic Party and the Liberal

Party, later to merge into the Liberal Democrats, and by organised critics of nuclear power like Friends of the Earth and citizens' initiatives now growing in the region. One initiative that played a significant role was the Severnside Campaign Against Radiation (SCAR), which developed after televised reports about heightened incidents of childhood leukaemia around Sellafield in 1984 put the issue on the national agenda.[358] Some critics even completed the circle and deemed it irrational to stick to nuclear power, arguing that those who connected the expansion of nuclear power with the economic future of the country, like Cottrell, were guilty of "scaremongering".[359]

Despite these heated elements surfacing in the public debates, however, the established emotional regime, which allowed for a broad variety of emotional statements in interaction with the factual development locally, did not dissolve. Most people still adhered to it, even when pressing local or regional issues came up. The most prominent example was a debate about leukaemia. After being discussed time and again in connection with nuclear power plants and nuclear waste transports around Oldbury since the 1970s,[360] it became a dominant subject in the regional debates after six cases of childhood leukaemia occurred in the nearby area of Lydney between 1979 and 1983. This was considered alarming, as only one case would have been expected. And because Lydney lay on the Severn's northern shore, opposite Berkeley, about 10 kilometres from Oldbury, nuclear power was soon considered a possible culprit.[361] These suspicions were met with fierce resistance by nuclear power's proponents – not because of their emotional character, however, but based on the conviction that there was no connection to the technology. This line of argument appeared in the *Gazette's* letters column in March 1984, when Alan Bryn Williams from Thornbury declared it "understandable that people in the district should be concerned" and even "understandable to some degree that those who are opposed to the use of nuclear power should assume that radiation is to blame", while insisting that "[s]uch an assumption can cause unnecessary alarm, because it is entirely without foundation". His conviction was based on the lack of knowledge about the reasons behind clusters of the illness, which occurred also in places – and at times – without nuclear power plants. He concluded: "Clearly more research is needed rather than the futile controversy we have had recently in the district."[362] His position was backed by the LLC, which pointed to radiation tests around the two regional power stations showing "that the radiological impact of the stations on the local community remained at a negligible level" for the last 26 years.[363] The *Gazette*, too, pointed out that "75,000 radiation measurements in the area around Berkeley" demonstrated "that Berkeley has not affected the environment in any way". Nevertheless, the newspaper remarked that "[i]t is right that nuclear safety should constantly be monitored and equally right that the public should be able to challenge and question the findings." At the same time, it criticised "those who are not prepared to accept the findings and choose instead to snipe at the scientists by constant innuendo and the belief that the experts are not to be trusted. What more must the nuclear people do to stop the sniping?"[364] The answer to that question, put forward by Lydney's Town Council and members of SCAR, was more research and transparency as a reaction to existing worries.[365] In line with the established emotional regime, this approach was supported by West Gloucestershire's Conservative Member of Parliament Paul Marland, who was "willing to help SCAR in its search for information" "to try to allay fears", although he was convinced that the existing measurements should be sufficient to show "that fears that the Berkeley and Oldbury nuclear power stations on Severnside are causing cancer are groundless". As a CEGB spokesperson put it: "In terms of known science the stations have had no recordable effect on the environment" meaning that "[i]n terms of scientific readings the stations do not exist". Adhering to the emotional regime, which welcomed expressions of emotions if those were construed as being in dialogue with factual developments locally, he added: "If people have got any worries at all, they can be quite happy about contacting the station manager or the CEGB in general. We have no secrets from them."[366]

The emotional regime according to which concerns or worries could and should be allayed by transparent studies and according to which emotions should be taken seriously as a reason to act was, in other words, still viable. This impression was further reinforced regarding the authorities' and the industry's willingness to be transparent, which was increased as a reaction to a surge of emotional statements following the catastrophe of Chernobyl. Up to 25 April 1986, the readiness to go beyond established measures when it came to allaying regional fears regarding leukaemia had been limited. The LLC, for example, was unwilling to set up "a sub-committee to investigate any links between low-level radiation from Severnside power stations and leukaemia in local children", which had been demanded in April 1985 by Chris Vidler from the Forest of Dean District, to which Lydney belonged.[367] One year later, with the topic still virulent, the situation was quite unchanged. The committee saw "no evidence at all" linking "low level radioactive emissions from the two sites with the cluster of leukaemia cases in children in Lydney". At the same time, councillor Bill Hobman from the Labour Party pointed out the lack of "any evidence to disprove the theory of a link", which was what many people suspecting nuclear energy to be the cause of the illnesses were looking for. In the words of the *Gazette*, summing up the argument that was presented to the committee: "Public anxiety about the subject would continue until some definite proof that the two were not linked was provided",[368] or, at least, more transparency about the topic established. As the Northavon District Council's Alliance group leader, councillor Dr George Conrad, put it,

> I deplore the way these figures have been reported . . . Why does the power station liaison committee meet in secret? Surely the whole object of such a committee is to improve the relations and trust between the public and the industry. This is a typical example of the obsessive secrecy which surrounds the nuclear industry. Until these meetings are

> open to the press and public they will be surrounded by the suspicion that they know something that they do not want us to know![369]

The latter demand gained ground after Chernobyl, against the background of "increased peoples [sic, CG] fears",[370] as Dan Norris, the prospective parliamentary candidate for Labour in Northavon, put it.[371] It was not only Norris who had the "hope" that more information about leukaemia would help "to protect Northavon people and reduce their fears".[372] The idea that transparency could "allay fears" – and that fears were a reason to act in the first place – was shared by the *Gazette*: "Fear will always persist, if secrecy is maintained, however good the intention, without the watchdog role of the press. For this reason alone, to say nothing of Chernobyl, then [sic, CG] liaison meetings should be open to the Press . . . Most of these fears would disappear with the simple expedient of 'open' meetings."[373] This reading of the situation seemed to prevail, with the CEGB soon declaring its willingness to support SCAR with advice for a planned monitoring project that was to be independent of the industry,[374] and with Keith Baddeley, Oldbury's station manager, convinced that: "By being secretive, people always assume the worst and we decided to tell them everything and come out of the long grass clearly and formally, warts and all."[375] Even Peter Edward Walker, the Energy Secretary, was now full of "hope that the revelations will allay any fears the public may have following the Chernobyl disaster".[376] It was in this vein that the government and the industry, in the aftermath of Chernobyl, initiated a number of measures to better inform the public about nuclear energy. These measures, often explicitly seen as a reaction to public worries, included open days at Oldbury,[377] the publication of "fortnightly news bulletins"[378] and "the setting up of an Emergency Planning Co-ordination Committee".[379] Also, the LLC was reformed and expanded.[380]

Obviously, the emotional regime in place in the debates in and around Oldbury had been challenged; the debate became more complex when it was connected to its more heated (trans-)national counterpart from the late 1970s onwards, with more information about nuclear power's downsides afloat and the issue being strongly intertwined with party-political conflict lines. The tendency not to accept the emotions of the respective other side, to equate them with irrationality or fearmongering, that was present here, had at least partly been taken up by regional voices. In the end, however, these more strident accusations did not replace the widespread acceptance of a wide range of emotions as an established part of the debate on par and in dialogue with factual arguments. And, when the Berkeley power plant was to be closed in 1989, the mostly pro-nuclear *Gazette* felt able to openly state its sadness: "What we can certainly say is that Berkeley has been a good neighbour to us and has had a significant role to play in all our lives. Its retirement is sad but Berkeley will not be forgotten, that is for sure."[381] The emotional regime that had been shaping the Oldbury debates from the start was still intact.

Rigidity or Reconciliation – Emotional Regimes Under Stress

Nuclear debates were emotive – and they were shaped by emotions. The analysis of the debates around Biblis and Oldbury confirmed this basic fact. And it revealed that the emotions intertwined with the "Peaceful Atom" were not constrained to hopes and fears, even though both played a large part. Early on, when the power plants were planned and constructed, their dimensions and novelty seemed fascinating. They were received with interest, with joy and pride, and with hope, which had also stood behind their construction in the first place. In the case of Biblis, it had been mayor Seib, motivated by his expectations for a better future for his municipality, who had striven to bring nuclear power to the small town in the Ried – and, in fact, continued to do so through his time in office. In Oldbury, no similar personal engagement stood out. Hopes, however, had been enshrined in the British nuclear power programme, and when the technology came to Oldbury – after being established in nearby Berkeley already – it was greeted with joy. Less prominently, other emotions also played a role, including forms of sadness, felt in Oldbury regarding the changes to come when the first power plant was built, but also by Biblis women and mothers when others were not ready to share their fears. Also among these emotions was anger. It was expressed not least when actors felt they were not being taken seriously, be it the critics of nuclear power in Biblis in the late 1980s, or Conservative parliamentarian Corfield, when the CEGB had made misleading statements about nuclear waste products. It resulted in communications and, quite possibly, contributed to more care being taken in explanations about the consequences of nuclear power plants. Also, courage and confidence were pointed out, mostly to demonstrate the absence or containment of one of the most prominent emotions in nuclear histories: fear. Fears were present manifold and from the start in both locales under consideration here. They came in the form of concerns and worries about the changes that power plants would bring to rural regions, and especially as fears about the dangers of radiation, most prominently so when accidents and incidents, locally as well as (trans-)nationally, drew attention to what could happen. While Chernobyl stood out, there were other events triggering fears, from accidents at Windscale or TMI to conspicuous numbers of leukaemia cases, to unexpected plans for expansion in both countries. Concerning the latter, it is quite striking that the following rise in fears occurred in Biblis in 1975, five years after the parallel development in Oldbury. This does not only imply that local developments could have strong impacts independent of the development of the (trans-)national debate. More importantly, it indicates once more that Britain cannot be considered to have experienced a less dynamic, quieter debate about nuclear power. There had been several conflicts and escalations, here, too. They were treated differently, however – rather leniently in Britain, restrictively in Germany – resulting in different outcomes. In both cases, nonetheless, fears, once triggered, motivated questions, protest or, most prominently and basically, participation, resulting in increasing public debates about atomic power as a "public technology"; shaping, in other words, the nuclear histories in question. And indeed, they were not only present on the side of nuclear power's critics. Fears had been felt, or at least recognised, by its supporters, too. Fears of the technology motivated safety measures, like the rules regarding distances between nuclear power plants and centres of population (cf. "Nuclear Debates"), but also the monitoring of radioactive emissions. One could even argue that they had motivated the German Atomic Energy Act's stipulation that nuclear installations always had to implement the most recent results of scientific research and technological developments – a rule that contributed to the steadily rising cost of atomic reactors and, thereby, facilitated the technology's downfall (cf. "Rational Atoms" and "Nuclear Futures"). Besides, fears for the atom had been experienced by nuclear power's proponents when the technology came under pressure in the public debates, most prominently in the case of Seib. Fears had not only been felt and motivated actions. There had also been many attempts to instil fears in others, be it *for* or *of* the atom.

How this was done and regarded in the public debates, how, when, and by whom emotions could be expressed in general, and with what kind of consequences, differed widely between the banks of the Rhine and those of the Severn. Here, the emotional regimes decisively came into play. They were anchored to the debates about the atom, they shaped their course and were, in turn, influenced by them. In Biblis, the ideal of being factual had been used early on by atomic energy's proponents to control the debate by disqualifying emotional statements critical of nuclear power, while ones supporting it were tolerated, even though they were often only made using some distancing techniques. The emotional regime thus helped to shape the early absence of debates in Biblis (cf. "Nuclear Debates"). When critics rose in larger numbers, they also employed this strategy, with the result that accusations about being emotional, or even trying to emotionalise others, became a prime tool in the dispute about nuclear power. That notwithstanding, respective attempts were numerous. This surely contributed to the attack on the emotional regime as such, which was in the main made by Biblis women and mothers in the wake of Chernobyl. While not unmaking the existing emotional regime, it resulted in changes, in that emotions gained value as a political argument, at least locally and regionally. Interestingly, this transformation indicated an inversion of the contents of the emotional regime, without changing its rather rigid style. In Oldbury, on the other hand, the emotional regime was different. It was more lenient and, in the end, more stable, less influenced by the dispute about nuclear power. Here, open emotional statements were accepted as valuable contributions to the assessment of nuclear power from the beginning, even if they were critical, and they were made by local authorities, too. In consequence, measures were taken explicitly because of them, combined with the expectation that, in view of these consequences, the emotions would also change. The radiation monitoring around the reactors is the most prominent example. As long as this kind of dialogue was kept open, emotions were taken seriously and emotional statements were welcomed. Deviations from this balance, however, met with resistance. When emotional statements seemed to be inflated or attempts to arouse emotions not based in the factual local situation were perceived, those responsible were reprimanded, as the example of Dr Fox demonstrated. This situation intensified from the late 1970s onwards, and several participants attempted to utilise emotions more in the way they were being used in Germany at the time. Connections to the (trans-)national debates came to the fore, and to the party-political fault lines now accompanying nuclear power. Despite this, however, as the debate about leukaemia cases demonstrated, the established emotional regime remained intact locally, and emotional expressions based in the local situation continued to be accepted and taken seriously as reasons to act.

The two emotional regimes found in Biblis and Oldbury, as well as their interactions with the debates they had been anchored to about nuclear power, were not specific to the two locales in focus here. They were rooted in the national debates, to which Biblis and Oldbury therefore provided exemplary insights. In Lingen, the factual was revered no less than in Biblis (cf. "Rational Atoms"), or, indeed, in Stade (cf. "Atomic Trust(s)"). And while the intensity of the fight for the right to have and express emotions and to have them considered in political decisions about nuclear power as it played out in and around Biblis was not matched in these two other German communities, it was not foreign there, either. Indeed, it was part of a shift that had been observable (trans-)nationally for years, and especially so in the context of TMI, the NATO Double-Track decision, and Chernobyl. Similarly, the rather open expression of emotions in the Oldbury debates, with them being taken seriously within the debates as long as a certain balance was kept, while not at all times found on the more "purely" national level, could also be observed in the other two British locales under consideration here. How to deal with fears of nuclear power, for example, was quite openly debated in Torness (cf. "Atomic Trust(s)"), as it had been in Hinkley Point (cf. "Nuclear Debates"). It is possible, therefore, to see, in the emotional regimes shaping the debates in Biblis and Oldbury, national ones. And while there even were some parallels beyond the national level – not least the rising intensity of emotions in the consequence of large-scale nuclear accidents, and especially Chernobyl – the question as to *how* these transnational trends were dealt with was obviously dependent upon said emotional regimes.

Some emotional expressions, however, and in all probability also the emotions behind them, seemed not to be influenced that much by the specific local situations or the national emotional regimes. Specifically, this held true for the forms of interest and joy, for the pride and awe, in short, the technological sublime observable whenever nuclear power plants were under construction. The way the rising technological artefacts were regarded was quite similar across the board, almost irrespective of time and place, pointing to the inclined temporal plane's relevance for the analysis. Corresponding articles were written in Biblis and Oldbury, but also in Stade (cf. "Nuclear Futures") and even

Lingen, where the construction of Emsland nuclear power plant took place during the 1980s. Here, like in the other places, the sheer dimensions of the plant, the masses of material used, and the numbers of workers active obviously inspired awe.[382] It was no different in Hinkley Point, where the employment of a "goliath-type crane" drew special attention during the 1950s,[383] as does its equivalent "Big Carl" in today's construction of Hinkley Point C.[384] There was little difference in Torness,[385] although a long-drawn-out conflict about the number of local workers employed here commanded most of the attention that the construction site received (cf. "Atomic Trust(s)"). Nuclear power plants on the rise, with the efforts made to contain and control atomic energy being visibly performed, it seems safe to assume, were (and are) able to evoke feelings of the technological sublime, quite independently of the specific time and place – while utterly dependent on the local artefact's situation, and its novelty. Once the construction was finished, taking the local power plant out of the realm of future promises where the sublime resides and into that of the present, or even past, once another, newer reactor came into being somewhere else, these emotions tended to fade away.[386] Certainly, this kind of tech-emotion was no exceptional feat of nuclear power.[387] It was more widespread – as were the emotional regimes that shaped the public debates beyond them. Neither did they start or end with the period under consideration here, nor were their constraints applied to one technology's history alone.

In sum, the close study of the emotional side of the nuclear histories of Biblis and Oldbury not only revealed very intense and telling relations between the "Peaceful Atom" and two communities that had, so far, mostly remained on the sidelines of researchers' interest. By highlighting the ways different emotions and the emotional regimes shaping their expression were anchored to atomic power, influencing the debates about it and being influenced by them in turn, it also provided a further part of the explanation for the different dynamics of the British and German nuclear debates, with more importance attached to control, facilitating escalating conflicts in the latter, and a greater openness and search for dialogue and compromise in the former.[388] Finally, it also showed that emotionality had nothing to do with irrationality, even though such a connection stood at the core of the German emotional regime. The issue of rationality, however, played an important role in the debates about nuclear power.

Notes

1 This chapter's introduction broadens the introductory considerations in Götter, Christian: Ein Herz fürs Atom. Die Stader Befürworter der Kernenergie, in: Heßler, Martina (Ed.): Technikemotionen. Paderborn 2020, pp. 60–82.
2 Press Release, 'Atoms for Peace' Speech, 8.12.1953, in: National Archives, Dwight D. Eisenhower Presidential Library, DDE's Papers as President, Speech Series, Box 5, United Nations Speech 12/8/53, pp. 7–8, www.eisenhowerlibrary.gov/sites/default/files/file/atoms_Binder13.pdf (23.1.2023).

3 Nye, David E.: American Technological Sublime. Cambridge 1996 (Original 1994), pp. 227, 234–235; for nuclear explosions as sublime: Hales, Peter B.: The Atomic Sublime, in: American Studies 32 (1991), 1, pp. 5–31; for the sublime of things that cause terror: Burke, Edmund: A Philosophical Enquiry into the Origin of Our Ideas of the Sublime and Beautiful. Ed. by Adam Phillips. Oxford 1990 (Original 1757), pp. 36, 121–122; for large technological systems: Hughes, American Genesis.

4 This is not unusual for new technologies: Sturken, Marita, Douglas Thomas: Introduction. Technological Visions and the Rhetoric of the New, in: Sturken, Marita, Douglas Thomas, Sandra J. Ball-Rokeach (Eds.): Technological Visions. The Hopes and Fears that Shape New Technologies. Philadelphia 2004, pp. 1–18, here pp. 1–3.

5 Kitschelt, Kernenergiepolitik, p. 176; Radkau, Joachim: Angst und Angstabwehr als Regulative der Technikgeschichte, in: Kerner, Max (Ed.): Technik und Angst. Zur Zukunft der industriellen Zivilisation. Aachen 1994, pp. 53–72, here p. 63; Schüring, Michael: Advertising the Nuclear Venture: The Rhetorical and Visual Public Relation Strategies of the German Nuclear Industry in the 1970s and 1980s, in: History and Technology 29 (2013), 4, pp. 369–398, here p. 391; Schüring, Die evangelischen Kirchen, pp. 235–240.

6 Weart, Nuclear Fear; cf. Weart, Rise.

7 Radkau, Angst und Angstabwehr, p. 55; cf. Deutsch, Karl W., Andrei S. Markovits: On Coping with Science as a Task of Policy: A Tentative Summary, in: Deutsch, Karl W., Andrei S. Markovits (Eds.): Fear of Science – Trust in Science. Conditions for Change in the Climate of Opinion. Cambridge 1980, pp. 223–235, especially pp. 224–225; Radkau, Angstabwehr; Jäger, Ludwig: Angstsemantik. Die Bedeutung von 'Technik' und 'Angst' in der Medienöffentlichkeit, in: Kerner, Max (Ed.): Technik und Angst. Zur Zukunft der industriellen Zivilisation. Aachen 1994, pp. 37–45; Nehring, Holger: Cold War, Apocalypse and Peaceful Atoms. Interpretations of Nuclear Energy in the British and West German Anti-Nuclear Weapons Movements, 1955–1964, in: Historical Social Research 29 (2004), 3, pp. 150–170; Hennig, Jochen: Der Atompilz: Ikone des Atomzeitalters in unterschiedlichen Konnotationen, in: Bigg, Charlotte, Jochen Hennig (Eds.): Atombilder. Ikonografie des Atoms in Wissenschaft und Öffentlichkeit des 20. Jahrhunderts. Göttingen 2009, pp. 145–148; Augustine, Dolores L.: Learning from War. Media Coverage of the Nuclear Age in the Two Germanies, in: van Lente, Dick (Ed.): The Nuclear Age in Popular Media. A Transnational History, 1945–1965. New York 2012, pp. 79–116, here pp. 94–95; Hogg, Family, p. 536; Laucht, Christoph: 'Dawn – or Dusk?' Britain's Picture Post Confronts Nuclear Energy, in: van Lente, Dick (Ed.): The Nuclear Age in Popular Media. A Transnational History, 1945–1965. New York 2012, pp. 117–148.

8 Radkau, Aufstieg und Krise, pp. 78–83; cf. Roßnagel, Spaltung, p. 175; Müller, Kernenergie. Vol. 2, pp. 28, 32.

9 Bloch, Ernst: Das Prinzip Hoffnung. In fünf Teilen. 2 Vols. Frankfurt/Main 1959; Jonas, Hans: Das Prinzip Verantwortung. Versuch einer Ethik für die technologische Zivilisation. Frankfurt/Main 1979.

10 Rusinek, Bernd-A.: 'Kernenergie, schöner Götterfunken!' Die 'umgekehrte Demontage'. Zur Kontextgeschichte der Atomeuphorie, in: Kultur & Technik 17 (1993), 4, pp. 15–21; Welsh, Mobilising, especially pp. 35–48; Schumacher, Frank: 'Atomkraft für den Frieden'. Eine amerikanische Kampagne zur emotionalen Kontrolle nuklearer Ängste, in: Sozialwissenschaftliche Informationen 30 (2001), 3, pp. 63–71; Metzler, Demokratisierung, pp. 270–271; Gelfort, Eike: Kernkraftwerke der I., II., III. und IV. Generation. Was heißt das für die Fortführung der Entwicklung?, in: atw 57 (2012), 7, pp. 486–494, here p. 490; Nehring, Holger: 'Atomzeitalter'. Die Debatten um Atomenergie in der Bundesrepublik

Deutschland der fünfziger Jahre, in: Erhardt, Hendrik, Thomas Kroll (Eds.): Energie in der modernen Gesellschaft. Zeithistorische Perspektiven. Göttingen 2012, pp. 223–243.

11 Heßler, Martina: Technikemotionen. Einleitende Überlegungen zur historischen Ko-Konstruktion von Technik und Emotionen, in: Heßler, Martina (Ed.): Technikemotionen. Paderborn 2020, pp. 1–36, here p. 7; cf. Heßler, Martina, Bettina Hitzer: Introduction. Tech-Fear. Histories of a Multifaceted Relationship, in: Technikgeschichte 86 (2019), 3, pp. 185–199; Kerner, Max (Ed.): Technik und Angst. Zur Zukunft der industriellen Zivilisation. Aachen 1994.

12 Heßler, Einleitende Überlegungen, p. 6; cf. Nehring, Internationalists; Nehring, Angst; Gassert, Philipp: Popularität der Apokalypse. Zur Nuklearangst seit 1945, in: Aus Politik und Zeitgeschichte 61 (2011), 46/47, pp. 48–54; Conze, Eckart, Martin Klimke, Jeremy Varon: Introduction. Between Accidental Armageddons and Winnable Wars. Nuclear Threats and Nuclear Fears in the 1980s, in: Conze, Eckart, Martin Klimke, Jeremy Varon (Eds.): Nuclear Threats, Nuclear Fear, and the Cold War of the 1980s. Cambridge 2017, pp. 1–23.

13 For a contemporary view: Jaenecke, Heinrich: Der Teufelspakt, in: Barthel, Wolfgang, Heinrich Jaenecke, Peter Thomsen: Der Unsichtbare Tod. Die Angst des Bürgers vorm Atom. Munich ^{2}1981, pp. 7–61, here p. 17; for their motivational role in the wake of nuclear accidents: Arndt, Melanie: Verunsicherung vor und nach der Katastrophe. Von der Anti-AKW-Bewegung zum Engagement für die 'Tschernobyl-Kinder', in: Zeithistorische Forschungen/Studies in Contemporary History 7 (2010), 2, pp. 240–258; for a call to consider not only fears of nuclear power, but also fears motivating its application: Kalmbach, Karena, Andreas Marklund, Anna Åberg: Crises and Technological Futures. Experiences, Emotion, and Action, in: Technology and Culture 61 (2020), 1, pp. 272–281, here pp. 276–277.

14 Radkau, Aufstieg und Krise, pp. 78–83, 417–418; Jäger, Angstsemantik; Gassert, Popularität, p. 54; Götter, Emotionen; cf., as a contemporary example: Bundesverband Bürgerinitiativen Umweltschutz: Atomenergie-Propaganda. Karlsruhe 1978.

15 Radkau, Joachim: Mythos German Angst. Zum neuesten Aufguss einer alten Denunziation der Umweltbewegung, in: Blätter für deutsche und internationale Politik 56 (2011), 5, pp. 73–82; Tompkins, Better Active, pp. 6–7; for a more nuanced view: Biess, Angst, especially pp. 450–452.

16 Radkau, Technik in Deutschland, pp. 26–27.

17 Radkau, Angstabwehr; Welsh, Mobilising, especially p. 43; Hogg, British; cf. Zeman, Scott C.: 'To See . . . Things Dangerous to Come to'. Life Magazine and the Atomic Age in the United States, 1945–1965, in: van Lente, Dick (Ed.): The Nuclear Age in Popular Media. A Transnational History, 1945–1965. New York 2012, pp. 53–77; van Lente, Introduction, p. 8; Schüring, Michael: Symbole der Hoffnung, Boten der Angst. Die Einstellung zum Atom hat sich verändert, in: Kultur & Technik (2013), 3, pp. 36–39.

18 Weisker, Emotion, pp. 205, 220–221; cf. Weisker, Expertenvertrauen; Cohen, I. Bernard: The Fear and Distrust of Science in Historical Perspective. Some First Thoughts, in: Deutsch, Karl W., Andrei S. Markovits (Eds.): Fear of Science – Trust in Science. Conditions for Change in the Climate of Opinion. Cambridge 1980, pp. 29–58.

19 Heßler, Einleitende Überlegungen, especially pp. 10–11, 25–26; for the general argument for taking emotions into account for historical analyses: Febvre, Lucien: Sensibilität und Geschichte. Zugänge zum Gefühlsleben früherer Epochen, in: Honegger, Claudia (Ed.): Schrift und Materie der Geschichte. Vorschläge zur systematischen Aneignung historischer Prozesse. Frankfurt/Main

1977, pp. 313–334; cf. Frevert, Ute: Was haben Gefühle in der Geschichte zu suchen?, in: Geschichte und Gesellschaft 35 (2009), 2, pp. 183–208; Plamper, Jan: Geschichte und Gefühl. Grundlagen der Emotionsgeschichte. Munich 2012; Frevert, Ute: Historicizing Emotions, in: Emotion Researcher. ISRE's Sourcebook for Research on Emotion and Affect. 2018, http://emotionresearcher.com/historicizing-emotions/ (23.1.2023).

20 Breach, Windscale, p. 80; Nehring, Cold War, p. 157.

21 Izard, Carroll Ellis: Basic Emotions, Natural Kinds, Emotion Schemas, and a New Paradigm, in: Perspectives on Psychological Science 2 (2007), 3, pp. 260–280, here p. 261; the model was developed throughout Izard's career and derived from his teacher, Silvan S. Tomkins. Cf. James, William: What Is an Emotion?, in: Mind 9 (1884), 34, pp. 188–205.

22 Izard, Basic Emotions, pp. 265–266; Izard, Carroll Ellis: Human Emotions. New York 1977, pp. 118, 125–126; regarding the latter aspect: Frevert, Ute: Angst vor Gefühlen? Die Geschichtsmächtigkeit von Emotionen im 20. Jahrhundert, in: Nolte, Paul et al. (Eds.): Perspektiven der Gesellschaftsgeschichte. Munich 2000, pp. 95–111, here p. 106; Kals, Elisabeth, Norbert Platz, Rainer Wimmer: Einleitung. Emotionen in der Umweltdiskussion, in: Kals, Elisabeth, Norbert Platz, Rainer Wimmer (Eds.): Emotionen in der Umweltdiskussion. Wiesbaden 2000, pp. 9–18, here p. 11; Frevert, Ute: Gefühle als geschichtsmächtige Kategorie. Ingrid Bauer und Christa Hämmerle im Gespräch mit Ute Frevert. (2013). L'Homme: Europäische Zeitschrift für feministische Geschichtswissenschaft, 24(1), pp. 109–118, here p. 112. Regarding their cultural dependence: note 19; Scheer, Monique: Are Emotions a Kind of Practice (And Is That What Makes Them Have a History)? A Bourdieuian Approach to Understanding Emotion, in: History and Theory 51 (2012), 2, pp. 193–220.

23 Averill, James R.: Intellectual Emotions, in: Harré, Rom, W. Gerrod Parrott (Eds.): The Emotions. Social, Cultural and Biological Dimensions. London 1996, pp. 24–38, especially pp. 25, 37.

24 Cf. Fredrickson, Barbara L.: The Broaden-and-Build Theory of Positive Emotions, in: Phil. Trans. Royal Society London B (2004), 359, pp. 1367–1377.

25 The original text describing the method is untraceable: Danner, Deborah D. et al.: Written Emotion Expression Code. Unpublished Manuscript, University of Kentucky, Lexington 2000; instead, shorter descriptions were used: Danner, Deborah D. et al.: Positive Emotions in Early Life and Longevity. Findings from the Nun Study, in: Journal of Personality and Social Psychology 80 (2001), pp. 804–813; cf. Scheer, Emotions, pp. 217–220.

26 The problems of this differentiation are pointed out by Schnell, Rüdiger: Haben Gefühle eine Geschichte? Aporien einer History of Emotions. Vol. 1. Göttingen 2015, pp. 17–19, 77, 90–94, 115–116, 259–261.

27 With a view to fears, this was also pointed out by: Jäger, Angstsemantik, pp. 41–42; cf., for the strategic thinking behind expressing emotions on the side of the critics: Jungk, Atomstaat, pp. x–xi; some of these strategies were analysed in: Götter, Emotionen. The importance of the analytical difference between emotions and communication of emotions was highlighted by: Schnell, Rüdiger: Haben Gefühle eine Geschichte? Aporien einer History of Emotions. Vol. 2. Göttingen 2015, pp. 654–656.

28 Cf. Mitchell, Robert C.: Rationality and Irrationality in the Public's Perception of Nuclear Power, in: Freudenburg, William R., Eugene A. Rosa (Eds.): Public Reactions to Nuclear Power. Are There Critical Masses? Boulder 1984, pp. 137–179, here p. 142.

29 For the concept of emotional regimes: Reddy, William M.: The Navigation of Feeling. A Framework for the History of Emotions. Cambridge 2001, especially

p. 129; Frevert, Gefühle in der Geschichte, pp. 205–207; cf. Parkinson, Anna M.: An Emotional State. The Politics of Emotion in Postwar West German Culture. Ann Arbor 2015, pp. 15–16; Biess, Angst, pp. 30–33.

30 Which, in general, they were not, cf. the argument made by psychologists specialised on emotion research like Carroll Ellis Izard and historians of emotion like Ute Frevert: Izard, Human Emotions, pp. 13, 43; Frevert, Gefühle in der Geschichte, pp. 189–191; cf. Greenberg, Michael R.: Nuclear Waste Management, Nuclear Power, and Energy Choices. Public Preferences, Perceptions, and Trust. London 2013, pp. xv–xvi; Biess, Frank, Daniel M. Gross: Introduction. Emotional Returns, in: Biess, Frank, Daniel M. Gross (Eds.): Science & Emotions after 1945. A Transatlantic Perspective. London 2014, pp. 1–38, here pp. 1–8; Parkinson, Emotional State, pp. 37–38. Nonetheless, as Joachim Radkau emphasised, to differentiate between rational and irrational forms of emotions like fear is problematic: Radkau, Angst und Angstabwehr, p. 57.

31 Cf. for the phenomenon's Western European dimension: Conway, Martin: Western Europe's Democratic Age, 1945–68. Princeton 2020, p. 88; for the argument of a nuclear phobia behind the critique against the technology: Bauer, Martin: 'Technophobia'. A Misleading Conception of Resistance to New Technology, in: Bauer, Martin (Ed.): Resistance to New Technology. Nuclear Power, Information Technology, and Biotechnology. Cambridge 1995, pp. 97–122, here p. 97; furthermore: Kalmbach, Afterlives, p. 5.

32 Cf. Goerke, Dietrich: Meinungsbildung in der Öffentlichkeit über energiepolitische Fragen, insbesondere im Bereich der Kernenergiesicherheit. Teil I. Ergebnisse und Bewertungen. Vechta 1978, especially p. 129; Heilmann, Klaus: Strahlenangst nach Tschernobyl. Was ist zu tun?, in: Krämer, Hermann (Ed.): Kernenergie heute. Die großen Themen. Bonn 1988, pp. 105–117; Staufenbiel, Rolf: Angst vor der Technik. Ein Anlaß zur Sorge?, in: Kerner, Max (Ed.): Technik und Angst. Zur Zukunft der industriellen Zivilisation. Aachen 1994, pp. 5–18, especially p. 16; Steinmetz, Willibald: Anbetung und Dämonisierung des 'Sachzwangs'. Zur Archäologie einer deutschen Redefigur, in: Jeismann, Michael (Ed.): Obsessionen. Beherrschende Gedanken im wissenschaftlichen Zeitalter. Frankfurt/Main 1995, pp. 293–333, especially pp. 331–332; Altenburg, Wandel, pp. 245–246; Schüring, Die evangelischen Kirchen, pp. 235–240; Frevert, Ute: Rationalität und Emotionalität im Jahrhundert der Extreme, in: Sabrow, Martin, Peter Ulrich Weiß (Eds.): Das 20. Jahrhundert vermessen. Signaturen eines vergangenen Zeitalters. Göttingen 2017, pp. 115–140, especially pp. 116–117, 123–125, 133–135; furthermore: Frevert, Ute, Anne Schmidt: Geschichte, Emotionen und die Macht der Bilder, in: Geschichte und Gesellschaft 37 (2011), pp. 5–25, here pp. 10–11, 18–19.

33 Nehring, Atomzeitalter, pp. 236–237.

34 Nehring, Angst, p. 451; Greiner, Bernd: Angst als Emotion und Instrument. Beobachtungen zu einem nervösen Zeitalter, in: Doering-Manteuffel, Anselm, Bernd Greiner, Oliver Lepsius (Eds.): Der Brokdorf-Beschluss des Bundesverfassungsgerichts 1985. Eine Veröffentlichung aus dem Arbeitskreis für Rechtswissenschaft und Zeitgeschichte an der Akademie der Wissenschaften und der Literatur, Mainz. Tübingen 2015, pp. 61–81, here pp. 67–68; cf. Engels, Naturpolitik, p. 331.

35 Jungk, Atomstaat, pp. x–xi; cf. Born, Nicolas: Rede in Gorleben, in: Haufs, Rolf (Ed.): Nicolas Born. Die Welt der Maschine. Aufsätze und Reden. Reinbek 1980, pp. 213–215; Jung, Öffentlichkeit, pp. 93–97.

36 Patterson, Walter Cram: The Fissile Society. Energy, Electricity and the Nuclear Option. London 1977, p. 83; Eiser, J. R., Joop van der Pligt: Beliefs and Values in the Nuclear Debate, in: Journal of Applied Social Psychology 9 (1979),

pp. 524–536; Irwin et al., Introduction, pp. 1–2; Welsh, Mobilising, especially pp. 208–209; Hogg, British, pp. 83, 90–91; 96–97; for the transnational context: Biess et al., Introduction; Erickson, Paul et al.: How Reason Almost Lost Its Mind. The Strange Career of Cold War Rationality. Chicago 2013; for the more fundamental argument: Weber, Max: Soziologie. Weltgeschichtliche Analysen. Politik. Ed.: Johannes Winckelmann. Stuttgart 1956; Giddens, Anthony: The Consequences of Modernity. Stanford 1990, pp. 38–40, 176; Fraunholz, Uwe, Anke Woschech (Eds.): Technology Fiction. Technische Visionen und Utopien in der Hochmoderne. Bielefeld 2012; cf. Gellner, Ernest: Nations and Nationalism. Oxford 1983, pp. 19–20.

37 Welsh, Mobilising, p. 44.

38 Heßler, Einleitende Überlegungen.

39 Cf. Nye, Sublime; Möser, Kurt: Technikfaszination und Technikkommunikation, in: Böhn, Andreas, Andreas Metzner-Szigeth (Eds.): Wissenschaftskommunikation, Utopien und Technikzukünfte. Karlsruhe 2018, pp. 183–200, here p. 188.

40 Gesetz über die friedliche Verwendung der Kernenergie, § 1.

41 Gesetz über die friedliche Verwendung der Kernenergie, § 7.

42 Cf. Weart, Rise, pp. 162–167.

43 Grundsatzprogramm der Sozialdemokratischen Partei Deutschlands. Beschlossen auf dem Außerordentlichen Parteitag der Sozialdemokratischen Partei Deutschlands in Bad Godesberg vom 13. bis 15. November 1959, p. 5.

44 RWE: Offener Brief. An die Einwohner von Biblis, Landkreis Bergstraße und Umgebung, in: *Mannheimer Morgen (MM)*, 22.6.1968.

45 RWE, Offener Brief.

46 Cf. 'Nuclear Debates'.

47 Landes, David S.: The Unbound Prometheus. Technological Change and Industrial Development in Western Europe from 1750 to the Present. Cambridge 1969, p. 21.

48 Cf. Roggenbuck, Klaus J.: RWE hält an Bibliser Kernkraftwerk fest. Nicht als Konkurrenzunternehmen zu Mannheimer Energieversorgung gedacht – Günstiger Standort entschied, in: *Südhessische Post (SP)*, 18.7.1968; Vier Mark pro Quadratmeter. Grundstückspreise für das Bibliser Atomreaktorgelände ausgehandelt, in: *SP*, 29.11.1968.

49 KR: Glänzende Kugel in der Rheinebene. Atomkraftwerk bei Biblis bereitet erhebliche Sorgen – Gefahr für den Fluß?, in: *Frankfurter Neue Presse*, 28.11.1969.

50 br: Im Sommer 1973 wird der Super-Reaktor geladen. Zwischenbilanz beim Kernkraftwerk Biblis – 'Für den Umweltschutz alles Erdenkliche getan', in: *Darmstädter Echo (DE)*, 13.3.1971.

51 Noack, Hans-Joachim: Ein Atommeiler kommt ins Dorf. Von Umweltverschmutzung wollen sie nichts hören, in: *Die Zeit*, 8.10.1971.

52 hz: Öffentlicher Anhörungstermin zum Bau eines Kernkraftwerks in Biblis. Bedenken jetzt ausgeräumt. Überdurchschnittliche Sicherheitsvorkehrungen – Fachvorträge klärten auf, in: *SP*, 19.1.1970.

53 sz: 'Wer bezahlt den Umzug bei Atomgefahr?' Öffentlicher Anhörungstermin über Einsprüche gegen Bau eines Kernkraftwerkes bei Biblis, in: *DE*, 16.1.1970.

54 Eirich, George: Erörterungstermin in Biblis. Plakette des TÜV für das Kernkraftwerk. Einwände gegen RWE-Projekt ohne Erfolg, in: *MM*, 16.1.1970.

55 hz: Öffentlicher Anhörungstermin zum Bau eines Kernkraftwerks in Biblis. Bedenken jetzt ausgeräumt. Überdurchschnittliche Sicherheitsvorkehrungen – Fachvorträge klärten auf, in: *SP*, 19.1.1970.

56 For the tendency to present concerns about and critique of nuclear power as rational, especially during large parts of the 1970s: Jung, Öffentlichkeit, pp. 90–97; Jung, Matthias: Emotionalität vs. Rationalität im Sprachgebrauch

der Umweltdiskussion, in: Kals, Elisabeth, Norbert Platz, Rainer Wimmer (Eds.): Emotionen in der Umweltdiskussion. Wiesbaden 2000, pp. 155–164, here pp. 156–159, arguing for 1977 as the year that this changed. In that year, Robert Jungk demanded to stop talking about the atom without being emotional: Jungk, Atomstaat, p. xi.

57 Ff: Umweltschützer verlangen Baustopp in Biblis. Resolution in der Hammer Kundgebung verfaßt. Auf Anhieb etwa 300 Unterschriften/Wissenschaftliche Beiträge gaben Einblick, in: *Wormser Zeitung (WZ)*, 22.1.1973 (date and newspaper deduced from context), Gemeindearchiv Biblis (GAB), Presseartikel KKW 1977.

58 k: Umweltschäden durch Atomenergie? Aktionsgemeinschaft Umweltschutz diskutierte mit RWE-Vertretern über Kernkraftwerk, in: *DE*, 31.1.1973.

59 os: Wie sicher ist nun ein Kernkraftwerk? Erstes Bibliser Kernkraftwerk-Gespräch – Fragerunde mit RWE-Verantwortlichen, in: *DE*, 26.5.1973.

60 os: Wie sicher ist nun ein Kernkraftwerk? Erstes Bibliser Kernkraftwerk-Gespräch – Fragerunde mit RWE-Verantwortlichen, in: *DE*, 26.5.1973.

61 Podiumsdiskussion über das Kraftwerk. Auch nach dem Pro und Contra blieben die Meinungen verschieden. Umweltschutz-Initiatoren bestritten die Feststellungen der Verantwortlichen, in: *MM*, 7.6.1973; for Geier, cf. Geier, Erna Maria, in: Hessische Biografie, www.lagis-hessen.de/pnd/1035014165 (23.1.2023).

62 Critics of nuclear power criticised the media for not being factual when they reported their position: Renner, Bauplatz, pp. 109–110.

63 Meyer, Jan-Henrik et al.: Bürgerdialog Kernenergie (1974–1983). Staatliches Handeln in der Auseinandersetzung um die nukleare Entsorgung und seine Bedeutung für das heutige Standortauswahlverfahren. Berlin 2024, http://nbn-resolving.de/urn:nbn:de:0221-2024040542658 (26.7.2024), p. 101; cf. Nelkin, Dorothy, Michael Pollak: Problems and Procedures in the Regulation of Technological Risk, in: Schwing, Richard C., Walter A. Albers Jr. (Eds.): Societal Risk Assessment. How Safe Is Safe Enough? New York 1980, pp. 233–248, here p. 242; Altenburg, Cornelia: Kernenergie und Politikberatung. Die Vermessung einer Kontroverse. Wiesbaden 2010, p. 62; Gassert, Philipp: 'Kernenergie akzeptabel für die Bevölkerung machen'. Bonn und der Anti-AKW-Protest, in: Lakeberg, Beata Dorota, Hans-Christian Pust (Eds.), Atom. Strom. Protest. 50 Jahre Wyhl und anderswo. Ostfildern 2023, pp. 116–129, here pp. 122–123.

64 For the rising fears of reactors in the early 1970s: Gassert, Popularität, p. 52; for Matthöfer: Rudzio, Wolfgang, Jiahn-Tsyr Yu: Matthöfer, Hans, in: Kempf, Udo, Hans-Georg Merz (Eds.): Kanzler und Minister 1949–1998. Biografisches Lexikon der deutschen Bundesregierungen. Wiesbaden 2001, pp. 470–474.

65 Der Bundesminister für Forschung und Technologie (Ed.): Kernenergie. Eine Bürgerinformation. Bonn 1975.

66 Cf. Matthöfer, Interviews; Röthlein, Kernenergie, pp. 84–95; Nelkin et al., Problems, p. 242; Nelkin et al., Besieged, pp. 171–173; Overhoff, Politisierung, p. 9; Rucht, Campaigns, p. 213; Flam, Political, pp. 335–336; Tiggemann, Achillesferse, pp. 443–450; for the information deficit model, cf. 'Nuclear Debates'.

67 lst: Wirtschaftsminister Friderichs: 'Wir brauchen mehr solche Beispiele wie Biblis'. Kernkraftwerk offiziell übergeben – Kritik des Ministers an 'rechtswidrigen Gewaltakten' – Beifall für Bürgermeister Seib, in: *DE*, 16.4.1975.

68 Bundesverband Bürgerinitiativen Umweltschutz, Atomenergie-Propaganda; Traube, umschalten, p. 239; Gloede, F.: Strukturelle Rahmenbedingungen und Vorgeschichte der Öffentlichkeitsbeteiligung zu GKN II, in: Blümel, Willi (Ed.): Die Öffentlichkeitsbeteiligung bei der Genehmigung von umweltrelevanten Großvorhaben. Zwischenergebnisse eines gemeinsamen Forschungsprojektes. Vol. 1. Speyer 1988, pp. 81–157, here pp. 88–89; Paul, Anti-AKW-Bewegung, p. 19; Hasenöhrl, Zivilgesellschaft, p. 407; Franke, BUND, p. 15; cf. Schluchter,

Wolfgang: Bürgerdialog und Partizipation. Zur Untersuchung des Battelle-Instituts in Frankfurt, in: Duve, Freimut (Ed.): Atomenergie/Brokdorf/Unterelbe. Reinbek 1976, pp. 84–98, here pp. 91, 96; Beker, Zumutbarkeit, pp. 34–35; Nowotny, Kernenergie, p. 71; Camilleri, State, pp. 118–119.

69 Friderichs, Dr. Hans: Kernenergie – Grundlage unseres Lebens. Nutzen größer als Risiken, in: *WZ*, 19.4.1975.

70 Siegelmann, Bert: 'Offener Brief' aus Essen an die Bürger: Kernkraftwerk Biblis bleibt Baustelle. RWE beantragt Reaktor-Block C und D. Gesamtleistung der vier Atom-Meiler rund 5100 Megawatt/Inbetriebnahme bis 1985?, in: *MM*, 14.5.1975; within the RWE, Biblis was deemed exemplary, technology-wise as well as with a view to the licensing process, and expected to contribute to a more rational perception of nuclear power plants: Das Kernkraftwerk Biblis des RWE, Ed. by: Energiewirtschaft und Technik Verlagsgesellschaft m.b.H., Munich 1975, especially pp. 11–12.

71 Cf., for an early example: haka: Stadtverordnete beraten 'heißes Eisen': Auch Lampertheimer SPD meint: Zwei Reaktoren in Biblis reichen. Umwelt bereits überlastet/Sitzung am Donnerstag mit acht Tagesordnungspunkten, in: *MM*, 14.5.1975.

72 'Keine Alternative zur Kernspaltung'. Die Junge Union verweist auf die Deckung des Energiebedarfs in den kommenden Jahren, in: *SP*, 30.6.1975; a similar call – albeit from a more critical perspective – was issued by the district's Social Democrats: Kernkraftwerk in der Diskussion: Das Pro und Kontra. SPD: Risikofaktoren sorgfältig abwägen. Die Parteien haben das Wort, in: *SP*, 30.6.1975.

73 dpa/lrs: Genscher soll Farbe bekennen. Aktionsgemeinschaft verlangt Auskunft über Atomkraftwerke, in: *MM*, 7.4.1973.

74 dpa/lrs: Genscher soll Farbe bekennen. Aktionsgemeinschaft verlangt Auskunft über Atomkraftwerke, in: *MM*, 7.4.1973.

75 rs: Energieversorgung – weltweites Problem. Öffentliche und private Studien sind jedoch weniger pessimistisch, in: *DE*, 17.7.1973; electricity shortages, more widespread in the years after the Second World War, partially returned in the winter of 1978/1979: Gerber, Sophie: Küche, Kühlschrank, Kilowatt. Zur Geschichte des privaten Energiekonsums in Deutschland, 1945–1990. Bielefeld 2015, here p. 272.

76 Schulze, Gerd-Peter: Wie sicher sind Atom-Kraftwerke? Der Mensch hat ein neues Feuer entfacht, um seinen Energiebedarf zu decken, in: *SP*, 27.2.1974.

77 Siegelmann, Bert: Umweltschützer diskutierten in der Riedhalle RWE-Baupläne. Bürger wollen objektivere Information. Buh-Rufe für Bibliser Ortsoberhaupt Seib. Gemeindevorstand will Ortsparlament-Beschluß überprüfen/Kritik an Katastrophenplänen, in: *MM*, 25.6.1975.

78 Karbiner, Hans: Stadtverordnetensitzung griff aktuelles Thema auf: Lampertheimer 'Nein' für weitere Reaktoren gründet sich auf Furcht vor Umweltschäden. Künstliche Panikmache abgelehnt/Kernkraftwerke dezentralisieren, in: *MM*, 20.5.1975.

79 os: Erst Erfahrungen sammeln. Neue Erklärung des SPD-Vorsitzenden zu Block C und D, in: *DE*, 9.6.1975.

80 'Keine Emotionen wecken'. Die CDU wirft der SPD Doppelstrategie vor, in: *SP*, 30.6.1975.

81 DE/os: 'Unverantwortliche Panikmache in der Bevölkerung'. Gespräch mit dem Bibliser Bürgermeister Josef Seib über die Pläne zur Kernkraftwerkerweiterung, in: *DE*, 7.7.1975.

82 DE/os: 'Unverantwortliche Panikmache in der Bevölkerung'. Gespräch mit dem Bibliser Bürgermeister Josef Seib über die Pläne zur Kernkraftwerkerweiterung, in: *DE*, 7.7.1975.

83 Cf. Kehl, Jürgen (Groß-Rohrheim): Ein 'Atom-Bürgermeister'. Betr.: Artikel 'RWE: Offener Brief an die Bürger von Biblis' in der 'Südhessischen Post' vom Freitag, 16. Mai 1975, in: *SP*, 22.5.1975.
84 DE/os: Unverantwortliche Panikmache in der Bevölkerung. Gespräch mit dem Bibliser Bürgermeister Josef Seib über die Pläne zur Kernkraftwerkerweiterung, in: *DE*, 7.7.1975.
85 Minutes, Meeting, Biblis Municipal Council, 18.6.1975, p. 8, in: GAB, Gemeindevertretungssitzungen.
86 Mössinger, Günter (Biblis-Nordheim): Biblis – ein Prüfstein der Demokratie, in: *WZ*, 5.1.1976.
87 Konkrete Bedenken gegen eine KKW-Erweiterung. SPD-Kreistagsfraktion befaßt sich mit der Problematik, in: *SP*, 28.1.1976.
88 Siegelmann, Bert: Experten standen in Biblis Rede und Antwort: Pläne für Block D ruhen vorerst. Reaktor-Massierung stellt Problem dar. Mandatsträger aus drei Gemeinden hörten Für und Wider zum aktuellen Thema, in: *MM*, 21.2.1976.
89 hez: Informationsveranstaltung der Kommunalen Arbeitsgemeinschaft 'Mittelzentrum Ried' über den weiteren Ausbau des Kernkraftwerkes Biblis, in: *Bibliser Blatt (BB)*, 27.2.1976.
90 Cf. Geschäftsordnung des Informationskreises Kernkraftwerk Biblis, in: GAB, Az.: 774–054.
91 hps: Informationskreis Kernkraftwerk Biblis. Ortspolitiker wollen Aufklärungsarbeit in der Bevölkerung verstärken, in: *DE*, 5.5.1976.
92 Seubert, Hans-Peter: Information statt Polemik, in: *DE*, 5.5.1976.
93 Rettweiler, Christian (Kernkraftwerksmeister, Betriebsratsmitglied, Sprecher der gewerkschaftlichen Vertrauensleute im KKW Biblis): Betr. Kernkraftwerk Biblis, in: *BB*, 12.8.1976.
94 Umhauer, Peter (Heppenheim): Die ganze Wahrheit, in: *SP*, 11.10.1976.
95 Lilge, Herbert: Biblis ist nicht Harrisburg, in: *DE*, 11.4.1979.
96 de: Willi Blodt: Aus Sorge um die Sicherheit, in: *DE*, 19.4.1979; for Blodt: Blodt, Willi, in: Hessische Biografie, www.lagis-hessen.de/pnd/1120556511 (23.1.2023).
97 Arndt, Verunsicherung, pp. 241–242.
98 hje: Hand bereits am Ausschaltknopf. Unsicherheit über die Sicherheitsfrage von Biblis/SPD im Konflikt mit Arbeitsplätzen, in: *WZ*, 21.5.1986.
99 Kilgus, Rudi: Keine Alternativen, in: *WZ*, 15.5.1986.
100 Zacharias, Christoph: Nach der sowjetischen Reaktorkatastrophe. Parteienstreit um Bibliser Kraftwerk. Bergsträßer Grüne fordern Stillegung. CDU fordert Überprüfung des Sicherheitsstandards/Sorge um Arbeitsplätze, in: *MM*, 21.5.1986.
101 BE: Die sanften Technologien weiter ausbauen. CDU-Arbeitskreis erörtert Tschernobyl und die Folgen – Unfall nicht verharmlosen, in: *Bergsträßer Echo (BE)*, 30.5.1986.
102 Minutes, Meeting, Bergstraße District Council 10/XI, 9.6.1986, in: GAB, Kreistag 3.
103 Minutes, Meeting, Bergstraße District Council 10/XI, 9.6.1986, in: GAB, Kreistag 3, p. 13.
104 Minutes, Meeting, Bergstraße District Council 10/XI, 9.6.1986, in: GAB, Kreistag 3, p. 15; for Kübler: Kübler, Klaus, in: Vierhaus, Handbuch. Vol. 1, pp. 464–465.
105 RWE-Information, in: *BB*, 15.5.1986.
106 hez: Biblis, drei Monate nach Tschernobyl, in: *BB*, 21.8.1986.
107 PB: Dr. Jürgen Dieter (SPD) fragt: Was geschieht mit den Arbeitsplätzen? Neuer Aspekt bei Diskussion um Kraftwerks-Stillegungen, in: *BE*, 24.5.1986.
108 Bezzenberger, Walter (Heppenheim): Besonnenheit ist das Gebot der Stunde, in: *BE*, 4.6.1986.

109 rhb: 'Patrioten' kämpfen mit Parolen für Atomenergie gegen 'Spinner'. In Biblis melden sich seltsame Verbündete zu Wort, in: *BE*, 1.8.1986.

110 Cf. Rettweiler, Christian: Presseinformation des Bibliser RWE Betriebsrates, in: *BB*, 29.5.1986; Rettweiler: Information des DGB Ortskartells Biblis, in: *BB*, 29.5.1986.

111 Rettweiler, Christian: Eine sachliche Information der Bibliser RWE-Betriebsräte. Risiko Kernenergie? Urteilen Sie selbst! GAB, Reaktorunfall Tschernobyl 1986 – Presse.

112 RWE: Offener Brief. An die Einwohner von Biblis, Landkreis Bergstraße und Umgebung, in: *MM*, 22.6.1968.

113 Wie wird das Kernkraftwerk aussehen? Dem Fortschritt kann man sich nicht mehr entgegenstemmen, in: *Bürstädter Zeitung (BZ)*, 19.7.1968.

114 ulf: Einzig in der Bundesrepublik. Bibliser Kernkraftwerk kostet 800 Millionen Mark, in: *WZ*, 10.4.1969.

115 hkh: Im Ried entsteht Europas größtes Kernkraftwerk. Das RWE baut bei Biblis/ Auftrag an die von AEG und Siemens gegründete Kraftwerk-Union vergeben, in: *MM*, 31.5.1969.

116 rn: Schwierigkeiten mit Schlamm. Auf dem Weg zum Atomkraftwerk Biblis rollen zuerst Bagger und Planierraupen. Vor dem Reaktorbau müssen die Straßen fertig sein – Überflutetes Gelände hemmt, in: *MM*, 1.9.1969.

117 rn: Großbaustelle am Rheinufer. Eine Wagen-Kette bohrt der RWE einen Weg. Schlamm und Matsch erschwerten die Arbeit. Ein Grundstücksbesitzer verzögert den Bau des Zufahrtsgeleises, in: *MM*, 19.2.1970.

118 Pk: Bagger wälzen Tausende Kubikmeter Erde. Bibliser Kernkraftwerk entsteht auf gleicher Höhe wie der Rheindamm. Riesige Kräne packen zu, in: *WZ*, 20.7.1970.

119 bert: Halbzeit auf der Großbaustelle. Kernkraftwerk wächst genau nach Zeitplan. Im Jahre 1974 wird der erste Strom geliefert. Rund 15 Prozent der Investitionskosten entstehen durch den Umweltschutz, in: *MM*, 10.3.1971.

120 Noack, Hans-Joachim: Ein Atommeiler kommt ins Dorf. Von Umweltverschmutzung wollen sie nichts hören, in: *Die Zeit*, 8.10.1971.

121 Nye, Sublime, p. 77; for aspects of the mathematical and the electrical sublime: Nye, Sublime, pp. 140, 143; for an interesting argument for the importance of size: Burke, Enquiry, pp. 124–125.

122 For the technological sublime's limited duration: Nye, Sublime, pp. 60–61.

123 ard: Kernkraftwerk Biblis als ein Arbeitsplatz für viele Spezialisten. Gemeindeparlament war Ende der sechziger Jahre Feuer und Flamme wegen der erwarteten Neuansiedlungen – Keine Ängste, in: *BE*, 24.5.1986.

124 Große Dinge kommen auf Biblis zu. Bürgermeister Seib berichtete von den Vorhaben der Gemeinde, in: *BZ*, 18.7.1968.

125 Grimm, Michael: 'Das ist keine Munitionsfabrik und keine Bombe'. Gespräch mit dem Bibliser Bürgermeister Josef Seib über Probleme mit Kernkraftwerken, in: *DE*, 19.3.1976.

126 Minutes, Meeting, Bergstraße District Council, 28.3.1980, TOP 1, Statement by Bergmann, pp. 6–7, Hessisches Staatsarchiv Darmstadt, H2 Bergstraße 9786.

127 Excerpt, Minutes, Meeting, Biblis Municipal Board, 28.6.1968, GAB, KKW Presseartikel 1968 bis 1970.

128 Grimm, Michael: Neuer Zugzwang?, in: *DE*, 24.1.1976.

129 ff: In Biblis heißt es umweltfreundlich CDU Lorsch taktiert jedoch anders, in: *WZ*, 18.10.1972.

130 An argument along this line was put forward in: Sicherheit bleibt erstes Thema. Mitglieder des Kreistages besichtigten Baustelle des Kraftwerkes, in: *MM*, 7.10.1970.

131 PS: Bescheidene Ergebnisse bei Kernenergieverwertung. Reaktorenentwicklung immer noch in Experimentierphase – Lange Reparatur- und Wartungszeiten, in: *DE*, 28.8.1968.
132 Noack, Hans-Joachim: Biblis bekommt den größten Reaktor. Von der Gurke zum Atom. Eine ländliche Gemeinde erwartet frohgemut die Kernspalter, in: *Süddeutsche Zeitung*, 10.11.1968.
133 Block A des Kernkraftwerkes Biblis offiziell übergeben, in: *Unidentified Newspaper*, 15.4.1975, GAB, KKW.
134 Die 'heile Welt' von Biblis . . ., in: *SP*, 17.4.1975.
135 Block A des Kernkraftwerkes Biblis offiziell übergeben, in: *Unidentified Newspaper*, 15.4.1975, GAB, KKW.
136 Kuhlo, Karl Ulrich: Wirtin Maria Höger: 'ANGST? Daß ich nicht lache!', in: *Bild am Sonntag*, 26.12.1976.
137 Similar statements were recorded with a view to further protest events against the Biblis power plants, which tended to be portrayed as events by and for outsiders: Beyeler, Susanne, Andreas Soschynski: Strahlende Zukunft. BRD 1977–1979, available as an attachment to: Baer, Willi, Karl-Heinz Dellwo (Eds.): Die Krebsfälle in der Elbmarsch/Der GAU in Fukushima. Hamburg 2012.
138 Kuhlo, Karl Ulrich: Wirtin Maria Höger: 'ANGST? Daß ich nicht lache!', in: *Bild am Sonntag*, 26.12.1976.
139 hez: Offizieller Antrittsbesuch von Bürgermeister Piegsa bei der Betriebsverwaltung des RWE in Biblis. Gute Zusammenarbeit zwischen RWE und Gemeinde soll fortgesetzt werden – Lob für die Kontaktaufnahme mit dem Betriebsrat, in: *BB*, 14.8.1984.
140 ehn: Bevölkerung im Schatten des Kernkraftwerks nicht verunsichert. Nur einer holte sich seine Jodpille bei der Gemeinde ab, in: *BE*, 2.5.1986.
141 PB: Landratsamt: 'Heißer Draht'. Telefonische Auskunft über die Strahlenbelastung, in: *BE*, 3.5.1986.
142 Cf., for 'tech-fear': Heßler et al., Introduction.
143 gz: Mehrheit gegen neue Atomreaktoren, in: *WZ*, 20.6.1975; DE/os: 'Unverantwortliche Panikmache in der Bevölkerung'. Gespräch mit dem Bibliser Bürgermeister Josef Seib über die Pläne zur Kernkraftwerkerweiterung, in: *DE*, 7.7.1975; mai: Bergsträßer Notizen, in: *DE*, 30.11.1978.
144 Seib, W., Bürgermeister Biblis: 'Unzutreffend und irreführend', in: *DE*, 7.11.1975.
145 cep: Keine Wahl. Wer langfristig die Versorgung der Menschheit sichern will, kann auf Kernenergie nicht verzichten/Absolute Sicherheit gibt es in keinem Bereich, in: *MM*, 18.2.1976.
146 cep: Keine Wahl. Wer langfristig die Versorgung der Menschheit sichern will, kann auf Kernenergie nicht verzichten/Absolute Sicherheit gibt es in keinem Bereich, in: *MM*, 18.2.1976; cf. Uekötter, Atomare, pp. 94–95.
147 Koch, Thilo: Kolumne. Büchse der Pandora, in: *DE*, 12.11.1976.
148 Rettweiler, Christian, et al.: Offener Brief an die Hessische Landesregierung, in: *BB*, 26.1.1984; Ro/zg: Energiepolitik in vernünftigen Bahnen. Bibliser RWE-Belegschaft schreibt Offenen Brief nach Wiesbaden, in: *MM*, 26.1.1984.
149 hez: Bürgermeister Piegsa warnt vor Ausstieg aus der Kernenergie. Es würden nicht nur Tausende von Arbeits- und Ausbildungsplätzen verloren gehen, sondern eine unverantwortliche weitere Luftverschmutzung wäre die Folge, in: *BB*, 24.1.1985.
150 Stellungnahme der CDU zum Grünen-Antrag auf Stilllegung des KKW Biblis, 27.8.1986, GAB, Gemeindevertretungssitzungen.
151 Stellungnahme der SPD zum Grünen-Antrag auf Stilllegung des KKW Biblis, 27.8.1986, p. 1–2, GAB, Gemeindevertretungssitzungen.
152 Ff: Umweltschützer verlangen Baustopp in Biblis. Resolution in der Hammer Kundgebung verfaßt. Auf Anhieb etwa 300 Unterschriften/Wissenschaftliche

Beiträge gaben Einblick, in: *WZ*, 22.1.1973 (date and newspaper deduced from context), GAB, Presseartikel KKW 1977.

153 Schulze, Gerd-Peter: Die Atomenergie nach Wyhl: Ratlos. Führen Bürgerproteste zur Krise beim Bau von weiteren Kernkraftwerken? – Alternativen sind nicht in Sicht, in: *DE*, 16.4.1975.

154 Schuvoir, Henry (dpa): Kernkraftwerke – ein Problem für die EG. Bericht von Hanna Walz sorgt für Gesprächsstoff im Europa-Parlament, in: *DE*, 28.8.1975; for the report by Hanna Walz cf. Gemeinsame Standortpolitik für Kernkraftwerke, in: Information (Europäisches Parlament) 1 (1976), pp. 8–12.

155 os: 'Wir wollen kein zweites Ruhrgebiet'. Auch Bibliser Gemeindevertretung fordert Aussetzung der Genehmigung für KKW, in: *DE*, 21.6.1975.

156 Twardzik, Werner (SPD-Vorsitzender): Antwort des SPD-Vorsitzenden in Biblis an Herrn Kernkraftwerkmeister Christian Rettweiler zur Stellungnahme Kernkraftwerk Biblis, in: *BB*, 19.8.1976.

157 os: SPD Biblis bei Osswald. Resolution gegen Kernkraftwerkserweiterung übergeben, in: *DE*, 28.11.1975.

158 Staab, Wolfgang: Katastrophenpläne lesen, in: *WZ*, 27.4.1974.

159 Bauer, L., Aktionsgemeinschaft Umweltschutz südl. Ried e. V.: Mitteilung der Aktionsgemeinschaft für Umweltschutz, in: *BB*, 12.8.1976.

160 Das RWE erklärt: Keine Gefährdung, in: *WZ*, 13.11.1974.

161 One example is the power plant's director Fred Meyer: Siegelmann, Bert: Wenn es im Kernkraftwerk brennt sollen Feuerwehren mit anpacken. Eigene Betriebsfeuerwehr aufgestellt/Biblis und Heppenheim werden direkt alarmiert, in: *MM*, 29.6.1974. For an in-depth analysis of the phenomenon of attribution of stronger emotions: Götter, Emotionen.

162 Rettig, Walter (Alsbach): 'Fortschritt' mit Todesverachtung, in: *DE*, 13.3.1976.

163 Cf. Hofmann, Matthias: Lernen aus Katastrophen. Nach den Unfällen von Harrisburg, Seveso und Sandoz. Berlin 2008, pp. 201–215.

164 Bauer, L., Aktionsgemeinschaft Umweltschutz südl. Ried e. V.: Mitteilung der Aktionsgemeinschaft für Umweltschutz, in: *BB*, 12.8.1976.

165 Bauer, L., Vorstand Aktionsgemeinschaft Umweltschutz südl. Ried e. V.: Konzentrierte Unvernunft, in: *BB*, 19.8.1976 (emphasis in the original).

166 Cf. Metzen, Margot (Worms): Das Atom-Ei von Biblis erregt Ärger, in: *WZ* (newspaper deduced from context), 16.5.1975, in: GAB, KKW.

167 os: Katastrophenabwehrpläne bleiben in der Schublade. Sechstes Bibliser Kernkraftwerksgespräch – Probebetrieb verläuft planmäßig, in: *DE*, 10.9.1974; for Wolter, who was called Rolf in the article: bert: SPD Biblis für sachliche Argumentation. Dem Gegner die Achtung nicht versagen. Thema Kernkraftwerk übergewichtig?/Rudolf Wolter wiedergewählt, in: *MM*, 19.2.1976.

168 Siegelmann, Bert: Unter der Lupe . . ., in: *MM*, 1.9.1974.

169 BS: Atom-Ei zieht Fische an. Kernkraftwerk Biblis wird bald auf Voll-Last gefahren, in: *MM*, 6.9.1974.

170 R.N.: Katastrophenschutz-Plan bleibt hinter Schloß und Riegel. Einblick für Bevölkerung höchstens teilweise, in: *MM*, 31.3.1977.

171 n: Bundesregierung befaßt sich mit Problemen der Kernenergie. Keine neue Genehmigung vor Lösung der Entsorgungsfrage, in: *DE*, 24.11.1976.

172 Some critics of nuclear power argued that the emotional assessment was providing the technology's ethic framework, cf. Bundesverband Bürgerinitiativen Umweltschutz: Atomenergie und die Zukunft unserer Gesellschaft. Karlsruhe 1978.

173 Cf. buck: Nuklearerhitzte Gemüter beruhigt. Informationskreis KKW trat kurzfristig mit RWE-Vertretern zusammen, in: *MM*, 9.5.1986.

174 AGU Biblis, Südl. Ried e. V., Grüne Liste Biblis: Harrisburg, Tschernobyl und dann Biblis? Versuch einer Stellungnahme, in: *BB*, 7.5.1986.

175 For the early Greens' relationship to emotions in politics: Raschke, Joachim: Die Grünen. Wie sie wurden, was sie sind. Cologne 1993, p. 83.
176 Aktionsgemeinschaft Umweltschutz südl. Ried e. V. Biblis. Sind unsere Atomkraftwerke sicher? Oder – was man durch Katastrophen lernen sollte!, in: *BB*, 15.5.1986.
177 Oberhollenzer, Dieter: 'Nachatomares Zeitalter angebrochen'. Biblis nach dem Reaktorunfall: Verunsicherung, Demonstration und Stillegungs-Diskussion, in: *WZ*, 17.5.1986.
178 For the change unfolding throughout the 1970s on the national level: Jung, Öffentlichkeit, pp. 93–94.
179 Nehring, Angst, pp. 452, 460–461. For Britain, cf. Patterson, Fissile, pp. 84–85.
180 Gassert, Popularität, p. 54; Greiner, Angst als Emotion, pp. 68–70, 76–78. For the temporary merging of peace- and anti-nuclear movements around the concept of a Nuclear Winter: Conze et al., Introduction, p. 13; Mausbach, Wilfried: Nuclear Winter. Prophecies of Doom and Images of Desolation During the Second Cold War, in: Conze, Eckart, Martin Klimke, Jeremy Varon (Eds.): Nuclear Threats, Nuclear Fear, and the Cold War of the 1980s. Cambridge 2017, pp. 27–54.
181 IG Frauen und Mütter: Was der Erde geschieht geschieht uns. Biblis 1967–1987, p. 10, GAB, Az.: 774–054; IG Frauen und Mütter aus Biblis trafen sich, in: *BB*, 26.1.1989.
182 In its wake, Cristina Perincioli had published a number of women's perspectives in a book that was well received by anti-nuclear groups transnationally: Perincioli, Cristina: Die Frauen von Harrisburg oder 'Wir lassen uns die Angst nicht ausreden'. Reinbek 1980; cf. Zaretsky, Atomic, pp. 56–57, 69–71.
183 Cf. Engels, Jens Ivo: Gender Roles and German Anti-Nuclear Protest. The Women of Wyhl, in: Bernhardt, Christoph, Geneviéve Massard-Guilbaud (Eds.): Le Démon Moderne. La pollution dans les sociétés urbaines et industrielles d'Europe. The Modern Demon. Pollution in Urban and Industrial European Societies. Clermont-Ferrand 2002, pp. 407–424, especially pp. 410–413, 420–421; Röhr, Ulrike, Dagmar Vinz: Frauen gegen Atomenergie. Die Auswirkungen von Tschernobyl auf das umwelt- und energiepolitische Engagement von Frauen, in: Mez, Lutz, Lars Gerhold, Gerhard de Haan (Eds.): Atomkraft als Risiko. Analysen und Konsequenzen nach Tschernobyl. Frankfurt/Main 2010, pp. 261–274; Summers, Sarah E.: 'Thinking Green!' (And Feminist). Female Activism and the Greens from Wyhl to Bonn, in: German Politics and Society 33 (2015), 4, pp. 40–52; Kirchhof, Astrid Mignon, Helmuth Trischler: The History behind West Germany's Nuclear Phase-Out, in: Kirchhof, Astrid Mignon (Ed.): Pathways into and Out of Nuclear Power in Western Europe. Austria, Denmark, Federal Republic of Germany, Italy, and Sweden. Munich 2020, pp. 124–169; for similar transformations during the 1960s: Hogg, British, pp. 114–119.
184 Excerpt from Brigitte, printed under: IG Frauen und Mütter: Eltern denken nach!, in: *BB*, 22.5.1986.
185 Nachdenken – Umdenken lohnt sich!, in: *BB*, 5.6.1986.
186 Minutes, Meeting, Bergstraße District Council 10/XI, 9.6.1986, in: GAB, Kreistag 3.
187 Minutes, Meeting, Bergstraße District Council 10/XI, 9.6.1986, in: GAB, Kreistag 3, p. 23.
188 flo: Frage nach Tschernobyl: Biblis A und B abschalten? Groß-Rohrheimer Parlament diskutiert Gefahren – SPD fordert internationalen Ausstieg aus der Kernenergie, in: *SP*, 11.6.1986.
189 Tschernobyl, in: *WZ*, 14.6.1986.
190 IG Frauen und Mütter: Gespräch mit Bürgermeister Piegsa und Pfarrer Muth brachte keine Wende!, in: *BB*, 19.6.1986.

191 IG Frauen und Mütter aus Biblis und Umgebung, in: *BB*, 18.9.1986.
192 IG Frauen und Mütter aus Biblis und Umgebung, in: *BB*, 9.10.1986.
193 Rettweiler, Christian: Mitteilung des RWE-Betriebsrates, in: *BB*, 23.10.1986.
194 Aktionsgemeinschaft Umweltschutz südl. Ried e. V. Biblis. Gedenktag, in: *BB*, 23.4.1987.
195 IG Frauen und Mütter: Was der Erde geschieht geschieht uns. Biblis 1967–1987, pp. 3–4, GAB, Az.: 774–054.
196 IG Frauen und Mütter: Was der Erde geschieht geschieht uns. Biblis 1967–1987, p. 5.
197 IG Frauen und Mütter: Was der Erde geschieht geschieht uns. Biblis 1967–1987, pp. 6–7.
198 IG Frauen und Mütter: Was der Erde geschieht geschieht uns. Biblis 1967–1987, p. 11.
199 IG Frauen und Mütter: Was der Erde geschieht geschieht uns. Biblis 1967–1987, p. 12–13.
200 IG Frauen und Mütter: Was der Erde geschieht geschieht uns. Biblis 1967–1987, p. 19.
201 'Muß ich mir wegen radioaktiver Strahlung um meine Familie Sorgen machen?', in: *BB*, 29.4.1987.
202 IG Frauen und Mütter aus Biblis und Umgebung. Tschernobyl ist immer noch überall . . ., in: *BB*, 14.5.1987.
203 C.H.: Aktionsgemeinschaft Umweltschutz südl. Ried e. V. Biblis, in: *BB*, 8.12.1988; cf. Radkau, RWE zwischen Kernenergie und Diversifizierung, p. 234.
204 Die Bibliser SPD nimmt Stellung zum Störfall im Kernkraftwerk Biblis, in: *BB*, 15.12.1988.
205 Sitzung der Gemeindevertretung vom 14.12.1988. Antrag der SPD-Fraktion zum Kernkraftwerk Biblis. Persönliche Erklärung, in: *BB*, 15.12.1988; for their declaration: Persönliche Erklärung von Karl-Heinz Brandt, Raimund Arnold und Rudolf Wolter in: Minutes, Meeting, Biblis Municipal Council, 14.12.1988, GAB, Gemeindevertretungssitzungen.
206 Begründung des Dringlichkeitsantrags der SPD, in: Minutes, Meeting, Biblis Municipal Council, 14.12.1988, p. 1, GAB, Gemeindevertretungssitzungen.
207 Begründung des Dringlichkeitsantrags der SPD, in: Minutes, Meeting, Biblis Municipal Council, 14.12.1988, p. 3.
208 Dringlichkeitsantrag der SPD Biblis, in: Minutes, Meeting, Biblis Municipal Council, 14.12.1988; Felber, Henri (Pressebeauftragter): Die Bibliser SPD nimmt Stellung zum Störfall im Kernkraftwerk Biblis, in: *BB*, 15.12.1988.
209 Cf. CDU-Information, in: *BB*, 29.12.1988.
210 Felber, Henri (SPD Biblis, Pressebeauftragter): SPD-Vorsitzender Twardzik kommentiert die Bibliser Gemeindevertretersitzung, in: *BB*, 22.12.1988.
211 C.H.: Grüne Liste Biblis, in: *BB*, 12.1.1989.
212 Letter to Piegsa, 10.3.1989, GAB, Az.: 774–054.
213 Before Chernobyl, 20 had been published.
214 C.H.: Grüne Liste Biblis. Gedanken zum Wahlausgang, in: *BB*, 16.3.1989.
215 Kissel, Raimund (Biblis), Ina Schott (Biblis): Leserbrief, in: *BB*, 30.3.1989.
216 Minutes, Meeting, Bergstraße District Council 3/XII, 10.7.1989, p. 13, GAB, Kreistag 1.
217 Minutes, Meeting, Bergstraße District Council 3/XII, 10.7.1989, p. 14.
218 Minutes, Meeting, Bergstraße District Council 3/XII, 10.7.1989, p. 20.
219 Kissel, Raymund (Biblis), Inge Würsching (Biblis): IG Frauen und Mütter informieren. Die Atomlobby und deren Wahnsinn, in: *BB*, 2.11.1989.
220 Cf. Heilmann, Strahlenangst, pp. 110–113; Jung, Öffentlichkeit, pp. 93–97; Gebauer, Annekatrin: Apokalyptik und Eschatologie, in: Archiv für Sozialgeschichte 43 (2003), pp. 405–420, here p. 417; Frevert, Rationalität, pp. 135–137.

221 Cf. RWE Presseinformation Kernkraftwerk Biblis. Informationsangebot genutzt. Kernkraftwerke erwarten den 5.000.000. Besucher, in: *BB*, 23.11.1989.
222 Cf. 'Nuclear Debates'.
223 A Programme of Nuclear Power, February 1955 (draft), in: National Archives (NA), CAB 129/73/c/55/31/31.
224 Ian Welsh argues that the leading British advocates of nuclear energy, gathered in the United Kingdom Atomic Energy Authority, were, via a virtual monopoly on sound statements regarding the atom, able and willing to generate enthusiasm for all things nuclear, to strengthen their own expert status: Welsh, NIMBY, pp. 18–22.
225 A Programme of Nuclear Power (cf. note 223), para. 1–3, p. 1.
226 A Programme of Nuclear Power (cf. note 223), para. 21, p. 5.
227 A Programme of Nuclear Power (cf. note 223), para. 36–37, pp. 8–9.
228 A Programme of Nuclear Power (cf. note 223), para. 53, p. 12.
229 Cf. Welsh, Nuclear Nation.
230 The First Heat Exchanger, in: *Thornbury Gazette (TG)*, 19.4.1958.
231 Nye, Sublime.
232 The paper was renamed in the *Northavon Gazette*, Thornbury Edition, from 1974 till 1985, when it returned to its original title, cf. Phillips, H. W.: The Gazette in Thornbury, www2.newsquest.co.uk/dursley/gazette_birthday/GAZETTE_BIRTHDAY25.html (23.1.2023); especially early on, another relevant newspaper for the regional debate was *The Citizen (TC)*.
233 Atomic Station, in: *TG*, 1.6.1957.
234 Power Station Reaches Half-Way Stage. Berkeley Shows What Can Be Done in 20 Months, in: *TG*, 23.8.1958.
235 Berkeley Is a Tonic to Our Industry, in: *TC*, 21.6.1958.
236 Power Station Work Is Up to Schedule. Two Reactors and Turbine Hall Almost Complete. Plant Installation Now in Progress, in: *TG*, 7.2.1959.
237 Cf. Bösch, Frank: European Media Events, in: European History Online (EGO), published by the Institute of European History (IEG), Mainz 3.12.2010, www.ieg-ego.eu/boeschf-2010-en (23.1.2023).
238 Cf. Welsh, Mobilising, p. 72.
239 Test Sites for Atomic Radiation, in: *TG*, 17.5.1958.
240 Radio-Activity on Plant and Animal Life. Research Assurance Sought by Farmers, in: *TG*, 26.4.1958.
241 According to Atomic Station Liaison. Exploratory Meeting to Be Held, in: *TG*, 13.12.1958.
242 Clean Air and Radiation. Two Subjects of Importance, in: *TG*, 22.11.1958.
243 A Programme of Nuclear Power (cf. note 223), Paragraphs 36 and 37, pp. 7–8.
244 Radio-Activity on Plant and Animal Life. Research Assurance Sought by Farmers, in: *TG*, 26.4.1958; Blair, A. R. (Stinchcombe): Radiation Risks From New Station, in: *TG*, 23.5.1959.
245 Test Sites for Atomic Radiation, in: *TG*, 17.5.1958; cf. Nicholson, Hayley: Oldbury Nuclear Power Station and Visitor Centre, in: Bradshaw, Jane, Joyce Tibbenham (Eds.): The Millennium Book of Oldbury on Severn. Clevedon 2001, pp. 78–80.
246 Minutes, Meeting, Public Health Committee, Thornbury Rural District Council, 4.12.1958, p. 275, Gloucestershire Archives (GlouA), DA38/100/32.
247 For a call to manage ones (negative) emotions regarding nuclear energy during the 1950s: Hogg, British, especially pp. 76–106.
248 Second Nuclear Station on Severn Bank. Application to Develop 175-Acre Site, in: *TG*, 28.3.1959.
249 Except in the Event of War. No Danger Envisaged from Radiation, in: *TG*, 11.4.1959.

250 Oldbury's Reaction to New Power Station. Farmers and Fishermen Perturbed but – 'It Has Got to Come', in: *TG*, 4.4.1959.
251 Radiation Hazards, in: *TG*, 9.5.1959.
252 Oldbury's Reaction to New Power Station. Farmers and Fishermen Perturbed But – 'It Has Got to Come', in: *TG*, 4.4.1959; for Fowler: A Radioactive Air Discharge from Oldbury 'B'?, in: *TG*, 25.4.1970.
253 Amor Patriae: Oldbury Power Station, in: *TG*, 11.4.1959.
254 Oldbury's Reaction to New Power Station. Farmers and Fishermen Perturbed But – 'It Has Got to Come', in: *TG*, 4.4.1959.
255 Oldbury's Reaction to New Power Station. Farmers and Fishermen Perturbed But – 'It Has Got to Come', in: *TG*, 4.4.1959; clearly, the coming of nuclear power endangered at least some livelihoods; cf. Presas i Puig, Albert, Jan-Henrik Meyer: One Movement or Many? The Diversity of Antinuclear Movements in Europe, in: Kaijser, Arne et al. (Eds.): Engaging the Atom. The History of Nuclear Energy and Society in Europe from the 1950s to the Present. Morgantown 2021, pp. 83–111, here p. 87.
256 Minutes, Meeting, Oldbury-on-Severn Parish Council, 21.4.1959, GlouA, P237a/PC/1/2, p. 368.
257 Minutes, Meeting, Oldbury-on-Severn Parish Council, 21.4.1959, GlouA, P237a/PC/1/2, pp. 368–369.
258 Minutes, Meeting, Oldbury-on-Severn Parish Council, 21.4.1959, GlouA, P237a/PC/1/2, p. 369.
259 Minutes, Meeting, Thornbury Rural District Council, 25.9.1959, p. 145, GlouA, DA38/100/33; Minutes, Meeting, Thornbury Rural District Council, 24.4.1959, p. 472, GlouA, DA38/100/32.
260 Nuclear Station. Council Told 90 Per Cent of Oldbury People Are Opposed, in: *TG*, 2.5.1959.
261 A Lone Voice, in: *TG*, 9.5.1959.
262 Blair, A. R. (Stinchcombe): Radiation Risks from New Station, in: *TG*, 23.5.1959.
263 Second Station. Hundreds Sign Petition, in: *TG*, 25.7.1959; cf. Unsigned minute, covering the Report of the Public Inquiry, NA, POWE 14/1248, p. 4; Minutes of Proceedings, NA, POWE 14/1249; Rough, Policy, p. 32; for Buchanan, in all probability Colin Buchanan: Hass-Klau, Carmen: The Pedestrian and the City. New York 2015, p. 87–94.
264 Report about the Public Inquiry by A.H.F. Linton and C.D. Buchanan, 4.7.1960, NA, POWE 14/1249.
265 Report about the Public Inquiry by A.H.F. Linton and C.D. Buchanan, 4.7.1960, NA, POWE 14/1249, p. 57.
266 Report about the Public Inquiry by A.H.F. Linton and C.D. Buchanan, 4.7.1960, NA, POWE 14/1249, p. 58.
267 Sheail, Environmental, pp. 103–109, 116–122, 132–137.
268 Report about the Public Inquiry by A.H.F. Linton and C.D. Buchanan, 4.7.1960, NA, POWE 14/1249, p. 60.
269 Report about the Public Inquiry by A.H.F. Linton and C.D. Buchanan, 4.7.1960, NA, POWE 14/1249, p. 62.
270 Report about the Public Inquiry by A.H.F. Linton and C.D. Buchanan, 4.7.1960, NA, POWE 14/1249, p. 27.
271 Report about the Public Inquiry by A.H.F. Linton and C.D. Buchanan, 4.7.1960, NA, POWE 14/1249, p. 71.
272 Minister Approves New Glo'shire Atom Station. Work Likely to Begin Soon at Oldbury, in: *TC*, 20.10.1960.
273 Topics Discussed by 'Double Glo'ster'. Exciting Decade, in: *TG*, 29.10.1960.
274 Corfield to Wood, 16.1.1961, NA, POWE 14/1248, p. 1.

275 Corfield to Wood, 16.1.1961, NA, POWE 14/1248, p. 1.
276 Corfield to Wood, 16.1.1961, NA, POWE 14/1248, pp. 2–3.
277 Corfield to Wood, 1.2.1961, NA, POWE 14/1248.
278 Goss to Bartlett, 6.2.1961, NA, POWE 14/1248.
279 Hinton to Corfield, 8.2.1961, NA, POWE 14/1248 (the day is hardly discernible); for Hinton: Gowing, Margaret: Obituary. Lord Hinton of Bankside, O.M., F.Eng., in: Biographical Memoirs of Fellows of the Royal Society 36 (1990), pp. 217–239, https://royalsocietypublishing.org/doi/10.1098/rsbm.1990.0031 (23.1.2023).
280 1955 and after, in: *TG*, 28.1.1961.
281 Topics Discussed by 'Double Glo'ster'. Unique Collection, in: *TG*, 12.8.1961.
282 Minutes, Meeting, Public Health Committee, Thornbury Rural District Council, 2.2.1961, p. 345, GlouA, DA38/100/34.
283 He was wrongly identified in the newspaper: Cooke, J. C. (Chairman, Thornbury Rural District Council): Good Wishes from Thornbury RDC, in: *TG*, 6.4.1963.
284 By Technology Minister, in: *TG*, 14.6.1969.
285 Minister of Power at Oldbury. Visit to Laboratories and The Nuclear Power Stations, in: *TG*, 18.9.1965; for Lee: Howell, David: Lee, Frederick, Baron Lee of Newton (1906–1984), Trade Unionist and Politician, in: Oxford Dictionary of National Biography, www.oxforddnb.com/view/10.1093/ref:odnb/9780198614128.001.0001/odnb-9780198614128-e-65951 (23.1.2023).
286 County Comment by Double Glo'ster: Nuclear Safety, in: *TG*, 10.2.1968.
287 Nuclear Power Station at Oldbury Fast Becoming a Reality, in: *TG*, 23.9.1961.
288 The Nuclear Power Race. Oldbury Contract Awarded, in: *TG*, 24.2.1962.
289 Nuclear Power Next Winter from Oldbury Station, in: *TG*, 12.2.1966; cf. Hill, Atomic Empire, p. 237.
290 Cf. Nye, Sublime, pp. 234–237; Jasanoff et al., Containing the Atom, pp. 125–128.
291 Good Progress Being Maintained. Oldbury Power Station Fast Taking Shape, in: *TG*, 7.9.1963.
292 Nuclear Power Next Winter from Oldbury Station, in: *TG*, 12.2.1966; cf. Piel, Helen, Rudolf Seising, Christian Götter: The Computer Metaphor. Anchoring the Fear of Human Obsolescence(s) since the 20th Century, in: Technikgeschichte 89 (2022), 1, pp. 87–118.
293 Power Station Progress at Oldbury. Project Is Approaching Half Way Stage, in: *TG*, 21.3.1964.
294 Oldbury Power Station Construction Prompts Awe and Wonder, in: *TG*, 6.2.1965.
295 By Technology Minister, in: *TG*, 14.6.1969; for Benn's increasingly critical view of nuclear power: Benn, Tony: The Benn Diaries. Selected, Abridged and Introduced by Ruth Winstone. London 1995, for example, 14.5.1969, pp. 205–207, 16.1.1977, pp. 389–390, and 28.4.1979, pp. 470–472; furthermore, Edgerton, Rise and Fall, p. 369.
296 Minutes, Meeting, Oldbury-on-Severn Parish Council, 4.8.1969, GlouA, P237a/PC/1/3, p. 118.
297 Nuclear Power. Stations in Estuary, in: *TG*, 14.2.1970.
298 Oldbury Power Station, in: *Northavon Gazette (NG)*, 29.7.1978.
299 Putting a question mark to the argument that the smaller British nuclear construction programme during the 1970s could explain the difference in the debates' dynamics, cf. 'Coming to Terms with Conflicts Anchored to the Atom: Methods'.
300 For the national debate: Patterson, Fissile, p. 83.
301 Scare Stories of Nuclear Tombs at Severnside, in: *TG*, 14.3.1970.

302 Scare Stories of Nuclear Tombs at Severnside, in: *TG*, 14.3.1970.
303 One exception was the Gazette itself, cf. A Radioactive Air Discharge from Oldbury 'B'?, in: *TG*, 25.4.1970.
304 One exception was the Gazette itself, cf. A Radioactive Air Discharge from Oldbury 'B'?, in: *TG*, 25.4.1970.
305 Nuclear Concern, in: *TG*, 30.5.1970.
306 Doctors Protest On Oldbury. South Glos Planners Told to Follow County Line, in: *TG*, 6.6.1970.
307 A Radioactive Air Discharge from Oldbury 'B'?, in: *TG*, 25.4.1970.
308 A Radioactive Air Discharge from Oldbury 'B'?, in: *TG*, 25.4.1970.
309 Worry Over Nuclear Build Up. Thornbury RDC to Press for Meeting with the Experts, in: *TG*, 30.5.1970.
310 Worry Over Nuclear Build Up. Thornbury RDC to Press for Meeting with the Experts, in: *TG*, 30.5.1970.
311 Nuclear Concern, in: *TG*, 30.5.1970.
312 Nuclear Concern, in: *TG*, 30.5.1970.
313 RDC and Nuclear Experts, in: *TG*, 6.6.1970.
314 Question and Answer about Nuclear Power Station Safety, in: *TG*, 20.6.1970.
315 Thornbury RDC Taking Lead in Joint Talks on Pollution, in: *TG*, 1.8.1970.
316 Criminal Folly of Pollution Menace, in: *TG*, 4.7.1970.
317 Minutes, Meeting, Oldbury-on-Severn Parish Council, 27.5.1970, GlouA, P237a/PC/1/3, pp. 134–135; this ongoing opposition demonstrates that mere closeness was no guarantee for "significantly positively shaped" "local perceptions", pointed out in: Konrad, Wilfried, Josep Espluga et al.: D4.2. Comparative cross-country analysis on preliminary identification of key factors underlying public perception and societal engagement with nuclear developments in different national contexts. Month 13. September 2016 (Update March 2018), p. 43, www.honest2020.eu/sites/all/themes/Porto_sub/downloads/deliverables/D4.2_revised_final.pdf (23.1.2023); the claims are based upon: Venables, Dan et al.: Living with Nuclear Power. A Q-Method Study of Local Community Risk Perceptions. Cardiff 2008 (=Social Contexts and Responses to Risk Network (SCARR). Working Paper 24-2008).
318 Oldbury Concern on Power Plan and Flooding, in: *TG*, 6.6.1970.
319 There had been an incident on August 24, 1970: Mahaffey, James: Atomic Accidents. A History of Nuclear Meltdowns and Disasters. From the Ozark Mountains to Fukushima. New York 2014, pp. 186–187.
320 Opposed to Second Station, in: *TG*, 19.9.1970.
321 Oldbury Protest Draws Assurance from Generating Board, in: *TG*, 28.11.1970.
322 Minutes, Meeting, Oldbury-on-Severn Parish Council, 7.9.1978, GlouA, P237a/PC/1/4, p. 29.
323 Open Day and Fete at Oldbury Nuclear Power Station, in: *TG*, 11.9.1971.
324 3,000 Attracted to Oldbury's Open Day, in: *TG*, 26.8.1972.
325 Cf. Pills to Combat Likely Radiation Danger from Nuclear Stations, in: *TG*, 1.5.1971; Assurances Wanted on Power Station Safety, in: *TG*, 1.5.1971; Risk Exists, But It Is Remote. CEGB Assurance to Thornbury RDC on Power Stations, in: *TG*, 5.6.1971.
326 Risk Exists, But It Is Remote. CEGB Assurance to Thornbury RDC on Power Stations, in: *TG*, 5.6.1971.
327 B., J. D. K. (Alveston): Pollution, in: *TG*, 25.9.1971.
328 Radio-Active Waste Fears Rejected by Council, in: *TG*, 2.10.1971.
329 Cf. Oldbury Power Station, in: *TG*, 25.3.1972.
330 Radio-Active Waste Fears Rejected by Council, in: *TG*, 2.10.1971. Strange, in all probability, worked at the power plant; cf. Radio-Active Waste Fears Rejected by Council, in: *TG*, 2.10.1971.

331 Pollution, in: *TG*, 2.10.1971.
332 Pollution, in: *TG*, 2.10.1971.
333 Nuclear Power, in: *NG*, 22.10.1977.
334 Cf. 'The National, the Spectacular, and the Complex Atom: State of Research'.
335 For the strategy: Welsh, Mobilising, pp. 141–142; cf. note 346.
336 Buckley, David (Dursley): Nuclear Power, in: *NG*, 22.10.1977.
337 Wheeler, J. E. (Wotton-under-Edge): Nuclear Power, in: *NG*, 29.10.1977.
338 Brown, Mary (Dursley): Nuclear Power, in: *NG*, 29.10.1977.
339 Buckley, David (Dursley): Nuclear Power, in: *NG*, 5.11.1977.
340 Buckley, David (Dursley): Nuclear Power, in: *NG*, 19.11.1977.
341 Brown, Peter (Dursley): Nuclear Power, in: *NG*, 26.11.1977.
342 Buckley, David (Dursley): Nuclear Power, in: *NG*, 5.12.1977.
343 Brown, Mary (Dursley): [untitled, CG], in: *NG*, 26.12.1977; this referenced a core element of the Atom State (cf., 'The National, the Spectacular, and the Complex Atom: State of Research').
344 Nuclear Waste Trains Stop at Stonehouse, in: *NG*, 12.4.1980; for Marjoram: Fenton, Susan: Longest Serving Councillor Calls Time, in: Green World, https://greenworld.org.uk/article/longest-serving-councillor-calls-time (23.1.2023).
345 Cf. 'Nuclear Debates'.
346 Welsh, NIMBY, p. 16; for Ridley: Cosgrave, Patrick: Ridley, Nicholas, Baron Ridley of Liddesdale (1929–1993), Politician, in: Oxford Dictionary of National Biography, www.oxforddnb.com/view/10.1093/ref:odnb/9780198614128.001.0001/odnb-9780198614128-e-53192 (23.1.2023).
347 For the adoption of German anti-nuclear protest strategies, cf. Welsh, Mobilising, pp. 160–165.
348 Euro MP Sees Nuclear Power as Essential, in: *NG*, 14.6.1980.
349 Cf. Patterson, Fissile, p. 83.
350 Cf. Green, L. (Miss) (Berkeley): Nuclear Waste, in: *NG*, 25.7.1981; Dellbridge, Olivia (Berkeley): Nuclear Wastes, in: *NG*, 18.7.1981; furthermore: Hamblin, Jacob Darwin: Poison in the Well. Radioactive Waste in the Oceans at the Dawn of the Nuclear Age. New Brunswick 2009.
351 Young people misled about our nuclear power, in: *NG*, 14.8.1982.
352 Patterson, Fissile, p. 83.
353 Taking Risks, in: *NG*, 21.3.1981.
354 Safety Check Finds Nothing over 25 Years of Activity, in: *NG*, 4.6.1983.
355 Better Safe . . ., in: *NG*, 18.6.1983.
356 N-Plant Safety Concern Creates 'Needless Fears', in: *TG*, 7.3.1986.
357 N-Plant Vote Is Welcomed, in: *TG*, 28.3.1986.
358 Cf. the description by SCAR's successor STAND (Severnside Together against Nuclear Development): Who Are We?, www.nuclearsevernside.co.uk/about.html (23.1.2023); Corner, John, Kay Richardson, Natalie Fenton: Nuclear Reactions. Form and Response in Public Issue Television. London 1990.
359 Conrad, Dr George (Pucklechurch): Coming Clean, in: *TG*, 28.3.1986.
360 Cf. the Gazette's letters column in February and March 1980.
361 Tylor, Dick, David Wilkie: Drawing the Line with Leukaemia, in: *New Scientist*, 21.7.1988, pp. 53–56, here p. 55; cf., for a short overview over different results by studies in Britain and Germany: Kaatsch, Peter et al.: Leukaemia in Young Children Living in the Vicinity of German Nuclear Power Plants, in: International Journal of Cancer 1220 (2008), pp. 721–726.
362 Williams, Alan Bryn (Thornbury): Nuclear Controversy Futile, in: *NG*, 16.3.1984.
363 N-Stations Have a 'Negligible Effect', in: *NG*, 27.4.1984.
364 Nuclear Sniping, in: *NG*, 21.9.1984.
365 Cash Appeal for N-Plant Investigation, in: *NG*, 22.6.1984; Leukaemia Link Probe, in: *NG*, 9.11.1984.

366 Hall, Richard: N-Plant Cancer Fears 'Groundless', in: *NG*, 22.3.1985; for Marland: Marland, Paul, in: Dod's Parliamentary Companion 158 (1990), p. 509.
367 No Probe into N-Plants, in: *NG*, 26.4.1985.
368 Nuclear Power Watchdogs Stay 'Secret', in: *TG*, 25.4.1986.
369 Nuclear Industry 'Too Secretive', in: *TG*, 25.4.1986.
370 Chernobyl 'Demands Honesty', in: *TG*, 16.5.1986; demonstrating heightened fears: Fleming, Pat (Gwent): Time to Call a Halt Now?, in: *TG*, 30.5.1986.
371 Chernobyl 'Demands Honesty', in: *TG*, 16.5.1986; cf. Candidates Condemn N-Plant Secrecy. Call to Reassure Public, in: *TG*, 30.5.1986.
372 Chernobyl 'Demands Honesty', in: *TG*, 16.5.1986.
373 Our Point of View, in: *TG*, 23.5.1986.
374 Call for Independent Radiation Monitoring, in: *TG*, 31.10.1986; cf. First Liaison Meeting Open to Press. Monitoring Service Is Welcomed, in: *TG*, 24.4.1987.
375 Allison, Hilary: Emergency Plans Open to Public. Power Stations' Bid to Dispel Secrecy, in: *TG*, 17.10.1986.
376 Allison, Hilary: Emergency Plans Open to Public. Power Stations' Bid to Dispel Secrecy, in: *TG*, 17.10.1986; for Walker: Garnett, Mark: Walker, Peter Edward, Baron Walker of Worcester (1932–2010), Businessman and Politician, in: Oxford Dictionary of National Biography, www.oxforddnb.com/view/10.1093/ref:odnb/9780198614128.001.0001/odnb-9780198614128-e-103315 (23.1.2023).
377 Pearce, Rosie: Thousands at Power Station Open Days, in: *TG*, 19.9.1986. Power plant visits were rather usual in both countries. The remarkable thing here was the way they were emphasised as opportunities.
378 Our Point of View, in: *TG*, 5.6.1987.
379 CEGB Plan Shake-Up in N-Plant Links, in: *TG*, 3.7.1987; for the new openness in the wake of Chernobyl overall: Nicholson, Oldbury.
380 Cf. Our Point of View, in: *TG*, 15.1.1988.
381 Our Point of View, in: *TG*, 15.1.1988.
382 Brinkmann, Hans: Lingen: Auf der Baustelle des Kernkraftwerks 1000 Emsländer tätig. An hiesige Firmen Aufträge von rund 180 Millionen DM. 'Gute Arbeit geleistet', in: *Lingener Tagespost (LT)*, 24.10.1984; Mit dem Baufortschritt weiter voll im Zeitplan. Schon viele Gebäude fertiggestellt – Kühlturm wird 1985 'hochgezogen', in: *LT*, 20.12.1984.
383 Will Be Biggest in the World. Contracts Placed for £60m. Hinkley Point Project, in: *Bridgwater Mercury*, 17.9.1957.
384 Robinson, James: Big Carl Takes the Strain: World's Largest Crane Lifts 1,600-Ton Steel Ring as Britain's Latest Nuclear Power Plant Takes Shape at Hinkley Point, in: *Mail Online (Daily Mail)*, 4.5.2021.
385 Cf. Campbell, Hamish: A Colony of Cranes, in: *The Scotsman*, 19.2.1982.
386 Nye, Sublime, pp. 60–61, 284; this can be gleaned from a movie presenting the construction of Hinkley Point in comparison to the already existing station at Calder Hall and the next one planned at Wylfa: Technical and Scientific Films Ltd.: Hinkley Point. The World's First 500,000-Kilowatt Atomic Power Station. Designed and Constructed for the Central Electricity Generating Board of Great Britain. UK, ca. 1960, www.youtube.com/watch?v=OREsNwYotP8 (23.1.2023).
387 Cf. Heßler, Einleitende Überlegungen.
388 Holger Nehring argued that this phenomenon could be explained with attempts to delimit the Federal Republic from National Socialist Germany: Nehring, Atomzeitalter; cf. 'Nuclear Democracy'.

3 Rational Atoms

Something changed with the "long 1970s". What it was, exactly, depends on the perspective taken. Ulrich Herbert saw the end of "High Modernity", which had befallen wide parts of Europe at the end of the 19th century, bringing with it near all-encompassing changes for the continent's population.[1] Environmental historians observed the dawning of an age of environmental protection or, at least, consciousness.[2] The economic and political dimensions of change were the focus of Anselm Doering-Manteuffel and Lutz Raphael, to whom it was the end of "a thirty years' boom" of an industrial society, spurred on by mass production and mass consumerism in the wake of the Second World War.[3] Uwe Fraunholz, Thomas Hänseroth and Anke Woschech emphasised the techno-utopian character of the age of high modernity, with two especially intensive phases around the fin de siècle and the 1960s.[4] Social scientists Daniel Bell and Ronald Inglehart, writing in the midst of the transformative decade, suggested that the western world underwent a transition from a material to a post-material age.[5] To their colleagues, not least Helga Nowotny, Peter Weingart, Ulrich Beck, and Anthony Giddens, analysing the change from the following decade's perspective, it had been centred upon the rational methods of science. To them, those methods had stood at the core of modernity since the 19th or even the 17th century, providing the critical tools that had brought an end to the traditional world by replacing tradition as a legitimation for political decisions with objective scientific findings.[6] These rational, analytical approaches had become nearly all-encompassing during the escalating decades of the Cold War,[7] and had finally been turned upon themselves. As Beck most famously put it, modernity had become "reflexive", or, with Giddens, "radical". Rational science's ability to deliver objective facts and truths eroded, with inner-scientific quarrels and interpretative ranges of statements by scientists coming to light. This, in turn, damaged the ability of scientific arguments to legitimise political programmes – while at the same time allowing a wider range of political positions to be backed up by scientific findings, therefore rendering its application as a legitimising strategy even more widespread. Weingart described these seemingly contrarian processes as the linked up "politization of science" and "scientization of policy".[8]

DOI: 10.4324/9781003583677-4

In this chapter, I analyse the connections between this erosion of rationality, in the sense of a science and technology-based "culture of knowledge" represented by scientists and technicians, in short, experts,[9] and the nuclear debates in Hinkley Point and Lingen (Ems). I analyse, in other words, an issue often studied from the perspective of science and scientists from a specific geographical area, and therefore from a more diverse perspective, which allows for more differentiated results.[10] Based upon this analysis, I argue that, *first*, indeed, the question of the role of rationality in the area of public political decision-making had become closely anchored to the debates about nuclear power.[11] Consequently, both issues influenced one another reciprocally. The local perspectives taken up here are outstandingly suited to demonstrating that the resulting debates were a prime example *for* and a crucial driver *of* the change described earlier: the erosion of the erstwhile almost unchallengeable status of objective science, scientists, and their rational arguments. Most importantly, however, *second*, I argue that the local perspectives also reveal that this change was mostly only an apparent one.[12] In fact, the decisive underlying transformation that occurred was a political emancipation process undergone by citizens critical of nuclear energy (cf. "Nuclear Democracy") transnationally. During the debates about atomic power, the technology's critics adopted a discursive strategy that had originally been applied by nuclear power's proponents and had allowed them to more or less steer the course of the debate: They had utilised experts as rationality's proponents to render their own political stances unchallengeable via a differentiation between the rational and the political. This, it must be remarked, should in no way imply that science and society were separate, or that said experts had not willingly taken up or even encouraged that role.[13] When the technology's critics adopted this strategy of utilising experts for their political goals, this not merely altered the development of the nuclear debates, but also crucially affected the way rationality was regarded. The result was a seemingly paradoxical situation: Rationality became simultaneously ubiquitous and irrelevant within the decision-making process.[14] *Third*, I argue that the main trigger behind this transformation was a combination of three elements around the 1970s: the transnational spread of alternative perspectives on nuclear power among growing numbers of experts, which demonstrated the interpretability of many formerly acclaimed facts about the technology; the active engagement of increasingly interconnected citizens with democratic institutions because of this spreading knowledge, which led to a growing realisation of their own political potential; and the emergence of issues provoking these citizens' all-out political opposition.[15] This requires a more detailed explanation.

In the years after its inception in the 1950s, the "Peaceful Atom's" advocates commanded overarching political majorities, backed in Britain, Germany and other industrialised countries by publicly renowned personalities from the fields of politics, media and, most importantly here, science and technology.[16] The corresponding lack of vocal opposition lent the proponents of nuclear power the ability to present the political decision to research and apply the atom's energy for peaceful purposes, under the label of progress, as one that was based upon reasons beyond the political sphere, from the realm of rationality.[17] The latter comprised arguments, knowledge and definitions based upon science and technology, which tended to be presented as facts by experts. These experts and their demeanour in the public debates, their way of communicating and constructing an argument, lent an aura of indefeasibility to the positions they stood for.[18] Once more indicating the different democratic cultures,[19] construing the realm of rationality as non-political was especially prevalent in Germany. Here, a corresponding tradition went back to the German Empire. It had been actively upheld by nuclear scientists after the Second World War, aiming to distance themselves and their work from the policies of National Socialism.[20] Nonetheless, this reading of rational science and its conclusions as unpolitical was not confined to Germany.[21] In both countries analysed here, the issue of nuclear power was so closely anchored to rationality from the start that even the most basic validity of statements made about the technology came to be seen as depending on their scientific and technical plausibility.[22] In the local debates, this plausibility was in practice taken to be guaranteed by an argument's presentation by experts from various fields, usually identified on the basis of their academic titles or their functions, which are, consequentially, highlighted in this chapter. A differentiation along scientific disciplines, in general, was of minor importance.[23] An exception to that rule were attempts to draw into question statements made by experts supporting an argument contradictory to one's own position: In the main, critics of nuclear energy accused the proponents of relying overly on physicists, even in questions concerning the biological effects of radiation, while the proponents accused experts criticising atomic power of being wanting in terms of physical or mathematical, and especially statistical, knowledge.[24] The elevated role of experts within the public debates can be gleaned from an interview with the German politician Franz Josef Strauß in October 1955, one day after his appointment as Federal Minister for Atomic Issues. Here, he stated his "conviction . . . that the exploitation of atomic energy . . . signifies an incision in human history comparable to the invention of fire". To prove the validity of his conviction, he added: "This view is shared by many experienced scientists."[25] In short, scientific or technical experts were charged with the ability to end any argument about nuclear energy, as they were held to simply be divulging facts which could not be argued with, which had to be accepted by all non-experts who, compared to them, were held to have an information deficit.[26] This, of course, was not so: In practice, most expert statements were rather interpretative than factual, as they themselves tended to point out repeatedly, especially in discourses with other experts.[27] The position ascribed to them within the public debates rather has to be understood as a discursive strategy, applied by nuclear energy's proponents to further their goals. That this was so, and that these goals were indeed political decisions and not ones dictated by rationality,[28] became visible when experts and rational arguments critical of nuclear power were not imbued with any comparable weight by nuclear power's advocates. In fact, an expert's affiliation with institutions formally involved in nuclear reactors' planning, licensing, construction, or oversight, the status, in short, of being an "official expert" – quite independent of even the greatest mere scientific status, up to and including a Nobel Prize – was a critical condition for their perspectives' consideration in the formal decision-making processes.[29] Rationality, to sum up, was utilised by the atom's proponents to immunise their political goals against critique.

For quite a long time, this strategy worked exceedingly well, clearly shaping the history of nuclear energy. The hurdle of rationality factually split the two societies regarding who could legitimately partake in the nuclear debates in that it allowed to largely exclude laymen and even experts critical of it from the core questions of nuclear energy, even though – or, more to the point, because – up to and in some cases even beyond the 1970s, the critics played along.[30] Time and again, they emphasised the rational worth of their arguments, concentrating on the debate's rational front. Nonetheless, their critique tended to be brushed aside – at least in fundamental questions while partially being accepted in rather peripheral ones. It was brushed aside, not because of shortcomings regarding its scientific-technical rationality, but because nuclear power's proponents, relying on their political dominance, were able to accept only those experts and arguments as rational and, therefore, valid, that concurred with their political goals. Others, including "critical experts", which increasingly appeared in the form of institutionalised groups during the 1970s,[31] could be cast out as lacking (official) expertise, even as irrational and, therefore, invalid. It was a clear demonstration of the primacy of the political over questions of rationality,[32] albeit the latter was publicly presented as decisive.[33]

Nuclear power's opponents were enabled to effectively counter this strategy once a combination of three factors came to play, which happened during the 1970s at most sites under consideration in this analysis.[34] *First*, the growing conscious realisation that scientific explanations were indeed, by and large, interpretations of facts, dependent on probabilities and perspectives, not unalterable facts themselves.[35] This insight spread via the transnational plane of debate in the wake of the inter-expert conflicts about the consequences of low-level radiation spurred on in the USA from 1969 onwards.[36] These went back to concerns about fallout in the wake of the USA's nuclear weapons tests in the early 1950s,[37] insights into the possibility of core meltdowns in light water reactors,[38] and considerations of a military "doomsday machine", rendering the planet uninhabitable, which in turn contributed to the conception of the planet as a system with "Limits to Growth".[39] In 1969, in any case, the physicist Ernest J. Sternglass published an article in *Esquire* blaming the rather high rate of child mortality in the USA's southern states on fallout from nuclear weapon testing in Nevada, downwind from which they lay. His claims gained credibility publicly, despite attempts by the

United States Atomic Energy Commission (AEC) to dispute them – up to the alleged predominant wind directions. Working for the latter, John Gofman and Arthur Tamplin found that Sternglass' arguments were, indeed, excessive. As a critic of the development of nuclear weapons, however, Tamplin pointed out that, even if the claims could not be proved, there was also no proof that there was no connection between radiation and children's deaths. Including reactors in the argument, Gofman and Tamplin publicly claimed that it could not be ruled out that their radioactive emissions during normal operations had harmful effects on somebody. This argument was quite successful; it was, in a way, a change of paradigms in the realm of nuclear sciences.[40] Soon, the consequences of uranium mining and nuclear waste had been attached to it, and it gained traction in the public debates.[41] Spreading via Sweden, the argument became a staple of the transnational debate about nuclear energy, acquiring relevance in many European national ones.[42] In Germany, it was introduced most forcefully by then-director of Friends of the Earth and former parliamentary expert Holger Strohm in "Friedlich in die Katastrophe".[43] Its impact can be gleaned from the fact that Tamplin even served as a speaker at the Brokdorf official hearing.[44] In Britain, the argument by Gofman and Tamplin had already been referred to in a critical local debate about plans to build a nuclear power plant at Stourport in 1970.[45] On the national level, the 1976 "Flowers Report" took up the dispute. While advising against foregoing nuclear fission, the report cautioned against the technology's unrestrained expansion because of the problems surrounding nuclear waste and those connected to a potential "plutonium economy". The report's presentation is held to have fuelled anti-nuclear positions in Britain.[46] The argument's spread was furthered by nuclear accidents like Three Mile Island (TMI) in 1979 and Chernobyl in 1986.[47] Both were widely considered as proof that (official) experts' relational statements about the likelihood of accidents were interpretations, and unreliable ones at that, as, in spite of being unlikely, large-scale accidents obviously did happen.[48] The argument's spread meant that the number of critics of nuclear power grew, helped them to recognise the discursive strategy employed by the advocates of atomic energy and provided them with arguments to use against nuclear power's proponents. The ability to make use of this realisation, *second*, hinged on the realisation of the political potential of the growing numbers of nuclear power's opponents. This awareness emerged in the context of a growing density of inter- and intra-societal communication once national debates had started and groups of critics and networks between them established, developing a rising willingness, besides new strategies, to actively engage in the political process.[49] This change was met somewhat more gracefully by the British than the German authorities.[50] An at least partial explanation for this difference can be found in the fact that, in Germany, the first generation born and raised in a democratic environment was confronting authorities who had grown up in less democratic surroundings.[51] *Third*, the resolution to make use of this political potential was furthered when issues arose that were outright rejected by local or regional critics of atomic energy,[52] motivating them to prioritise their political stance regarding the technology vis-à-vis considerations from the realm of rationality. Mostly, these were issues popularised by the spreading inter-expert conflicts, namely the consequences of low-level radiation connected with normal reactor operations, nuclear waste, and large-scale accidents. In sum, via their anchoring to the issue of nuclear power, the issue of rationality's role within the public debates not only shaped atomic energy's history in societies split over the technology, but apparently underwent dramatic changes itself.

On the (trans-)national plane, these changes tend to be seen as an endpoint of high modernity, a moment where technological utopias and widespread trust in experts collapsed. As the local perspective reveals, however, such a break did not quite occur within the conflict about nuclear energy; or only as long as the focus remained firmly on the highest level of the debate and the experts employed by atomic power's proponents.[53] Broadening the view and taking in lower strata of the debate and the expertise of nuclear power's opponents, it becomes clear that, instead, experts and the rational culture of knowledge they represented were at the same time devalued *and* upheld as gatekeepers for the validity of any contribution to the debate.

On the one hand, critics of nuclear energy copied the strategy of the technology's proponents, utilising experts within the debate according to their political goals and subordinating their acceptance under those goals. Thus, rationality as a criterion for decision-making seemed to be devalued. In a way, certainties gave way to probabilities, allowing the uncertainty principle to take hold in societal debates:[54] To many critics of nuclear power, for example, climate change, prominently debated in the context of the North Atlantic Treaty Organization in view of concepts of environmental warfare in 1969, after being taken up in many states of the Eastern Block,[55] was a distant bogey. To the atom's advocates, on the other hand, it was real enough to use it as an argument for their preferred technology, for example, in Germany's Bürgerdialog Kernenergie (Citizens' Dialogue on Nuclear Energy, cf. "Emotional Atoms").[56] Deciding which way to go meant juggling with unknowable quantities. To some extent, this devaluation of rationality went so far as to blur the border between fact and interpretation. Arguments were openly judged depending on the side that used them, up to the point where they were accepted when useful for one's own position but rejected when they were useful for the other side. This kind of tribalised relativisation spread to topics beyond nuclear energy, suggesting the twilight of a more rational era.[57]

On the other hand, this devaluation of rationality was only an apparent one. For one, rationality's superelevation had primarily been a discursive strategy in the first place. What is more, scientific-technical experts, their knowledge, and arguments, did remain utterly relevant. This pillar of high modernity even gained in importance and outreach, as experts were now routinely employed by proponents as well as critics of nuclear power, both ready and willing to fill the general public's assumed information deficit.[58]

Experts continued to be rather ritualistically treated as proof of the validity of arguments by both sides. Indeed, they were now utilised in that way more than ever, especially so as they came to be employed to back up nearly every perspective and interpretation brought up within the debates.[59] Beyond that, more and more people appropriated the expertise used, even addressing its epistemological limitations.[60] The experts' culture of knowledge, in other words, partially extended into that of the laypeople, evening out the distance between them and, at the same time, expressing and furthering higher educational standards (which, it should be remarked, was partially necessitated by the rationalisation processes driven by the expansion of nuclear energy).[61] Rationality, in other words, went from being brought into play by a few experts to a widespread commodity in the supposedly emerging "knowledge society".[62]

When seen in combination, these two opposing trends show that the debates about nuclear energy functioned as a motor *and* a roadblock for the role played by rationality within societal debates. They spurred on the bloom of this pillar of high modernity during the 1950s and 1960s and contributed to its seeming collapse in the 1970s and 1980s, as well as its simultaneous, ultimately utilitarian, continuity. Scientisation was writ large – albeit in a cynical way[63] – with consequences permeating the time and topic under consideration here.

While this process clearly influenced and was influenced by the (trans-) national debates about nuclear energy, its sometimes subtle shifts can best be analysed on the local level, as the examples of the Emsland nuclear power plant in Lingen and the ones at Hinkley Point show. These two locales are very different in that they represent one of the longest and one of the shortest debates investigated,[64] but they are also similar in that they contain the two countries' last reactor projects within the period under investigation here; successful in Lingen, (initially) unsuccessful in Hinkley Point. This combination of similarities and differences, especially regarding the inclined temporal plane, helps to illustrate the significance of the transnational level for the aforementioned change, and to highlight different aspects of it. The shorter, more compressed debate in Lingen lends itself to a close analysis of the dynamics of change, and is exemplary of the tactics applied by nuclear power's opponents in their attempt to handle the hurdle of rationality. As the Lingen debates only took off at the end of the 1970s, however, the longer lines in the development of those approaches are less visible here. These are prominent in the debates about Hinkley Point. Their long duration renders it possible to emphasise the sometimes slow and subtle shifts and detailed positions within the dispute and to highlight the significance of the debate's national and transnational planes for the local and regional ones. Taken together, the two rather dissimilar sub-chapters help to demonstrate the importance of the three central reasons behind the change in the debate: the emergence of "critical" counter-expertise, the realisation of the political potential of nuclear energy's opponents, and the rise of aspects of the technology rejected outright by those critics. The two case studies also help to illustrate the paradoxical character of the change the debate underwent. Thus, Lingen and Hinkley Point offer an insight into the reciprocal significance of the public debates on nuclear energy and the shifts in the public role of science and rationality associated with the upheavals of the 1970s. I begin with the British example, to provide a foundation for the larger development of changes and continuities playing out around that hinge decade of the twentieth century's second half.

Hinkley Point – The Slow Spread of a Discursive Strategy

Hinkley Point in Somerset was one of the first sites chosen for the construction of a commercial nuclear power plant in Britain, with the public debate starting in 1956. It was also the site of the one nuclear power reactor planned, but put on hold because of a moratorium taking effect in Britain in 1989. Later, it came to be the site of the first reactor to be built in the country after the turn of the century. The site itself was a rural place on the Bristol Channel's southern coast, opposite Cardiff, within a few hundred metres' distance from the hamlets of Knighton, Burton, Shurton, Wick and Stolford, in Stogursey Parish, about 15 kilometres from Bridgwater and 25 from Taunton, the seat of Somerset County Council. The most important voices within Hinkley Point's public debate were the electricity industry, namely the Central Electricity Generating Board (CEGB) and its forerunner, the Central Electricity Authority (CEA), and the area's local authorities. These comprised Somerset County Council and its officers, Stogursey Parish Council and Williton Rural District Council, to which Stogursey Parish belonged (it was reformed into West Somerset District Council in 1974), and the Bridgwater Rural District (merged into Sedgemoor District in 1974), which almost surrounded the site. Also, groups like the local branches of the National Farmers' Union (NFU), and individual participants played a vocal role in the debates, as did the *Bridgwater Mercury*, widely read in the locale, and quite a few individuals living in and around the area (cf. "Nuclear Debates").

To analyse the effects of expert-backed rationality on the atomic debates and the, at times subtle, changes its role underwent within the public discussions around Hinkley Point, this sub-chapter's three sections highlight three phases of the debate. First, pro-nuclear experts and their arguments were installed and widely accepted as decisive at the core of the debate by the pro-nuclear political majority. They were framed as the only ones possessing the necessary experience for rational decisions on the topic.[65] Opposing voices – mostly laymen, even if they were, at times, able to refer to experts – were largely taken to be unable to meaningfully contribute to basic discussions on nuclear energy. The technology's proponents were thus able to accomplish their plans to a large degree. Second, from the 1970s onwards, the local and regional critics of nuclear power received a double boost from the (trans-) national plane of the debate, in the form of spreading critical expertise and growing political potential, which was used in the early 1980s, when nuclear waste turned out to be an issue unacceptable for many on political grounds. The third section focuses on the debates about a third local power station, Hinkley Point C. It serves as a textbook example of the seemingly paradoxical parallel of the clear political utilisation of and ostensible reverence for experts, and helps to demonstrate the enduring dominance of the political dimension over the rational or scientific one within the debate.

The Seeming Dominance of Experts, Politically Backed Up

The importance ascribed to experts and arguments from the realm of rationality in the first months of the public debates around Hinkley Point can be primarily demonstrated in three core areas: economic efficiency, site selection, and nuclear safety. Of course, these were not the only topics that were discussed. Other questions, however, such as the necessity of industry in rural areas, the local labour market, infrastructure, or the consequences for landscape and tourism – questions in which expertise was widespread and at least partly founded upon local people and local knowledge[66] – were characterised less by references to experts and their arguments and more by negotiation processes between different opinions and perspectives (cf. "Nuclear Debates"). Regarding the issues of economic efficiency, site selection and nuclear safety, however, three things can be shown. First, that experts and their judgements based upon a rational culture of knowledge were generally regarded as decisive for the acceptance of arguments as valid foundations for political decisions; second, that the crucial element regarding the decisions in question was the political course of the government; third, that the latter was reflected in the selection of the experts and perspectives ultimately considered.

The question of economic efficiency was the one that was least discussed in public initially. After all, with the 1955 "Programme of Nuclear Power", the decision in favour of nuclear energy had already been taken.[67] Locally, it hardly played a role. One instance where it was mentioned by the *Bridgwater Mercury*, however, is a perfect example of the extensive deference to the rationale of experts in this respect. When the topic was brought up, the paper simply stated that it "can best be left to the experts to explain".[68] No attempt was made to question the arguments presented, which were deemed rather complex. Therefore, the underlying political decision was not publicly discussed at the time (I return to this aspect later in this chapter).

The issue of site selection was debated much more intensively, especially considering its links to questions of the specific design of the buildings, their exact placement on site, its greening, or even the fundamental conflict over the transformation of rural areas and the need for electricity as such. At heart, the question was why Hinkley Point, of all places, was chosen for a nuclear power plant. After all, as a part of Bridgwater Bay, the area had been categorised as "agricultural" in the county development plan just four years earlier.[69] To now accept the building of a power plant here, the county authorities demanded a solid justification.[70] To provide it, the CEA put forward "four principal technical requirements for power station sites for the economic exploitation of nuclear energy", or, in other words, four criteria to rank potential power plant sites in an ostensibly rational, non-political and therefore hardly disputable way. They were:

> 1. Proximity to areas of demand for electricity not already satisfied by local generation and where generation from coal would be costly, owing to distance from coal fields. 2. Foundation conditions suitable for supporting the unusually heavy loads of the reactors and their shielding. 3. Access to large volumes of cold water for condenser cooling. 4. Low population density in accordance with the policy of siting these stations at a distance from centres of population until more experience of safe operation has been obtained.[71]

According to these criteria, Hinkley Point was suitable "to an exceptional degree", as the CEA's experts never tired of emphasising.[72] The way these arguments were taken up in the public debates clearly shows the dominance of the official experts and their perspectives – and the ultimately political character of the decisions on nuclear energy. The former became visible in attempts to bring other perspectives to the debate, undertaken not the least by one of the most important early critics of the project, Colin Mackenzie of Stolford. To him, the site chosen was "valuable agricultural land".[73] While his local expertise on the topic was backed by other critics,[74] the argument went under quite quickly, as it could be second-guessed locally, in the parish council.[75] It could not, in other words, axiomatically be established as an incontestable argument on par with those brought by the CEA and its experts.[76] The ultimately political character of the decision and the difference in the valuation of the local and national perspectives were rendered visible in the county council's attitude to its own planning for Hinkley Point. It withdrew to the position that its own considerations had to take a back seat to the arguments put forward by the CEA's experts, representing the government's political will, as the council lacked the expertise to judge the relative weight of the arguments presented for and against the choice of site. As its clerk Ernest Stanley Rickards put it: "The County Council feel that although they have been told of all the points in favour of the site there is a national interest involved and that they have not the knowledge of the detailed evidence to decide how that national interest should override planning consideration. 'In their view it is for the Ministry to make that decision and not for the County Council to advise them one way or another.'"[77] Unable to judge the arguments made, the council accepted the pro-nuclear ones and the national political considerations behind them as being decisive.

That the government's political course dominated the decisions and that its experts, arguments, and perspectives were chosen accordingly – in other words, that the debate was not a politically "neutral" competition for a scientifically quantifiable optimum to be found by rational arguments alone, even if they and their presentation by experts were obviously necessary for a position's legitimation – became even more clear when it came to the issue of nuclear safety. This was a topic of paramount importance for the local debates (cf. "Nuclear Futures"). Nonetheless, it was rapidly delegated to the experts officially tasked with implementing the "Programme of Nuclear Power", while doubts and potential problems were largely sidelined. One could say that, because of the decision being taken in Westminster, those living in the area had little choice except to trust in the principle of hope to frame their lack of influence as acceptable: When questions about the plant's safety were raised at a conference between local authorities and members of the CEA on 11 December 1956, the latter stated that the reactor would be "inherently safe". It could not explode and "shuts down automatically" in the case of malfunctions. The normal radioactive emissions would be limited and quite harmless, the estuary would not be negatively affected and the area's amenities would be preserved, no dangers would arise from transports of fuel elements to or from the station and it would not be used militarily. The participants seemed to rely on the experts' statements,[78] even though, in the following days, some councillors wrote to "seek assurances" on those points from the CEA.[79] The county council later on was even satisfied that "it had been possible to announce publicly that the Authority had stated categorically that the degree of radio activity in the air and in the sea emanating from the station would be minute and unharmful".[80] The forceful statement by official experts, in other words, was accepted and presented as proof for the installation's extensive – albeit not absolute – safety. The *Mercury*, accordingly, gave quite a lot of room to experts explaining it. Quoting the CEA's station engineer Donald Clark, who was "mainly responsible for the siting and building of nuclear power stations", as well as project engineer W. H. C. Pilling after a press conference in March 1957, the paper was confident that, based upon their statements, "safeguards are being taken in every conceivable form". While there were "certain hazards, . . . such had been the progress in handling it [nuclear power, CG] that there was only a million to one chance of anything ever going wrong. Indeed, so stringent had been the precautions taken that never once in the 10 years that the Atomic Energy Commission has been operating has there ever been a fatal accident."[81]

Quite in line with the reverence for experts expressed by nuclear energy's proponents, even open doubts regarding nuclear safety, raised by individual critics, tended to be either based upon experts' statements or aiming for the inclusion of further experts' opinions. This indicated that the latter were quite universally regarded as necessary guarantors of the rationality and validity of arguments and positions. K. Margaret Page, for example, concerned about "the long term effects of not one, but several power stations in one area", wondered "[w]ho has estimated" them and complained that information stemmed from the CEA, not from "independent scientists",[82] thereby opening up a line of argument that was to become central for the conflict about rationality in the dispute over nuclear energy. Others went further. A "Bridgwater District Resident" added to these questions a quote from an article in New Zealand's *The Evening Post* about "a Dundee doctor", Harry R. Lillie, who claimed that the current statements about long-term effects of radioactivity on humans were based on short-term experiments with animals. He concluded: "None of the atomic experts know anything at all about the level of radio-activity which is harmful to human and to other life. All their statements are conjectures and about as accurate as the weather forecasts . . . I feel that nothing is known about the effect a major accident might have on, say, Bridgwater and the surrounding districts."[83] A farther-reaching critique regarding the role of experts was rarely put forward, except for Newton A. Rowe. The outspoken critic of the reactor project highlighted the possible consequences of radioactive emissions. In his view, the local population was to be used as test animals by official experts whom he quite frankly mistrusted: "We have got to a point where, in the name of officialdom and progress, any mob who care to sound off can do practically anything they like. We would be better off with less power than this sort of thing."[84]

Although doubts were expressed with references to expert statements, they did not prevail. The decisive factor was not an argument's epistemological validity or scientific quality, but its political support, which was reflected in which experts and interpretations were accepted by the Minister of Power. In the case of the Hinkley Point nuclear power plant, these were the arguments and expert opinions put forward by the CEA, which was presented at the public inquiry in Bridgwater, on 14 and 15 May 1957, as the most experienced body regarding nuclear power plants.[85] It, in turn, leaned upon the United Kingdom Atomic Energy Authority (AEA), which was referenced as the only reliable group of experts for nuclear phenomena.[86] Explicitly, S. B. R. Cooke from the CEA hedged any doubts regarding the CEA's expertise by referring to the help of "the Atomic Energy Authority, a very experienced body in these matters".[87] The AEA's chief safety officer Frank Reginald Farmer emphasised that it "was the only organisation with the necessary experience to give technical advice on nuclear plant [sic, CG], and they have made arrangements with the Central Electricity Authority to make that experience available to them".[88] Although critics like Geoffrey Knowles and Dorothy Irving-Bell questioned the safety assurances, not least by pointing to accidents which had already happened in nuclear installations and the biological consequences of radioactive emissions,[89] the Authority's official experts prevailed. Their rejection of these arguments was, according to the inspectors Howard Walker Grimmitt and H. R. Wardill, outstandingly effective: "Mr. Farmer did allay the objectors [sic, CG] fears regarding the safety aspect, to such an extent, that some suggested that the policy of placing these stations in isolated places should be relaxed."[90]

This was overly positivistic. While, indeed, many doubts and much critique were mitigated or resolved by compromises (cf. "Nuclear Debates"), others lingered on. This holds true especially regarding the area of nuclear safety, where conflicting expert opinions and even more fundamental critique regarding the basic expertise behind nuclear power had been pointed out during the debate, without substantially being acknowledged during the official decision-making process. During the following two decades, these doubts were renewed and grew whenever incidents around nuclear technology seemed to falsify the statements and forecasts made by the CEA's experts during the early debates. This happened most prominently on nuclear safety's sub-topics of radioactive emissions and nuclear waste. For the time being, however, the corresponding critique focused the rational level of scientific and technical arguments, concentrated on expertise and experts, attempted to draw into question those arguing for nuclear power, or pointed to others who were sceptical about it. The political basis of the decisions in question, and, thereby, the political character of the utilisation of experts and rational arguments meanwhile was rarely commented upon – or even put to use.

One exceptional example for such a utilisation was the area's NFU in its reaction to the Windscale Fire in October 1957 (cf. "Nuclear Debates"). The farmers saw themselves and their livelihood endangered if something akin to this accident should happen at Hinkley Point. In view of the potential losses through the destruction of milk or a decreased marketability of animals, the NFU wrote to the CEA, asking for reassurances. In its calculation, an assurance from Westminster would improve the farmers' position for demanding compensation in case of need.[91] Their request was fulfilled a few months later, when the group received the assurance "that there was no possibility of a similar mishap to the one at Windscale" because of the differences in the reactors – albeit learning that "[t]his cannot of course be taken to mean that in no conceivable circumstances could an accident of any sort ever happen at Hinkley".[92] With this, the farmers had got what they wanted: a reassurance that handed them an argument for negotiations should the worst come to the worst. They, in other words, had utilised the official experts in a way not dissimilar from the way they were employed by the proponents of nuclear energy – although in a field rather on the sidelines of energy policy.

The critics of nuclear power who had focused on the safety aspect reacted to the Windscale Fire in a less utilitarian way. Although they were not blind to the fact that rational arguments and the experts presenting them were, in the end, simply fulfilling a supporting role in a politically dominated debate, they kept their arguments mostly to its rational front at that point in time, attempting to highlight the errors and fallibilities of the official experts or pointing to the plausibility of arguments and experts criticising nuclear energy. One reader of the *Mercury* highlighted the epistemic limits of the available expertise by emphasising that all reassurances by experts about the local reactors would be worthless, as similar assurances would have been given, in the weeks before the Windscale Fire, to people in that area – to no avail.[93] Another reader, who spoke up during the following months, simply called pro-nuclear positions "rubbish" and "[p]ropaganda", demanding "that we should be suspicious of every utterance that the scientists make today in respect to nuclear discoveries".[94] Rowe, however, who can be seen as an indicator for the "mainstream" of critics because of his pronounced activity and rather prominent position among them, did not reject expertise as such. On the contrary, he pointed to "the report of the Fleck Committee on the Windscale mishap, recently published, which has special reference to the power stations now building, and gives serious warnings". In other words, he took rational arguments and experts very seriously, just not necessarily the ones prioritised by the atom's proponents, retrospectively wishing there had been "expert evidence" available at the public inquiry, where, without it, he had been unable to counter assurances made by official experts. "One can see plainly now what a killing could have been made at the Hinkley inquiry, with the aid of expert evidence. It is true that the inquiry was conducted against a background in the countryside where everything was over bar the shouting, which was the reverse of democratic; the official reason being that time might be saved."[95] This statement exemplifies two things: It shows that Rowe had recognised that the political decision about nuclear power had had the primacy, with the experts' statements serving for little more than public justification. It also shows that, nonetheless, his critique focused on that very front of expertise, accepting the conflict's framing as a rational one to be settled by scientific, technical arguments. Some months later, in view of a government decision to modify Hinkley Point to enable it to produce plutonium for military purposes, Rowe once more focused on this rational plane, voicing his mistrust in the official experts, the "sensible, practical men with no nonsense about them", who could err, as, to him, a changed situation regarding coal supplies and the development of the prices of nuclear energy showed.[96]

This focus on the realm of rationality was shared by most proponents of nuclear power. At least, it was displayed by default in the *Mercury*, which sought out the CEA's experts for reassurances. Talking to Hinkley Point's resident engineer, D. M. Longman, the newspaper learned that "[t]here is no possible danger of any mishap occurring at Hinkley Point similar to that which occurred at Windscale". The rationale behind his conviction was that the reactor fulfilled a different purpose and was not cooled by outside air but by CO_2. This would be kept inside the reactor. But even if it did escape, "there would be no harmful effects since its addition to the normal CO_2 content in the atmosphere would still result in a negligible percentage of that element in the air."[97] The editors seemed satisfied with this statement, as they continued the practice of turning to official experts when doubts regarding the local power plant's safety recurred some years later.[98] In any case, it was true for a reader who emphasised that "we can put our faith in our British scientists".[99] In other words, experts still were imbued with the power to end debates.

After the Windscale Fire, the public debate around Hinkley Point quietened down. Nonetheless, time and again it flared up, with the topic of radioactive emissions being a rather steady catalyst. This helps to demonstrate the continuity of the equilibrium that had been reached in the late 1950s, with experts and rational arguments widely taken to be the decisive plane of a dispute, the political character of which was not unknown but hardly focused on by the opponents of nuclear power. In that vein, in early 1963, with the power station slowly nearing completion, the county council's health committee agreed to the "proposal to discharge liquid and gaseous radioactive waste from Hinkley Point nuclear power station into the Bristol Channel" made by the CEGB because of the arguments presented by its experts. Specifically, the "senior radiochemical inspector" of the responsible Ministry of Housing and Local Government had convinced them "that there would be no hazard to public health, or to fish, animals or crops; nor would there be any danger of contamination of the foreshore of the Bristol Channel".[100] Another example can be found in the reactions of multiple councils to radioactive emissions during the early 1970s. Regarding the discharge of radioactive liquids into the Bristol Channel during tests for the Hinkley Point B power station, Williton Rural District Council vowed to "take it [the discharge, CG] seriously" – "[u]nless we get a full assurance that there will be no danger". The council relied, in other words, on the estimates of the CEGB's experts to render acceptable an action they had doubts about. The same held true for Stogursey Parish Council, whose chairman, Norman Jones, explained that it planned to "ask the Central Electricity Generating Board to give an assurance that everything will be absolutely safe".[101] Similarly, Bridgwater Rural District Council accepted an increase in tritium emissions in 1973, based upon the expertise provided by the Secretary of State for the Environment and the Ministry of Agriculture, Fisheries and Food, although councillors did not understand the matter. In the end, they refrained from comment, as long as the council would receive biannual reports showing that the newly set limits were kept.[102] Two years later, West Somerset District Council's concerns about waste treatment dissolved after a talk with the station's personnel.[103] This kind of reliance on experts and their rational explanations, indeed, was the rule during most of the decade and it could be found far beyond the region's governing politicians. In 1977, for example, the "Bridgwater and District Business and Professional Women's Club" visited Hinkley Point and, afterwards, was impressed by the "size and efficiency" of the plant, the control room of which reminded the members of "Star Trek", with the difference that "the cast were calm, British scientists, completely in control, and they weren't wearing plastic ears or zany costumes!"[104]

Parallel to this deference to official experts and their culture of knowledge, critics continued to exist and to speak up. In some cases, they even based their critique on expertise contrary to that which the authorities' decisions were based upon, sometimes challenging the latter. Already in 1960, one reader of the *Mercury* had criticised that nuclear power was employed at all, as it was – he was convinced based upon, not least, an article in *The New Scientist* – uneconomical, and produced waste products like plutonium.[105] Three years later, H. J. Lederman, who earlier on had argued against nuclear weapons testing because of the consequences of radioactive fallout as described by Nobel laureate chemist Linus Pauling,[106] complained about "the apparent indifference on the part of the authorities in connection with the disposal of radio-active waste from Hinkley Point". To him, assurances of safety had to be understood as limited by the extent of the experts' knowledge (which they themselves freely conceded):

> The plain fact is that we only know by experience. We only know now that air pollution is a risk to health, after the damage has been done. Statistics are available for those who want the truth . . . I wonder how many people though, realise that the maximum permissible amount of medical X-ray exposure is only 15 per cent today of what was considered absolutely safe 10 years or so ago . . . Must we wait for irreparable damage to be done and suffering to be caused, simply because money cannot be found to protect us from the follies of our rulers?[107]

Lederman developed, in other words, a methodologically based critique regarding the experts' limits of knowledge and, therefore, their ability to predict the power plant's consequences, drawing into question political decisions referencing these kinds of calculations. It was a concern he repeated regarding the actual emissions once the power plant had taken up its operations.[108]

This kind of critique, however, focused on technical and scientific arguments and the experts presenting them and was quite routinely rebuked or even ignored by the proponents of nuclear power whenever it was expressed, as they could refer to official experts who presented different interpretations of nuclear power and its consequences. This situation began to change once the critics, motivated by an issue they absolutely rejected, put into practice the fact that, as necessary as experts and rationality obviously were for the acceptance of arguments in the debates about nuclear power, it was a political debate, dominated by political decisions taken by governments elected by a democratic majority. In other words, the critics managed to demonstrate that they not only had scientific expertise to point to, with interpretations of nuclear power differing from the ones presented by the technology's backers, but that they also had a significant number of the voting population supporting their opposition. This combination of factors occurred when nuclear waste, namely used fuel elements, rose as a topic of regional importance.

The Realisation of the Debate's Political Foundation

During the 1970s, critique of nuclear power took off (trans-)nationally. In the British case, public critique was fuelled by considerations to import American pressurised water reactors (PWR),[109] combined with the consequences of an incident at Windscale in 1973. Here, a radioactive leak led to the shutdown of the oxide fuel Head-End Plant, and to plans to build a large Thermal Oxide Reprocessing Plant (THORP), which would reprocess British, but also foreign nuclear waste.[110] In view of these occurrences, the standing Royal Commission on Environmental Pollution decided, in early 1974, to investigate the subject of nuclear energy,[111] asking members of the public to submit observations.[112] While its investigations were under way, the British branch of Friends of the Earth began to criticise the ideas behind THORP in early 1975. Once these were more widely publicised later that year – after an article in the *Daily Mirror* on 21 October, headlined "Plan to Make Britain World's Nuclear Dustbin" – a heated public debate ensued, including several questions asked in Parliament.[113] The critics' arguments were substantiated by the "Flowers Report" in 1976. In this, its famous sixth report, named after its chairman, Sir Brian Flowers, who had been a part of the British-Canadian nuclear energy research team during the Second World War,[114] the royal commission advised against giving up using nuclear fission as a source of power, because its risks did not seem to be out of scale. On the other hand, it cautioned against a large expansion of the existing capacities, counselling research into alternative power sources and measures to conserve energy instead. Behind this stance stood two main lines of argument: One was concerned with the security aspects of plutonium and the implicated societal consequences of a "plutonium economy", the other was the open question of the treatment of nuclear waste.[115] Both of these aspects fitted well into the ongoing critique of THORP, which resulted in a public inquiry far beyond site-specific questions, investigating the national nuclear waste policy.[116] Beyond that, both remained important pillars of the nuclear critique in Britain. They were taken up in local and regional debates, which continued to escalate after the Windscale inquiry, not least as the report resulting from it was seen, as Walter Cram Patterson put it, as "a heavy-handed nuclear apologia, so clumsily one-sided as to provoke unease even among many Britons previously unmoved by the issue".[117] This national development facilitated the formation and connection of local groups of critics, as a large number of people became actively involved in the nuclear debates.[118] In a democracy, this meant that critics could muster more political self-efficacy than had been the case up to then.

In the debates around Hinkley Point, this change took effect in combination with another result of the "Flowers Report": the search for an underground storage facility for nuclear waste. Up to the end of the 1970s, the issue of used fuel elements had largely been put aside in the Hinkley Point area by pointing out that they were not the power plant's business, but that of British Nuclear Fuels, who were processing them at Windscale.[119] Framed in this way, despite the dual use implications, the highly radioactive materials and their treatment were largely accepted, although some critical voices existed. Margaret Rees, for example, chairwoman of Sedgemoor's environmental health committee, pointed out in 1975 that accidents, including the "lethally-toxic plutonium" could happen independently from experts' expectations, as Windscale had shown. Therefore, regularly transporting used fuel elements through Bridgwater, to her, was not an acceptable practice. Most people, however, kept quiet on the topic.[120] It was no different in 1978, when

Lederman once more wrote to the *Mercury* regarding a recent incident in Aldermaston, where working women had been contaminated with plutonium. Pointing to the arguments from 15 years earlier, he restated that every level of radiation was dangerous, that accidents could never be excluded, and that "[t]he opinions of nuclear physicists differ widely and cannot be trusted", which meant transports of plutonium through Bridgwater and on sea had to be stopped.[121]

Around the end of the decade, the critics of nuclear power gained buoyancy in the debate when they were able to form interconnected groups and began prioritising the debate's political character over the role played in it by rational arguments and experts concerning the issue of nuclear waste storage – which several people flat out rejected, unwilling to argue. Examples of early groups around Hinkley Point and active in this debate are the Somerset and Dorset Nuclear Reactor Vigilantes and the Ecology Party's Bridgwater chapter. Both were critical of a potential third power station at Hinkley Point (see later), demanding that such a plan, at least, be publicly "discussed",[122] thereby forming a hinge between the national and local level of the debates with a view to prioritising a political decision over experts' arguments. That is not to say that they would have dismissed rationality: Explicitly, the Ecology Party's Bridgwater secretary described the group as "reasonable people" who "try to make up our minds on the evidence".[123]

A slight change became clear when, in October 1979, an umbrella organisation, spanning both groups and Friends of the Earth Somerset and the Taunton Anti-Nuclear Group, was formed with the Somerset Anti-Nuclear Alliance (SANA).[124] SANA was critical of nuclear energy as such, demanding a stop to new reactor constructions until the problem of nuclear waste had been solved,[125] and explicitly rejected a third power plant at Hinkley Point and "the dumping of nuclear waste" in Somerset.[126] While it also presented rational arguments,[127] SANA mainly sought to act on a political plane, with "mass petitions and the lobbying of MPs [Members of Parliament, CG] and local councils".[128] Soon, it protested against nuclear waste transports through Bridgwater, sea dumping, and deep burials in Sedgemoor. It also challenged industry claims that these practices were harmless.[129] For example, in July, a member of Hinkley Point's staff described the additional level of radioactivity inserted into the naturally active water as "totally insignificant". In response, the group's Danielle Grunberg insisted that it was "dangerous waste",[130] and that it was "totally immoral to dump this waste at sea".[131] In other words, she not only challenged the experts' assessment of the situation but explicitly included a wider societal dimension by pointing to the question of the practice's moral acceptability. To her, the treatment of nuclear waste clearly was a political question, to be decided not by experts in terms of technical feasibility but by the people and their moral compass. Therefore, when the plan to drill about 30 boreholes in the Sedgemoor area to test the geological feasibility of an underground storage site was publicly debated in late summer 1980,[132] one of her first points of critique was the way the information about this had been spread – or, rather, had not been spread. To her, the authorities wanted to prevent people's participation. To counter that, she planned an independent meeting.[133] From here on, while rational arguments were not left out, the main dispute was over the fundamental question of whether the handling of nuclear waste should be decided by politically "neutral" criteria established by experts and based on their scientific and technical culture of knowledge, or whether it needed to be negotiated politically in the first place.

In contrast, the proponents of nuclear energy mostly argued that the question was not a political one, but one the "correct" answer to which could only be found via scientific experiments and methods, applied by experts in a rational way. As Thomas Jeremy King, Bridgwater's conservative Member of Parliament, who was to briefly become Secretary of State for the Environment in 1983,[134] pointed out, they tended to hint at the royal commission, which had called for more research into the treatment of nuclear waste.[135] Among the possible options, the plan to store nuclear waste underground was, to Mike A. Oldfield, Hinkley Point's health physicist,[136] a positive example for all industries of how to treat the environment. Opponents of test drilling, to him, were opposing "research" out of "very little understanding of general science and a superficial knowledge of this particular problem". He claimed they had "got hold of a few half-truths and cannot, or do not wish to analyse the problem in a proper manner. For example, do they realise that a coal station causes a greater radiation exposure to the public than a nuclear one?" He elaborated further on this point, referencing a paper by P. T. Manning, and ending with the call: "Please 'environmentalists' open your minds and be objective."[137] At later opportunities, he called for "some sanity in this debate!"[138] R. W. Halder, who protested against calling plutonium a waste product and demanded it be treated as a resource instead, also called for critics of nuclear power to "stick to the facts and stop trying to scare the good people of Somerset with works of fiction based on fundamental misconceptions".[139] To him, the research-based view of the AEA that underground storage facilities were the safest method available to store nuclear waste was a fact.[140] Also, a number of Sedgemoor's district councillors backed the collection of information,[141] and the *Mercury* continued to highlight opinions of experts from the pro-nuclear side, for example, "an Atomic Authority spokesman", who explained that, as only small boreholes would be necessary, there would not be "a wholesale despoliation of the area",[142] or Harwell's Dr John Lewis, who argued that nuclear power was less deadly than coal-burning, and that research would be able to establish the best way to treat the technology's waste products.[143]

On the other hand, the critics of the plan to look for a geologically suitable site for the storage of nuclear waste in the neighbourhood of Hinkley Point certainly did not forego highlighting expertise and rational arguments. They questioned, for example, the pro-nuclear arguments presented by

Oldfield,[144] Halder and others, with one critic, Robert H. Jervis, explicitly calling claims as to the safety of underground storage facilities "totally irrational and unscientific", as any kind of tests or proof were found wanting.[145] They also emphasised the existence of other options for nuclear waste storage,[146] pointed out the epistemological inability to guarantee the long-term safety of underground storage sites,[147] specific problems discussed by various scientists,[148] or argued for overground storage, where future scientific insights could still be applied to the stored waste,[149] and for which others had already devised arguments, for example, Patterson in "Nuclear Power".[150] Most importantly, however, they argued that this was not a question to be decided in the realm of rationality, but politically.[151] In the words of the Ecology Party's Jonathon Porritt, nuclear power was a technology that "helped shape society" – and that, as a first, "[t]he needs and values of society" should be taken into account,[152] instead of questions of scientific or technical feasibility. In a similar vein, P. A. White emphasised that the protesters' aim was not "obstructing research" but to prevent "the proposed burial of toxic radioactive waste" – a position that required no "understanding of general science". To them, the drilling was not "research", but "a cut and dried plan", and all attempts at shielding the nuclear material did not matter, as they did not take away its radioactivity. On a side note, White also questioned Oldfield's statements regarding the radioactive emissions of coal-burning, not least regarding the volume and composition of the emissions.[153] Even some local councillors like Mick Briscombe argued that, in the end, the waste would go to a place where research had not been opposed, so that allowing research was, in the end, a preliminary decision in a political sense, and not a scientific question.[154] Along this line, after a poll in the city, the Bridgwater Anti-Nuclear Group (BANG) concluded that the majority rejected an underground storage facility in the area, so that any collection of information regarding its suitability was unnecessary.[155]

Against this background, SANA and Grunberg organised meetings, protests, and petitions,[156] the People Against Nuclear Dumping in Sedgemoor (PANDIS) was formed,[157] a group of about 40 mothers opposed to the mere idea of underground storage in populated areas,[158] and, in December 1980, Sedgemoor District Council, in light of the public upheaval, positioned itself against the drilling, calling upon the county council to do the same so as to force a public inquiry.[159] Indeed, following protests by Sedgemoor District Council, ten parish councils, groups like Friends of the Earth and 126 individual opponents as well as 2,500 protest signatures, the county council's planning and transportation committee, in early 1981, referred the question of drilling back to the Department of the Environment, demanding a special inquiry.[160] This would, according to two Labour councillors, Angela Sneddon and Kathleen Phelps, who were opposed to the drilling, facilitate a broad debate.[161] To BANG, the decision was "a great victory".[162] On the topic of an underground storage site for nuclear waste, the primary political character of the nuclear debates had thus been clearly revealed.[163]

This did not mean, however, that the nationally governing Conservatives, in general backing nuclear power, accepted the shift. In what Grunberg called "a complete disregard for what people think" – the wording being another hint at the predominance given to the political level of contesting opinions rather than to oppositional scientific arguments presented by differing groups of experts – King opposed a fundamental debate in the context of a "special inquiry".[164] To him, the political decision to research underground storage had already been taken on the national level, when the royal commission had considered waste storage options.[165] Indeed, the requested special inquiry was not set up. Consequently, the county council did not back the drilling plans, which meant that a "local inquiry" would have to be held, focused on local environmental and technical questions.[166]

What looked like a victory for those advocates of nuclear energy counting on the debate-ending capability of rational arguments presented by official experts was, in fact, rather short-lived. Less than three months later, the government pondered the idea of above-ground storage for up to one century, pointing out that experts had named it the best option in the second annual report of the radioactive waste management advisory committee. To BANG, this was a decision brought about, at least in part, by the widespread critique voiced against the concept.[167] In other words, a political victory was reached by moving the topic of nuclear waste treatment out of the seemingly unpolitical sphere of rationality. At the year's end, even King buried the concept of underground storage,[168] which, in the eyes of BANG, positively proved that the topic had been a political, not a scientific, question from the very beginning. His change of opinion came without any drilling having taken place – without, that is, any new scientific data having been collected.[169]

The conflict did not end here. It comprised aspects like sea dumping; emissions during normal operations;[170] the safety of waste transports, a stop for which was demanded not least by the Campaign Against Nuclear Transport (CANT);[171] and renewed worries about final repositories in the area in 1983 and 1987.[172] However, it had clearly moved into the political sphere. Therefore, in 1987, when a map was presented by the Nuclear Industry Radioactive Waste Executive (NIREX), showing potential places for a final storage facility, which included sites in Somerset, its "head of information" quickly explained that the places named should not imply planned facilities, only locations where they were technically possible, ensuring that, before a decision was to be taken, there would be "full consultations".[173] Experts and their culture of knowledge were, in other words, not excluded from the debate. They continued to play an important role, pointing out and explaining facts. The interpretations of the latter, though, and the consequences derived from them were now openly treated as a political debate, wherein experts and their arguments were mostly limited to a legitimising function.

This change did not remain confined to the topic of nuclear waste storage. Rather, it permeated the whole range of nuclear issues. In the power plant's Local Liaison Committee (cf. "Nuclear Debates"), for example, near the end of 1982, county councillor Phelps demanded that emission reports be changed. Up to then, identical formulations had been used over years to interpret the meaning of the measurements, although the concrete numbers varied. To her, this meant that "the true facts" were hidden.[174] While she did not claim that the numbers presented were false, she demanded that the experts compiling the reports explain them in a way that enabled the laymen councillors to explain them, in turn, to the public.[175] She demanded, in other words, that the interpretation of the information was moved outside the black box of expertise into the public sphere. This demand was backed by county scientific adviser Wilf Cassidy, who, like Phelps, was convinced that it was wrong to expect councillors to "take everything on trust" because they "aren't scientists" and therefore "can't understand". Rather, experts had to explain their interpretation of the facts to make the latter transparent – and, thereby, criticisable – to the public.[176]

Some critics of nuclear power even attempted to drive farther this "demotion" of experts within the debate. In 1987, for instance, Bridgwater Labour councillor Bob Brookes claimed to know about faulty welds in the second Hinkley Point power station, in the construction of which he had been involved, demanding an "independent investigation" by the Nuclear Installations Inspectorate (NII).[177] The latter had been formed in 1960, in the wake of the Windscale Fire, to licence and supervise commercial nuclear reactors not directly controlled by the AEA.[178] When the NII rebutted his claims, pointing out, among other aspects, that the parts in question "had been replaced over the years", he called it "a whitewash" and accused the CEGB and NII "of complacency to the point of stupidity".[179] He, in other words, rendered his acceptance of facts presented by experts conditional on the conformity to his interpretation. Such a complete devaluation of experts and the facts presented by them certainly remained rather rare in the debates around Hinkley Point.[180] In this specific case, the dispute slowly settled after Brookes lost his seat in the council elections.[181] Calls for a renewed investigation stalled in Somerset's public protection committee.[182] The debate was, in other words, ended politically.

By and large, the now open primacy of the political plane of the conflict did not mean that the role of experts and rational arguments within it had been dissolved completely. On the contrary, in a new equilibrium, both sides relied on them to understand questions regarding the complex matter of nuclear energy and to validate the points they each wanted to emphasise – even though they used different experts, highlighting diverse perspectives that fit their conflicting interpretations. The important difference was that experts and rationality were now openly utilised by both sides in the debate to back up their respective political goals, not just by one. This was a substantial change to the way the nuclear debates and, with them, nuclear history developed.

Political Decisions, Expertly Backed Up

This new equilibrium can best be demonstrated by examining the conflict about Hinkley Point C. Here, both sides relied heavily on scientific and technical expertise regarding various aspects of nuclear power, for example, the question as to whether the normal operations of reactors induced leukaemia. The corresponding debate shows that experts and their culture of knowledge were held up as decisive by all participants without being seen as sufficient to come to a conclusion. Conclusions were only attainable by political majorities, as is demonstrated by the issue of nuclear power's economy, concerning which the critics presented more convincing arguments during the public inquiry. Nonetheless, the CEGB carried the day with reference to the government's insistence on using nuclear power as part of the country's energy-political setup – at least until the end of the inquiry. Then, in a turnabout, it became clear that the private companies about to take over Britain's electricity industry under Margaret Thatcher's government would not bear the technology's financial risks. Rating the plan for privatisation more highly than the endorsement of atomic power, its expansion was held up, clearly showing that nuclear energy was a political energy, with political decisions determining its fate, and experts and rational arguments merely providing legitimacy after the fact.[183]

A third power station at Hinkley Point had been the topic of local debates time and again since the 1970s. From the beginning, there had been scepticism, especially as an American PWR was considered by the CEGB as part of a pondered general switch to this technology.[184] Critics believed these reactors responsible for rising leukaemia cases in surrounding areas, among other problems.[185] Soon, in relation to this, the question of the public's dependence on experts[186] and their political dependencies[187] was raised, with critics demanding more transparency from the official experts as a solution.[188] At the same time, the latter continued to command most people's trust.[189] The debate tapered off for the time being in 1975, with the Department of Trade and Industry denying any current plans for a third station at Hinkley Point and King assuring that, ere another power station was built, "very considerable discussions" would take place.[190] At that point in time, however, this assurance rang rather hollow.

In fact, in 1981, the government claimed that the possible construction of Hinkley Point C was not a political decision, but one taken according to scientific criteria by the CEGB.[191] But while the board's experts emphasised that the site was geologically suitable,[192] its resolve to go forward with the project, announced in the following August,[193] was based upon the Thatcher government's determination to ensure for Britain "a strong and larger nuclear component", as Parliamentary Under-Secretary of State for Energy Norman Stewart Hughson Lamont put it in Parliament on 26 January 1981.[194] This, while explained not least by the goal of averting climate change,[195] was a political decision by the Conservatives,[196] opposed by Labour and the

Ecology Party, and numerous citizens' initiatives, most importantly Stop Hinkley Expansion (SHE), which developed out of the Alliance against Hinkley C.[197] Consequently, the oppositional groups planned to gather political backing for their position, within the local area and beyond. Explicitly, Wendy Hudson declared that BANG's strategy would be "to gauge public opinion and campaign to stop the proposed pressurised water reactor at Sizewell", as a success there would help to prevent a new power plant at Hinkley Point.[198] Besides Sizewell, the group also looked to Cornwall, where the plan to build a nuclear power plant had faltered.[199] Taken together, a growing network of atomic opponents was forming, undergirding the impression that the upcoming public inquiry, though ostensibly fought with rational arguments and experts, was at the same time considered a political contest,[200] where backing had to be organised for one's positions, and arguments and interpretations picked accordingly.

It took about six years from the CEGB's announcement of its resolve to build Hinkley Point C in August 1982, until the public inquiry into the proposal started under Michael Barnes on 4 October 1988.[201] Before going forward, the board wanted to wait for the end of the public inquiry into Sizewell B, where the first PWR in the United Kingdom was under scrutiny, to decide on the technology to be used at Hinkley Point. After the report on Sizewell B had been published in 1987, the board went ahead, planning for a PWR – and the opponents of this plan also consolidated. There were more than 20 councils, from Stogursey Parish Council to Somerset County Council, and some from adjacent areas in Wales and southwestern England. With the exception of Sedgemoor District Council, which backed the planned extension, they formed the Consortium of Opposing Local Authorities (COLA).[202] The group relied on "expert witnesses" to present its case at the inquiry and produced the *Hinkley Headlines* to promote their perspective and their arguments beyond it.[203] In addition, there were groups like SHE, a self-described "grass roots organisation based near Hinkley Point nuclear power station . . . campaigning since 1983" that consisted of about 450 members and 90 groups,[204] which published the biweekly *Hinkley Inquirer* to cover the proceedings.[205] In total, about 90 organisations, including the Campaign for Nuclear Disarmament (cf. "Nuclear Debates"),[206] and more than 1,500 individuals objected the plan directly, with about 20,000 signatures backing them up,[207] stemming from the local area, but also from places like Cambridge,[208] Plymouth,[209] and Kildare in Ireland.[210] The CEGB, on the other hand, received the backing of a group called the Association of Power for Good, consisting of "highly skilled professional people from industry and the universities",[211] besides the Conservatives, who controlled the national government and provided vocal backing during the inquiry. Especially active in that regard was the regional Member of the European Parliament Richard J. Cottrell, who insisted that nuclear power would help with "reducing the CO_2 burden on the atmosphere".[212] This claim was contested, for example, by Friends of the Earth's Michael Birkin,[213] although the group was very willing to point to the problem of rising sea levels to dispute the siting choice for Hinkley Point C later on during the debate, in an exceptionally clear example for the level the utilisation of arguments had reached at that point in time.[214] Altogether, the inquiry, called "another Battle of Sedgemoor" by the *Mercury*,[215] saw over 600 people giving evidence on more than 180 days in session, spread over 14 months, with 4,500 documents being submitted.[216] The aspects considered ranged from electricity demand and the cost of nuclear power to its safety. This included questions of emissions during normal operations and leukaemia, accidents with a special focus on Chernobyl, emergency planning, the need to control the spread of plutonium, the treatment of nuclear waste, and questions of decommissioning. Also under consideration were ways to conserve energy or generate it via alternative means, like wind, waste, tidal, or solar energy, and the issue of fossil fuels and the greenhouse effect, besides consequences for the local economy, infrastructure, and landscape.[217]

To make their point regarding these various fields, both sides referred to experts and rational arguments, for example, regarding the question as to whether nuclear power stations during normal operations induced leukaemia.[218] On that issue, the public inquiry resumed a debate that had erupted in the area in 1984, in the wake of a television report about leukaemia around Sellafield (cf. "Emotional Atoms"). At the time, the county's community physician, Dr Cameron Bowie, had analysed 11 years of cases and found that there was no leukaemia cluster around Hinkley Point,[219] although case numbers had been rising over the years. Whether this finding represented a real increase in cases or better registration or diagnostics had been controversial.[220] The dispute had been renewed in 1986, when Under-Secretary of State, Health and Social Security Raymond William Whitney stated that, while there was no connection to nearby nuclear installations, the incidence of cancer deaths in Sedgemoor was twice the national average. As a counter, SHE's Crispin Aubrey, a well-known journalist who had moved to Somerset in the late 1970s, saw "mounting evidence" for a connection to low-level radiation, although "it's very difficult to prove these links", while Bowie once more pointed to his study.[221] In the following year, a study by the Medical Research Council and Imperial Cancer Research Fund (ICRF) showed that ten miles around Hinkley Point, the number of deaths caused by cancer of the liver were twice the national average from 1965 to 1980. Reasons for this finding remained unclear, although similar findings were available for other reactor sites.[222] A report based upon these data later showed that children living around old reactors were four times as likely to contract leukaemia than those living elsewhere.[223]

During the public inquiry, against this background, the CEGB's health physicist David Jeremy Weston argued that "humans have evolved in a radioactive world. Radioactivity occurs naturally. It follows, therefore, that the mere presence of radioactivity does not in itself imply a hazard. Compared with natural background, the effect of nuclear power is totally insignificant – less than 2 per cent even for people close to Hinkley Point."[224] Referring to the ICRF's findings, the board emphasised that there was no heightened number of leukaemia cases around its power plants; the number of cancer deaths was even lower than elsewhere.[225] On the side of the critics, the Severnside Campaign Against Radiation (SCAR)[226] argued that there was "accumulating scientific evidence" pointing to leukaemia clusters around reactors.[227] The facts and arguments presented with the aim of questioning a causal link, to them, were not enough to cast doubt on their assumptions of a connection.[228] They claimed that

> the public is not only rational in its analysis of risk but in doing so parallels the scientific form of analysis. Essentially the public's concern arises from a gap between what the nuclear industry predicts and what is observed . . . The public is faced with two hypotheses – one is "low risk" generated by "experts" [sic, CG] models, the other a high risk concept based on observation. The more the observations of accidents or unusual health models occur the more the public inclines to the high risk hypothesis.[229]

SCAR, in other words, argued for the validity of their argument by pointing out its rational, scientific character, drawing into question that of the CEGB's, as they called them, "experts", whose calculations they saw falsified by observations. Additionally, they presented an analysis by statistician Dr Michael K. Palmer, who opined that leukaemia clusters were rare outside areas near nuclear installations,[230] reducing the likelihood of other causal explanations.

The topic was highlighted once more in late 1988, when Bowie and Dr Paul Ewings presented a study that pointed to a high number of cases around Hinkley Point between 1969 and 1973 without an unambiguous explanation.[231] This was the core problem of this conflict: There were grave questions that could not be conclusively and definitively answered. In this case, Dr Richard Taylor, head of health and safety at the CEGB, argued that the below-average number of cases in the power plant's direct vicinity strongly indicated that it was not the cause,[232] and insisted that "research had failed to show any logical link between low radioactive discharges and the incidence of the disease in surrounding communities. He said clusters of leukaemia cases had been recorded in areas far from nuclear installations and even before the technology was launched."[233] Weston added that the regular monitoring that had started before the first reactors commenced operation showed "no evidence of any sustained build-up of radioactivity in the environment. In most cases, the results indicated that the presence of the stations made virtually no contribution to overall background levels."[234] Critics like Aubrey, on the other hand, thought that normal "low-level" emissions could be responsible,[235] as, according to Dr Alice Mary Stewart from Birmingham University, the officially used models for estimating the consequences of radiation were wrong, being based solely on the fittest of human beings.[236] According to

David Gilliam from the Irish Sea Project citizens' initiative, the "current cancer risk estimates could underestimate the risks from radiation by a factor of 100" because "certain pathways had been ignored and the resulting human exposure under-estimated".[237] In the eyes of SCAR, the conflicting interpretations between official and critical experts meant that "the burden of proof has now shifted – the leukaemia cases must be assumed to be connected with the nuclear plants until proven otherwise."[238]

This claim regarding "the burden of proof" and the widespread reliance on expert witnesses demonstrate the importance attached to expertise and rational modes of argumentation – as does the widely shared conclusion that, in the end, it was a positive thing that the dispute had triggered research into the causes for leukaemia.[239] Phelps said it best when she demanded during the course of the inquiry: "[L]et us be genuinely scientific and try to face all the facts, good and bad, of this very controversial issue."[240] Scientific and technical data, arguments and experts were clearly revered by both sides of the dispute. The question, though, as to what to do in view of, as Barnes called it, the "considerable disagreement between experts",[241] could not be solved in this way. It required weighing the arguments presented from a "higher" point of view. Taking up such a position in his report, Barnes thought that the evidence presented did not speak against allowing the use of nuclear power with its various benefits.[242] To SHE, obviously, this was the wrong conclusion.[243] Experts, in other words, while revered by both sides of the dispute, could not solve it. It remained, as it had been from the beginning, a political one, wherein rational arguments and experts played a merely utilitarian role.

This was made especially clear when it came to the question of nuclear power's economics. At first glance, here, too, experts and their arguments dominated the dispute. For SHE, Professor Jim Jeffrey claimed at the public inquiry that electricity from nuclear power would cost 4.7 pence per kilowatt-hour generated, almost double the cost for that from coal-fired plants. The CEGB, on the other hand, claimed that the price would lie at 2.2 pence.[244] However, after Barnes demanded to see concrete numbers for both kinds of power stations,[245] Frank Jenkins, the CEGB's Director of Strategic Studies, had to concede that electricity from coal was cheaper than that from nuclear energy, albeit dependent on coal prices and rates of return.[246] As it turned out, nuclear electricity cost about 6.25 pence, if the cost for dismantling disused power plants was excluded![247] In Britain, therefore, a rather widespread observation, pointedly stated for the USA by James Mahaffey, was quite fitting: "Economics is what killed the expansion".[248] To Aubrey, at least, Jenkins' concession was "a defeat" for the CEGB, enabling "a real debate about the comparative costs of nuclear power and other means of generating electricity".[249] In the summer of 1989, then, the public inquiry was put on hold, after the rising costs of Sizewell B had drawn into question the foundation of the economic arguments for nuclear power even in the eyes of the private entities designated to take over the British electricity industry. None of them wanted to accept the oldest stations and the impending cost of dismantling them. On 9 November 1989, the government had to desist from demanding an expansion of nuclear energy as part of the country's energy mix. Effectively, Energy Secretary John Wakeham pronounced a moratorium on new nuclear power plants until a review of the technology in 1994, when the completion of Sizewell B was expected. Until then, no new nuclear power plant would be built – including Hinkley Point C.[250] Britain, in other words, often held to have seen less resistance against nuclear energy, saw an official moratorium in the 1980s, while in Germany, there was a dispute over this decision until the turn of the century. To critics of nuclear power in Somerset and Britain overall, the "massive case against the proposed PWR" presented by them had helped force the government's hands,[251] clearly showing that, to them, nuclear power was indeed a "public technology" par excellence. They celebrated a great triumph, as they had "succeeded in proving that a PWR at Hinkley is uneconomic, unnecessary and potentially unsafe".[252]

This, however, was not quite what had happened. Nuclear power was not given up for all time by the British government because its economic troubles had been scientifically proven, for all to see. Its expansion was put on hold because the privatisation of the electricity industry ranked more highly on the government's political agenda than the expansion of nuclear energy. Taking up an idea of then Secretary of State for Energy Nigel Lawson from 1982, a government White Paper in 1988 had initiated the plan to privatise the electricity industry. After calculating costs, private companies had not been interested in acquiring nuclear power reactors, especially the old Magnox ones, with decommissioning looming. In the end, most nuclear reactors had to be kept in the state's hands.[253] This did not mean that the government would have been convinced in any way that the technology was not usable at all. This was illustrated once Barnes, in his report, recommended the power station's construction after weighing all arguments. To him – and the government – nuclear power remained a non-fossil-fuel energy source that would be useful with a view to limiting CO_2 emissions, and, in turn, "global warming". If this goal was prioritised politically,[254] then, it would be rational to grant consent to the nuclear power plant, even if this consent could not result in any practical steps for now, in view of the moratorium established until 1994.[255] In fact, Hinkley Point C remained dormant until 2008. In the 1990s, the British government was not interested in financing it, and British firms were not interested in building another reactor. Existing ones soon even ended up under the control of the French Électricité de France.[256] Then, however, the Labour government of Gordon Brown reassessed the technology in view of climate change, prioritising the arguments for nuclear power regarding this topic over the technology's well-known problems.[257] Today, the third power station at Hinkley Point is under construction, and its costs are rising.

Obviously, the arguments regarding nuclear energy's economic side presented by its critics during the "battle of Sedgemoor" had been correct. No less obviously, however, the experts and rational arguments that both sides had relied on to present their cases had not been decisive. Whether or not to use nuclear power was not a question that could be answered by a contest of scientific and technical arguments presented by experts. Even though experts were essential for understanding, conceiving, and using the technology, and even though they were revered and utilised by both sides, the conflict was decided by political considerations. Rationality, in other words, was essential but not decisive at all. It was a tool within the debate, and its utilisation had been monopolised early on by the technology's proponents, enabling them to subdue unwanted critique within the debate. During the 1970s, this monopoly had begun to waver. It faltered in the light of the debate about underground storage of nuclear waste, when the growing group of the atom's opponents had found an issue they rejected independently of all rational arguments and started to openly treat nuclear power as a political question that was decided by majorities, not scientific prowess, with experts providing argumentative ammunition for the conflict. The resulting new equilibrium had been visible in the debates about Hinkley Point C.

The changes demonstrated here had already started to unfold when the public discussions about a new nuclear power plant took off in Lingen. There, the change in the debate played out in an accelerated form, highlighting some of the steps taken by nuclear power's opponents to adapt to their antagonists' utilisation of rationality more clearly.

Lingen – Elements of an Adoption

Lingen at the river Ems was the largest city of the Emsland District, created in 1977. It was situated in the west of Lower Saxony, about 25 kilometres from the border between Germany and the Netherlands. In the very year the Emsland District took form, a public debate ensued about a new nuclear power plant for Lingen, commanding an improved version of the PWR used at Biblis, with an output of about 1.3 gigawatts of electricity.[258] Initially, the debate had even been about four such reactors, two at Lingen, to be operated by the Vereinigte Elektrizitätswerke Westfalen AG (VEW), and two at Meppen, the Emsland's district town, about 20 kilometres to the north of its bigger neighbour, which were to be built for the Rheinisch-Westfälisches Elektrizitätswerk AG (RWE).[259] In the ensuing debate, that number shrank fast, and in the end only one reactor at Lingen was realised as the Kernkraftwerk Emsland (KKE). It was run by the Kernkraftwerke Lippe-Ems GmbH, which was controlled by the VEW, with the Elektromark from Hagen as a junior partner, supplanting a joint project the two had initially planned for Hamm in 1975 but abandoned in view of the nuclear-critical stance adopted by the government of North Rhine-Westphalia.[260]

The timing of the debate's start means that the emergence of critical expertise on the (trans-)national plane of debate, as well as the realisation of the political potential of the "Peaceful Atom's" opponents, had already occurred. Therefore, the focus in this sub-chapter rests on the analysis of the parallel existence of a public reverence for experts as signifiers of the rational, working as a discursive strategy towards the political goals of the technology's advocates, and the growing readiness of its critics to undercut this strategy. The latter approach was modified, in the wake of the catastrophe of Chernobyl, from attempts at defiance within the realm of rationality to an open political challenge of the pro-nuclear ones. This development can be exemplified by five episodes of the debate, highlighting different topics. The first three unfolded partially parallel to one another, while the final two took off subsequently.

First, the conflict about building a new nuclear power plant at all exemplifies how atomic power's proponents staged rational arguments and facts (re-)presented by experts, as the indispensable foundation of the debate about nuclear power, which was accepted by the technology's opponents. At the same time, the dispute reveals that the decisive criterion for an argument to be accepted was not to be found in its rational merits but its political leaning. Second, a dispute about a possible cluster of leukaemia cases shows how opponents of nuclear power attempted to establish the perspective of "critical" experts within the debate, which failed because the atom's proponents were able to deconstruct the critics' scientific expertise and standing. Parallel to these two episodes, a conflict about a cooling water reservoir for the power plant developed. It reveals how the opponents of nuclear power attempted to undermine the official experts by proving them wrong or, at least, fallible, in the end even attacking the debate's rational framework. In a fourth section, the debates after the catastrophe of Chernobyl demonstrate how the critics, to overcome their opponents' dominance, came to openly prioritise experts' political stances, mirroring the level of utilisation practised by nuclear energy's advocates. Finally, a debate about the decontamination of whey offers an insight into the new equilibrium, where both sides used their political prowess to fight for their positions, but still referred to experts, expertise, and rational arguments as a means of debate, demonstrating that rationality, while being subordinated to political decisions, remained a sine qua non for the acceptance of arguments as valid within the public debates about the atom.

Experts Enabling and Ending Arguments – The Debate About a New Reactor for Lingen

Between 1978 and 1982, Lingen's public debates revolved around the new nuclear power plant plans. The debates reveal the fundamental importance attached to rationality by atomic energy's proponents as well as its critics. Regarding the proponents, this can be gleaned from the fact that, in the summer of 1978, a few weeks before the VEW applied for the license for the planned Lingen power plant on 17 August,[261] Lingen's city councillors and its chief city director Karl-Heinz Vehring visited the company's headquarters in Dortmund to be educated about the project. Here, the chairman of the VEW's executive board, Professor Dr.-Ing. Klaus Knizia, marked as an expert by his academic titles, highlighted in the newspaper coverage of the visit, laid out the main arguments for a new nuclear power plant at Lingen, putting special emphasis on the advantages of nuclear energy over fossil fuels: "The climate of the earth could deteriorate incurably" if the latter were unabatedly used to generate electricity, he pointed out, possibly costing "this world . . . its viability for us". His argument was not only a tip of the hat to ideas of environmental protection in general; it was aimed at those councillors of Lingen's Social Democratic Party of Germany (SPD) who did not oppose the idea of a new local power plant as such but preferred a coal-fired one. The argument was outlined as a rational one, based on scientific expertise and presented by an expert, so Knizia and those among the Lingen delegation supporting a new reactor expected it to be accepted. At least, Knizia pointed out that his aim was "that the citizens attain a clear understanding of what is done" – meaning that they should come to the same pro-nuclear conclusions.[262] Whenever this did not happen, when critics remained unconvinced, although "about no new technology, as much information had been given" as was the case with nuclear energy,[263] the reason was sought on the side of the opponents in question. This had already been demonstrated by the technical director of the older Lingen demonstration power plant,[264] Dr Ottmar Deublein, in 1978, when he had emphasised nuclear energy's safety by pointing out that there was no "accumulation of radioactive materials in the vicinity of nuclear power plants". Claims to the opposite effect by critics of nuclear power, to him, were mere signs of a "catastrophe hysteria, which testifies to an ignorance of technical processes and contexts".[265] The chairman of the workers' council at Lingen nuclear power plant, Heinrich Kösters, went even further and deemed critics of nuclear power enemies "of the political landscape or the Federal Republic's social fabric". To him, such a fundamental, political motive was the only conceivable reason behind any opposition to nuclear reactors, as radioactivity's dangers were widely accepted for lesser reasons than satisfying the electricity demand: "Whoever carries a luminous watch confronts oneself with three times the radiation dose that people in the vicinity of a nuclear power plant receive."[266] To oppose nuclear reactors was, from this perspective, irrational and, therefore, unacceptable.

Lower Saxony's government argued along similar lines. At a talk in Geeste, just north of Lingen, in May 1978, the State's Minister for Science and Art, Professor Dr Eduard Christian Kurt Pestel, designated as an expert via his identification as one of the founding members of the Club of Rome,[267] explained how he, albeit arguing for the intensified use of coal until some years ago, had come to change his opinion: It had been "scientists all over the world", pointing to the consequences of burning coal, namely the rising level of CO_2 in the atmosphere, leading to a "greenhouse effect" and resulting in rising sea levels and a "karstification" of central Europe, that convinced him to embrace nuclear energy. Compared to that, he argued, the comparatively low risks of atomic power, including the solvable problem of nuclear waste, had to be preferable when trying to meet the rising demand for energy.[268] Pestel thereby presented himself as an expert who had learned from pro-nuclear experts, as was to be expected of every rational person. In the words of State Minister for Education – later to become Minister of the Environment – Dr Werner Remmers, who stemmed from the Emsland and was the chairman of the Christian Democratic Union of Germany's (CDU) Osnabrück-Emsland group, those who resisted experts' arguments were surely "[n]ew luddites", bent on "a militant fight against" nuclear energy. In an exhaustive 1980 text in the *Lingener Tagespost*, he painted as irrational the SPD members following nuclear critic Erhard Eppler, who, in 1979, had presented an energy-political concept bent on conservation, "although his energy scenario had been reduced to absurdity point by point by experts".[269]

In a similar way, the heads of the Emsland District and of the city administrations of Meppen and Lingen used experts and their rational considerations as "last arguments", to which one was expected to give in. Regarding his own area of expertise, the local impact, chief district director Karl-Heinz Brümmer saw a new reactor as a sound foundation for further industrial settlements. Therefore, he counted on "the common sense of the people" to use this "chance . . . for gaining new jobs" and "to deal constructively with the profound structural crisis".[270] To him, the rational and therefore only acceptable solution would be to back the new power plant, especially as it was deemed safe for its surroundings – safer not less than a comparable coal-fired plant, as he emphasised with reference to numbers produced by the Physikalisch-Technische Bundesanstalt (PTB). According to those, he pointed out, the latter would emit radioactivity in the range of 100 times that emitted by the planned nuclear reactor.[271] Vehring argued along similar lines, adding that reactors "increased" "the natural radioactivity of 120 millirem . . . by one percent", while "in coal mines people die, are injured and fall ill with dust lung in rows at the age of 45 to 50 years".[272]

Even the *Tagespost* portrayed those unwilling to give in to the pro-nuclear arguments as suspicious, at least. In early 1979, Lingen's SPD youth organisation, the Working Group of Young Socialists in the SPD (Jusos), expressed their categorical opposition to the construction of nuclear power plants because of "occurrences in the atomic power plants Brunsbüttel, Gundremmingen, Karlsruhe, Biblis, Ohu and others, not least the atomic power plant Lingen I" and many of the well-known arguments against nuclear power. Without explicitly pointing out experts' reports, however,[273] their position was sharply criticised by the *Tagespost's* Alois Dickopp as "carrying the features of a dangerous tactic of destruction and insecurity, which is reflected in blanket statements without evidence", while "verifiable experiences are consciously or unconsciously regarded as non-existent by the Lingen Jusos".[274] To him, by not explicitly referencing experts, the Jusos had violated a core rule of the debate.

In fact, most critics of nuclear power did regard experts and their culture of knowledge as a necessary foundation of the debate. Lingen's Bürger gegen Umweltvergiftung (BgU), for example, one of the region's citizens' initiatives, argued that energy conservation measures had to be prioritised, and "conventional technologies" used to meet the remaining demand. Their environmental problems, they insisted, "can and have to be reduced to a minimum by using the most modern and, if necessary, costly technology".[275] To undergird this position, the initiative invited then-CDU Member of Parliament Herbert Gruhl. Gruhl, who left the party two weeks after the talk to found the Grüne Aktion Zukunft, had been one of the CDU's most prominent advocates of environmental protection. In 1975, he had published the bestselling "Ein Planet wird geplündert", discussing mankind's options on a planet Earth as seen from a cybernetics-systems perspective,[276] prominent since the 1972 publication of "The Limits to Growth".[277] He had also been one of the first members and, up to 1977, the chairman of the Bund für Umwelt und Naturschutz Deutschland e. V. (BUND) (appearing, in its first year, with an inverted placing of Natur and Umwelt, as BNUD). This group, founded in 1975 as a national voice of environmental protection, quickly established itself as an interlocutor for leading politicians such as Minister for Research and Technology Hans Matthöfer and Federal President Walter Scheel, and, in 1989, aiming to acquire transnational presence, joined Friends of the Earth.[278] In June 1978, however, in Geeste, Gruhl presented a number of arguments against the utilisation of nuclear power in principle, not least its facilitation of a policy of growth that was doomed to fail as the planet's resources were limited, and its effect on the jobs situation in the meantime, as the electricity it provided advanced rationalisation.[279] Besides inviting experts, citizens' initiatives also referred to their arguments, often in the form of their publications. A prime example for this is the Bürgerinitiativen Emsland gegen Atomanlagen (BEgA), an organisation combining citizens' initiatives from Lingen, Meppen, Geeste, and other places in the region. In May 1979, it sent Brümmer and the Emsland's district administrator Klaus Stricker copies of the German translation of US physicist Amory B. Lovins' book on "Soft Energy Paths" from 1977,[280] recommending its use as "food for thought". Against that background, the group wanted the nuclear reactor built where the energy was needed, in North Rhine-Westphalia, not in the Emsland, where it was unnecessary.[281]

Similarly, the region's SPD referred to experts' rationale in their critique of a new Lingen reactor, not least arguing along the lines sketched out by Eppler and by questioning nuclear power's necessity,[282] its economic validity, and the creation of more nuclear waste, besides emphasising that a new reactor would endanger thousands of jobs in coal mines in nearby Ibbenbüren,[283] preferring the construction of a coal-fired plant instead.[284] This position, as Martin Braese, the chairman of Lingen City Council's SPD group explicitly pointed out, was taken up by a majority after "a long debate and careful consideration of the pros and cons", reaffirming the councillors' claim at partaking in the debate in a distinctly rational way – and, therefore, at being taken seriously.[285] The Emsland District's SPD argued along similar lines and explicitly pointed out that it knew very well the problems of coal-fired power plants, tipping its hat to the arguments regularly made by the atom's advocates.[286] Reverence for the realm of rationality was even more visible in the surrounding region. In early 1979 in Nordhorn, Lingen's westward neighbour and district town of the Grafschaft Bentheim District, the SPD planned "a weekend event for the summer, during which expert speakers will present the problems of nuclear energy".[287] In fact, many local councils and administrations in this district explicitly presented themselves as unable to handle the licensing process without the help of experts, thereby affirming the assumption that these and their expertise were a fundamental requirement for valid political decisions. The administration of Schüttorf, for example, while sceptical of the project's implications for the municipality, in June 1979, "consider[ed] itself unable to draw up an opinion on the safety report for the planned nuclear power plant . . . Therefore, it want[ed] to hire an expert to inform council and administration"; to validate, in other words, their critical opinion. To this end, the council resolved "to hire a 'free expert'", as the SPD had suggested, not "an expert from the Federal Ministry of Research", as had been the CDU's plan.[288] Other councils, not the least Nordhorn and Bentheim District, planned to follow suit.[289] When the radio-ecological study by Professors Hellmut Glubrecht and Wilhelm Kühn, jointly commissioned by the councillors of Nordhorn, Bad Bentheim, Schüttorf, Neuenhaus, Emlichheim, and Uelsen arrived,[290] it served as confirmation and legitimation for councillors opposing the power plant.[291]

This differentiation between experts with ties to the authorities administering the licensing process and those outside of it already points to the difference between "official" and "critical" experts, with the former dominant in the formal proceedings. It was this situation that allowed the proponents of nuclear energy to control the debate's course while coating their wielding of majority powers with the widely accepted irresistibility of rational arguments and expertise. This became obvious in Lingen City Council's dealings with the topic. In the wake of the TMI catastrophe, for example, often held to have demonstrated that large-scale accidents were not as unlikely as they were deemed to be by pro-nuclear experts,[292] the city council once more turned to experts in the industry and the state government to explain the reasons for and safety of a new reactor. Here, the council learned, besides electricity, steam could be provided for an existing industrial area with sufficient connections to the grid and transport infrastructure; experienced local personnel could be employed; and the region would profit economically from a modern reactor protected against crashing military airplanes, exploding gas clouds, quakes, or sabotage. Beyond that, reports by neutral experts on siting, meteorology, radioecology, and other aspects were promised,[293] with "[t]he TÜV [Technischer Überwachungsverein (Technical Inspection Association), CG] Hanover . . . commissioned to produce a main report on reactor safety",[294] to clear up all open questions. These outstanding reports, then, allowed the councillors backing the project to defer the debate and a final decision until the pro-nuclear expertise was at hand – while further preparatory steps could be undertaken until that point in time. This strategy worked because experts' arguments were accepted as a precondition for a rational and therefore valid debate by nearly all participants in it.

The strategy was applied again on 9 October 1979, when the council's SPD group proposed postponing a decision about the development plan necessary to move forward in the licensing process until there had been a political debate about the power plant that included the public.[295] This was a clear attempt to openly prioritise the political decision-making process. The CDU's speaker Georg Wenzel now insisted that the decision had to be taken at once, as this was the only way the licensing process would be able to move forward – up to the point where the expected experts' reports were presented, which in turn were a precondition for a political debate in a "factual atmosphere of discussion".[296] He was backed by Vehring, who argued that, without the reports, any debate would take place "in a vacuum",[297] and by the *Tagespost's* Johann Hinrich Derr, who emphasised: "Indeed, factual discussions on this topic are only possible if concrete statements on the local level about the establishment of the planned Lingen nuclear power plant can be made on the basis of expert opinions."[298] By applying this strategy, the licensing process was moved forward and the status of experts – of pro-nuclear experts who were part of the licensing process, who furthered the political goals of the pro-nuclear majorities on the federal, state, and local levels – and their arguments above the "uneducated" opinion of laymen secured.

Again, this strategy was repeated in later steps of the procedure,[299] until, in February and March 1982, results were presented and the council's majority satisfied with the experts' evaluations.[300] Even an SPD councillor now lauded the "factual, very informative debate . . . All have learned something . . . regardless of how one regards the nuclear power plant in principle."[301] One week after this March meeting, the council approved the development plan necessary for the power plant's construction. Once more, Wenzel emphasised: "The city council has to rely on the judgment of others, though, who are appointed in the lawful procedure to give their expert opinion and to check and approve within their area of responsibility."[302]

Clearly, the dominance of pro-nuclear experts and expert opinions was ingrained into the structure of the nuclear licensing process and the debates about nuclear power in Germany. While experts were not taking political decisions – most explicitly so[303] – they and their examinations and evaluations were presented as the unavoidable arbiters of said decisions on the complex field of nuclear energy. Even more so, understanding the topic in the first place was deemed impossible without them. Which experts were considered relevant during the licensing process, however, was ultimately determined through the licensing procedure, wherein experts from the industry conferred with those from the authorities, together seeking a way to build and operate the planned nuclear power plants in conformity with applicable laws. Other, "critical", experts who had an interest in preventing a reactor's construction, inevitably remained outsiders in this procedure. This had been exemplified at the official hearing for the KKE, which took place in Lingen from 12 to 15 May 1981, under the auspices of head of section Dr Wilhelm Knollmann from the Ministry for Social Affairs (cf. Nuclear Democracy). Here, detailed objections brought forth by critics, sometimes experts themselves, sometimes referring to them, were presented to the panel of experts from the industry and the authorities. In the end, however, the latter decided when and if a question had been answered satisfactorily and when and if an objection had been upheld or disproved. Disproving them and their arguments was impossible under these conditions, especially for non-experts. This was rather inadvertently pointed out by Knollman concerning the question as to whether laypeople could ever hope to understand the complex matter. To him, this was "certainly impossible".[304]

When the construction of a new nuclear power plant for Lingen was debated, in other words, experts and their culture of knowledge were deemed a necessary precondition for a valid, rational argument to be presented by advocates as well as opponents of the project. Whether or not an expert argument was taken seriously within the licensing process, however, was decided by those controlling the process; in this case, supporters of nuclear energy. Their experts' opinions dominated the official side of the debate, while critics remained at the fringes. This situation was rather strongly, albeit ineffectively, criticised by the opponents of the new nuclear power plant in 1984, when another hearing had to be held because of changes to the reactor's safety concept. This was largely boycotted by the project's opponents.[305] The *Tagespost's* Hans Brinkmann commented that the critics, "instead of providing evidence that the changes pose unacceptable risks, . . . turned their backs on the experts after a blanket statement . . . By avoiding an open discussion . . ., the citizens' initiatives apparently want to avoid publicly falling behind in their argumentation, and to ensure that power plant builders and authorities appear as a 'clique'."[306] In fact, this appearance could hardly be disputed (cf. "Nuclear Democracy"). And as there was no option to forgo experts altogether either, the critics attempted two other strategies. The first of them, discussed in the following section, was to find experts for questions that shed a critical light on nuclear power and which could be proven relevant, albeit outside the official licensing process. Such an attempt was undertaken regarding leukaemia cases, seen by nuclear power's critics as an unavoidable consequence of the low-level radiation emitted by every reactor in operation.

Critical Experts as an Attempted Answer – The Debate About Leukaemia

In 1978, well-known right-wing nuclear critic Walther Soyka, a former member of the National Socialists' "Schutzstaffel"[307] who had studied political science in Vienna and was now working at the University of Bremen,[308] besides having founded, together with Roland Bohlinger,[309] the Institut für biologische Sicherheit (IbS), claimed to have found, in an 80-kilometre area around Lingen, a dramatically heightened number of childhood leukaemia cases for the years 1968 to 1978. Significantly, these were the years since the old reactor there had become operational. His data stemmed from replies to advertisements in Bremen's *Weser Kurier* and Oldenburg's *Nordwest-Zeitung* in August and September 1978, calling for information about cases from 1945 to 1978, and from conversations with people from the region. His findings were publicised – although Soyka himself, at that point in time, thought the material unready for publication – by an article in the *Niedersächsisches Ärzteblatt*, written by Dr Hermann Kater, a physician from Hameln. The latter knew Soyka from protests against the Grohnde nuclear power plant. Having met him again at Geeste, he had learned about the study and then written about it – albeit remaining sceptical of the IbS.[310] In the wake of the publication, a heated debate began.

Quickly, broad criticism regarding the study's methods, validity, and its authors' expert status as such was voiced by proponents of nuclear power to deny the critics' claim to a valid, rational basis for their critique. In September, Soyka's and Bohlinger's claims about a causal connection between the power plant and raised numbers of leukaemia cases had already been attacked by the VEW and the authorities as being "without scientific merits" and "dubious".[311] After Kater's report had popularised the topic, the criticism sharpened in tone. Franz Holtgreve, managing director of the Public Services, Transport and Traffic Union's (ÖTV) Emsland district, called the study "irresponsible panic-mongering with scientific coating", by the IbS, which had been "founded specifically to strike against nuclear energy".[312] Brümmer emphasised that it was run by "one historian and one social scientist, without biologists, and was financed by opponents of atomic power plants", besides pointing out methodological weaknesses in the study.[313] Georg Hiemann, a reader of the *Meppener Tagespost*, furiously described the latter, pointing out that a member of the institute had called a citizen of Meppen, wrongly claiming that this citizen's daughter had leukaemia, and asking for additional information. Concerning Soyka personally, he added that he "is calling himself Absolv. rer. pol. [meaning that he studied political sciences, CG] and has, probably, as little contact with the field of biology as every normal citizen. Therefore, doubts regarding the scientific qualification of this two-man institute should be understandable."[314]

Beyond criticising Soyka's qualifications and study, the Ministry for Social Affairs emphasised that, since July, it was working on its own report, based on official death registers, and that, so far, it had found no indications for an increase in leukaemia cases.[315] During the following days, this information was reinforced. On 31 October, the *Tagespost* reported that the number of leukaemia cases in Lingen was well within the average rate of about 37 cases per one million inhabitants.[316] Hiemann in his letter, referred to "official sources", which showed that "there was no raised level of leukaemia cases during the last years. As long as birth rates were rising, so were – with a lag of about two years – the numbers of leukaemia cases in children. As the birth rates have been going down since 1972/73, so are the statistic numbers for cancer in children."[317] On 7 November, Brümmer pointed to leukaemia researcher Professor Landbeck, who had observed a steady level of the illness among children throughout the last two decades. This, he added, corresponded with the findings of the Emsland Health Authority.[318] Two days later, the *Tagespost* emphasised that, according to state and federal governments, the number of leukaemia cases in Lingen was within the normal range.[319] To the *Tagespost's* Dieter Sommer, this was a sign that "[w]hat has been said and written about the alleged rise in deadly leukaemia cases around the Lingen nuclear power plant turns out to be less and less sustainable."[320]

By then, however, Willi Wolf, SPD councillor and chairman of Lingen City Council's environmental committee, had – quite in the tradition of Tamplin – attempted to use the faltering study to reverse the burden of proof regarding the acceptability of a local nuclear reactor by arguing that even "well-intentioned and well-founded counterstatements" were "of little help for reducing the insecurities that have arisen among the citizens", "[a]s long as the claims by the 'Bremer Institute for Biological Safety' about a cluster of leukaemia cases in the area of the Lingen nuclear power plant are not faultlessly disproved".[321] Unwilling to wait for the "'official' investigations" by the government, he demanded that his committee take up the question and organise a public hearing.[322] This demand, however, was taken out of his hands by the CDU's majority in the council, which insisted on a town hall meeting – after the report the Ministry for Social Affairs was working on had been published,[323] thereby attempting to rein in the topic raised by antinuclear experts.

In the meantime, the situation remained, as the *Tagespost's* Dickopp put it, "inscrutable". This became obvious at a conference on the topic organised by a regional Catholic educational institution, the Ludwig-Windthorst-Haus, in late November.[324] Here, Bohlinger contested the numbers given by the Ministry for Social Affairs of "30 deadly leukaemia cases in the Emsland" between 1968 and 1977 and claimed there had been 43 to 50. Based thereupon, Soyka argued that it was "very, very unlikely" that mere chance was behind these numbers, as they were higher for years after 1968, when the nuclear power plant had become operational.[325] On the other hand, as the *Tagespost* put it, "scientists denied his [Soyka's, CG] study any claim on scientific value", with a physicist from the TÜV emphasising that the emissions of the Lingen power plant in 1976 had equalled the dose one would have received during one flight to Mallorca. Nonetheless, as the directors of two university clinics from Hanover and Münster pointed out, the consequences of low-level radiation remained unclear. No wonder, then, that the participants agreed that the state government should start a cancer registry.[326] Dickopp assented. To him, too, it was high time "[t]o create, with scientific accuracy, the tools to give citizens more security, trust and confidence".[327] The conference, in other words, strengthened the position of the official experts, as they were the ones now generally looked to for a definite evaluation of Soyka's findings.

In February 1979, a panel discussion chaired by Vehring took place in Lingen. Here, Bohlinger was framed by Dr Günther Keding from the Ministry for Social Affairs; the professors W. Seelentag for the Federal Ministry of the Interior, Günther Schellong, and Christian Streffer, a radiation biologist from Essen; the Gesellschaft für Reaktorsicherheit's chartered physicist L. Ferdinand Franzen; the physician and chemist Dr F. Portheine from Nordhorn; and Dr.-Ing. Joachim Adams from the VEW, who was to become director of Lingen's nuclear power plant. In front of a 300-strong audience, the phalanx of experts refuted Soyka's findings.[328] Franzen and Seelentag explained why it was unlikely to expect a measurable increase in cases in the first place, because of the level of emissions of a nuclear power plant in relation to the wide range of background radiation in which humans lived, and why it would be quite impossible to link an increase, if one was to be found, to one single cause.[329] More specifically on the point in question, Seelentag described how, due to the lack of a German cancer register, "80 international cancer registries" had been studied by the authorities' researchers, leading to the assumption that seven leukaemia cases per year and 100,000 people were to be expected as an average number, and Schellong detailed that in Britain, Sweden, or the USA, about 4 out of 100,000 children would contract leukaemia. He concluded that there were "no indications for an increased number of children suffering from leukaemia in the Lingen area. An increase like the one claimed by Soyka, Bohlinger and Dr Kater would have been reflected in our clinic." Keding even emphasised that the Emsland District's "mortality of children with leukaemia lies below the State's average". Beyond disputing Soyka's findings and interpretation, his study's validity was once more dissected, and his expert and personal character attacked. Seelentag pointed out the study's methodological problems and said that it was unacceptable to "play number games with the terrible fate of cancer patients. Mr Soyka's conclusions lack any evidential value!" Schellong confirmed that Soyka's studies "prove nothing, but have caused unrest nevertheless" and Portheine criticised Soyka's statistical numbers as too small, while describing himself as "shocked . . . by the advertisements of W. Soyka in the newspapers", adding: "With those, you have put yourself on the sidelines. That is panic-mongering." In view of this broad attack, Bohlinger "defended himself against accusations that the Soyka Institute was imitating science and was panic-mongering and was acting negligently", but conceded that Kater had published unfinished results. He then claimed that the study's most important result had been to bring attention to the question of "radiation and leukaemia". Indeed, Keding argued that "Soyka and his institute should be thanked for having brought about a public discussion" of a topic that "[i]n expert circles was deemed over for a year" because "[e]xtensive studies and debates have shown that, at the current state of research, there are no indications for a rise in the incidence of leukaemia in the area of nuclear power plants".[330]

Obviously, with one of the fathers of the study retreating, the attempt to establish an expert-based argument against nuclear power outside of the formal licensing process had failed.[331] This result was reinforced one year later, when the study promised by the Ministry for Social Affairs was presented, based upon death certificates and interviews with physicians and clinics. It showed that, while the number of children in Lingen was above the State's average, the number of those dying from leukaemia was below it.[332] It was utilised by Meppen's former deputy chief district director and now CDU Member of the State Parliament Winfried Hartmann to once more emphasise that the IbS's study had been unqualified from a scientific standpoint and, in the end, "panic-mongering". This, he added, was not unusual for "critical" institutions, explicitly pointing to Freiburg's famous Eco-Institute,[333] which had been founded to provide scientific expertise to citizens' initiatives.[334]

Critics of nuclear energy made some rather inconsequential attempts during the following years, especially after the catastrophe of Chernobyl, to return to the question and criticise the methodology of the counter-studies presented by the authorities. They also criticised the way the media had treated their own contributions, for example, by Bentheim's Bürgerinitiative Umweltschutz,[335] and, once more, attempted to reverse the burden of proof by pointing out that the counter-studies had not disproved the possibility of a connection between leukaemia cases and the nuclear power plant.[336] However, an essential lesson was seemingly drawn by the opponents of nuclear energy: It was possible to deconstruct experts on the basis of methodological doubts, and subsequently largely ignore their statements and the questions posed by them. In other words, while it still seemed necessary to listen and react to experts and their statements, accepting their validity was something else. Consequentially, the atom's opponents attempted, as a second strategy to gain an upper hand in the debates, to publicly draw into question the experts and arguments presented by nuclear power's proponents outside of the core of the licensing procedure, that is, on a field of expertise outside the realm of nuclear power, thereby implicitly paving the way to a more general questioning of experts and the rationality embodied by them.

Unconvincing Experts? The Debate About a Cooling Water Reservoir

It had been clear early on that the planned KKE would need a reservoir to supply it with cooling water during dry months. Corresponding plans had been hatched since 1970, although, in the end, the reservoir's size was reduced to fit two instead of the four reactors originally planned. That implied an aboveground, artificial lake of about 1.6 square kilometres, with dams about 13 metres high. It was to be sited near Lingen's northerly neighbour Geeste.[337] Geeste's citizens were less than happy with the proposal because, as its district councillor Hermann Muke put it, in the event of a breach in the dam, "our village would be razed to the ground".[338]

In view of this local assessment, initially, offers from the VEW's and RWE's experts to come to Geeste and explain the artificial lake, from which "no risks whatsoever" would arise, as the VEW's Dr.-Ing. Ulrich Täubert emphasised in the district council's construction and planning committee had not been taken up.[339] After a few weeks, however, in a further example of the importance attached to the recognition of rational arguments and those presenting them within the dispute, Geeste welcomed the experts. In February 1978, its population, politicians, and members of citizens' initiatives listened to Täubert explaining the planned construction, and to the RWE's Dr G. Ortner's assurances regarding its safety. There was, he pointed out, nothing special from the perspective of hydraulic engineering. Dozens of similar installations existed throughout the Federal Republic of Germany (FRG), without larger accidents having occurred. Experts from both companies emphasised that recent dam breaks at the Elbeseitenkanal and at Kirchheim resulted from the transition from earth to concrete, which would not occur here, and from neglect of maintenance, which would be avoided through continuous controls. Worries about a destruction of the dams were described as groundless, with below-sea-level provinces of the Netherlands serving as proof.[340]

Nonetheless, Geeste remained sceptical and its municipal council, for now, rejected the reservoir.[341] A reader from Biene, Lingen's northernmost quarter, which was to border the reservoir's southern end directly, explained that those having to live closest to it also opposed the construction and doubted that it could be known that the dams would hold. Consequently, he asked politicians to back their voters instead of the industry,[342] or, in other words, the laymen's doubts instead of the experts' calculations. These doubts were fuelled by Lingen's SPD, whose Harald Höhne was convinced that the reservoir, in combination with the planned power plants, would "endanger the population of the Emsland and the neighbouring districts" because recent dam breaks, including the Elbeseitenkanal's, "reduce[d] the faith in the planned reservoir's safety to zero". His party colleague Hans-Joachim Wiedorn added: "There is not a single insurance company in Germany willing to assume the risk for residents. One must think about that."[343] To them, as to anti-nuclear citizens' initiatives like Lingen's BgU, the artificial lake was not only a danger in itself but also a precursor of the opposed nuclear reactors. As such, it was a "risk for the population and the whole ecology of the area", as the citizens' initiative put it.[344]

No wonder, then, that attempts were made to stop the reservoir's construction. First, an ecological report that had been commissioned,[345] and presented as a necessary step "according to the water act",[346] was attacked by Geeste's Bürgerinitiative Umweltschutz gegen Atomenergie. It wrote to the President of the Regional Council, Weser-Ems, protesting against the reservoir and demanding that the expert commissioned to produce the ecological report should be replaced, as he allegedly had been asked to come to a positive result.[347] This attempt to derail the process by undermining the credibility of the official expert failed. The report was finalised and presented at a special meeting of Lingen's City Council and Geeste's Municipal Council in May 1980,[348] in the wake of which the majority accepted the reservoir.[349] Two years later, when the council had to pass the necessary land use and development plans, Wenzel stated that the reservoir had to be accepted, as "[t]he experts convincingly showed that the safety is guaranteed up to twice or two and a half times the legal requirements." Also, additional demands, for example, regarding the creation of a biotope and the generation of recreational value would be met.[350] Not everybody had been brought round, though. People from Biene remained sceptical. Nonetheless, its municipal council accepted the plan, as it was convinced of its safety.[351] In other words, the official experts functioned in the expected way, enabling decisions even in view of unabated scepticism about the projects in question. Thus, the first attempt to derail the process by questioning the official experts had failed. On the other hand, the fact that scepticism remained indicated that, while fulfilling their role within the lawful proceedings, the experts and their arguments had been unable to sway those opposed to the project.

Consequently, the reservoir came under attack again.[352] The most significant attempt to stop it, and, with it, the nuclear power plant, was made by the chairman of the Green Party group in the district council, Werner Bußmann, in December 1984. Together with Johann Stockum, a critic of nuclear power marked as an expert by his engineering diploma,[353] he presented accusations by two former employees of one of the companies working at the reservoir. According to them, rootstocks, instead of being burned, had been "ploughed under" and were now residing in the reservoir's floor and in the dams. Bußmann and Stockum claimed: "On an average of 100 square metres, there are probably two stumps to be found" and added that the former employees, whom they deemed trustworthy, "could even now name the places where there are rootstocks in the dam body". Therefore, they demanded that the construction site be shut down to search for the rootstocks, because: "Stumps that rot can cause subsidence, which later leads to cracks in the sealing and in the long run to a dam burst."[354]

This expert-backed attempt to disprove the pro-nuclear experts' safety assurances by hinting at practical mistakes undermining their calculations and, at the same time, aiming to discredit their reliability in view of their alleged lack of control over the construction process provoked a massive reaction. In the words of the *Tagespost's* Brinkmann, "this accusation by the Greens gets down to the nitty-gritty. It touches directly upon the nerve of safety, because if rootstocks really rest in the supporting body of the mighty earth wall, then subsidence could occur, resulting in the threat of cracks in the seal and dam bursts."[355] That his description of the possible consequences corresponded almost verbatim with the one presented by Bußmann and Stockum indicated the importance attached to the expert's assessment of the situation. Nonetheless, Brinkmann expressed doubts, which spoke of the high regard in which the authority's and industry's experts responsible for the construction were held:

> The question is whether these claims also accurately correspond with the truth. In view of the control system described by the VEW, it must

> be impossible, actually, that entire stumps in the dam and floor remain undetected. On the other hand, the Greens insist they can produce highly credible key witnesses. Since these are not just gewgaw, but core questions of safety, there is now only one consequence: Create absolute clarity, as quickly as possible! The two informants from the Greens must show precisely whether and how they can prove the serious accusation.[356]

On the very next day, Lingen's city councillors aimed for just that clarification, demanding explanations from 14 experts from the VEW, the Meppen water management office, Essen's earthworks laboratory, the companies Siemens and Strabag and a company group active at the construction site, because, in the words of Wenzel, "[t]he accusations made in the press are so serious".[357] During the meeting, the phalanx of experts insisted that, as the VEW's Täubert put it, because of "the way the dam is built and . . . controls" "the installation of rootstocks" was "to be ruled out", as the dam was built up in layers 20 to 30 centimetres thick, in which rootstocks could not be overlooked – especially so as the every layer's construction was checked by the construction companies, by the VEW, and by the authorities. Nonetheless, Täubert urged the latter to investigate the matter.[358] His estimation was backed up by the experts from the water management office, who stated that "no foreign objects which could endanger the safety are or were integrated into the security sensitive areas of the installation."[359] In the end, Wenzel and the SPD's Robert Koop demanded that the Greens prove their accusations.[360] In a press statement, the VEW followed suit, and the construction manager emphasised that "[e]very substantiated hint . . . will be taken serious, because the reservoir's safety has priority".[361]

Soon, it became clear that the attempt to undermine the assurances made by the industry's experts had failed. The VEW and the construction companies publicly pondered criminal charges.[362] In an open letter, employees of the construction companies emphasised their expertise and reliability, for which they claimed to be known nationwide, a reputation which they were not willing to risk, especially in their home region.[363] In the district council, the CDU group's chairman Heinrich Hövelmann and the SPD's Hermann Proske, who was also a Member of the State Parliament, were "furious", calling it unacceptable "that the Greens constantly unsettle the population with unproven claims". The council's Green members Bußmann and Ruth Jacobsen, however, doubled down, with the latter, in view of the massive attacks against them, claiming that "the messenger was to be beheaded because he brings bad news". "She [Jacobsen, CG] also expressed the conviction that the dam of the reservoir would break in the long run: 'maybe not immediately, but in 20 years'."[364] And while the accusations soon faltered completely,[365] the latter statement is quite telling for the role that experts and their rational arguments now played in the debates: It indicates how at least some of the opponents of nuclear power had come to see fit to openly call into question all kinds of experts and scientific or technical expertise that were not explicitly anti-nuclear, demonstrating how rationality's standing had deteriorated, even beyond the field of nuclear energy. It became increasingly clear that, while experts remained important for accessing the public debate in the first place, the way experts' statements were regarded was most dependent on the side they took in the conflict about atomic power. This situation intensified in the wake of Chernobyl.

Indispensable Yet Insignificant – Experts in the Wake of Chernobyl

In the wake of the Chernobyl catastrophe on 26 April 1986, the rational side of nuclear energy came to the fore again. Once more, it became all too clear that experts were needed to handle such a complex topic, from the basic measurements and descriptions of what was going on, to translating available data into laymen's terms, to the interpretation of what that, in turn, meant. It became equally conspicuous, however, that these interpretations were not confined to the realm of rationality but highly political, as was the decision over whose interpretation to consult as a foundation for political programmes – and that this fact, dominating the debates about nuclear power from the beginning, had been recognised by its critics. This became especially obvious when, regarding the state elections of 1986, under its frontrunner Gerhard Schröder, who would later become Federal Chancellor, Lower Saxony's SPD openly declared its determination to adhere to the nuclear critical experts once the elections were won.[366]

First, however, after Chernobyl, the importance of experts and their expertise was rendered glaringly obvious once more when, suddenly, the newspapers were full of measurements of radioactivity in air or rainwater and of the degree of contamination in various foodstuffs.[367] Of course, while there was a large amount of information, that did not mean that exact data for every place and time were available. In the Emsland District, measurements were primarily provided from the power plant site, where the necessary sensors were installed.[368] Also, the timing of publication was rather problematic, as information about peaks in radioactivity often became available only once they had passed, if at all, as some data was simply not published.[369] Still, a wide variety of measurements was publicised. The meaning of the various numbers and units, though, often remained rather unclear. Obviously, experts were needed to explain them.[370] As the *Tagespost* put it: "It is certainly advisable not to simply put measurements out there, but to provide them, if possible, with an explanation of how they are to be classified and what effects they could have. Anyway, a 'normal mortal' cannot find his way through the tangle of terms like Becquerel, rad, rem, Curie, and Sieverts."[371]

Regarding the interpretation of the measurements, it became obvious once more that there were different groups of experts, some sceptical of nuclear power, some relaxed – and that the decision about whom to believe, whose interpretation to translate into action, was a highly political one. On 30 April, one and a half days after the catastrophe had become public, the *Tagespost* documented this split by pointing out that the federal government saw no cause for concern in the heightened measurements of radioactivity, while, for Green politicians, they were exactly that.[372] During the following weeks, the difference between political groups adhering to different groups of experts, interpreting the existing measurements differently, remained glaringly obvious, for example, concerning recommendations made regarding milk. Here, the Commission on Radiological Protection stated that up to 500 Becquerel (Bq)[373] per litre were acceptable, while other experts, adhered to by the Greens and some German states like nuclear critical Hamburg, saw a more reasonable limit value at 50 Bq, with a view to keeping children safe, or were advising pregnant women and children to avoid fresh milk completely.[374] In early May, while there was deemed to be "no acute danger", even Lower Saxony's pro-nuclear state government recommended using ultra-high temperature milk or dried milk for the time being.[375] A few days later, while still emphasising that dairy cows should not be allowed to graze in the open, it argued that milk from dairies could now be consumed.[376] Locally, the district administration mirrored the recommendations from the federal and state level,[377] while the Emsland Greens compiled their own recommendations, "warning, among other things, . . . against the consumption of fresh milk and fresh dairy products".[378] Besides, further experts' recommendations were published, for example, by two children's physicians from Meppen.[379]

The situation, in short, was quite confusing, for political parties,[380] the district administration under district director Dr Walter Kolck,[381] the *Tagespost*,[382] and especially for individual citizens. This can be gleaned from a letter to the editor written by a reader from Orthermersch, about 25 kilometres east of Lingen:

> It is a reassuring feeling to know that the radioactive contamination in the air is decreasing. Was it high? How high? "Normal"? What is that? Is there no danger to the public? Why can't we have fresh milk? Why can't cows graze? Why are children allowed outside? At what point is radiation harmful? . . . We note that to date there is no satisfactory information on the situation, let alone instructions on actions (except: nobody is at risk).[383]

In fact, this reader was not the only one wondering what to do and whom to believe. The same applied to a group of women and mothers who founded the Frauen gegen Atomkraft some weeks later. Looking back, they described the situation right after the nuclear accident:

> [A]uthorities, politicians and scientists contributed to the insecurity with a wide variety of information, and helplessness became widespread among many women and mothers. Should we let our children play outside? Should we stay at home when it rains? What should we

> buy? Questions that nobody could answer clearly. Questions that affect women in particular, who have to find answers and make decisions every day.[384]

Although this kind of confusion – in the latter case also carrying a certain anti-elitist streak complaining about those who were too detached from daily lives to recognise the questions occupying those in the middle of if – was far from general, as surveys in the inner cities of Lingen and Meppen showed,[385] it was a clear sign of the parallel existence of two groups of experts and participants in the public debate following them. This was a result of the fact that the reverence for scientific and technical experts and their expertise, for rationality, was not beyond political positions, as nuclear energy's proponents and critics each had their "own" experts on whose expertise and arguments they based their positions.

The conflict between these two groups markedly intensified in the weeks after Chernobyl, with the respective experts and their reliability taking centre stage. To many, the reactor catastrophe meant that the credibility of the experts most outside the group of nuclear critics had relied on had been damaged. As the *Tagespost's* Alfred Brugger had put it on 30 April:

> Something bad happened. We still do not know exactly what is going on in Chernobyl . . . Only this much is known: Something that, according to experts' assurances, happens only once in ten thousand years, has happened. The reactor core of an atomic power plant has melted, the abbreviation GAU became a terrible reality . . . in the 32nd year after the first Soviet nuclear reactor commenced operations. This Saturday has changed the world, it is the beginning of the end of seemingly complete safety . . . We are at the wind's mercy . . . A catastrophe of these dimensions is as uncontrollable as the technology that brought it about . . . That will have consequences. The first is that the assurances of scientists are turned into conciliations, fostering fears rather than suppressing them.[386]

His was not a single voice, as indeed trust in experts – pro-nuclear experts, that is – suffered, a situation that was widely exploited by the technology's opponents to publicly expand upon their doubts in them and the rationale behind nuclear power represented by them. Soon, the district council's Greens pointed out that "the 'German atom industry'" had claimed in 1983 that reactors of the Chernobyl type were safe.[387] To the Greens, as well as the area's citizens' initiatives, the catastrophe now "proved true" that atomic power was "not controlled and once an accident occurred, wide parts of the planet are radioactively contaminated".[388] The argument that "once in 10,000 years" could also mean "tomorrow" was widespread,[389] and Bußmann pointedly summed up: "[T]o pretend that German nuclear power plants are protected against catastrophes of the same magnitude can only be done if one has lost all contact with reality."[390]

As, on the other hand, experts were still clearly necessary to navigate complex questions like nuclear energy, the solution seemed obvious: One had to rely on critical experts, like the Greens' Stockum,[391] or, as recommended by Wilfried Telkämper, who was about to enter the European Parliament for the party, "renowned atomic scientist Klaus Traube" who had stated: "Whoever says that such a catastrophe can be ruled out in our country either does not understand the subject matter or deliberately tells an untruth."[392] With the Greens was the SPD, whose Michael Weidekamp in Chernobyl saw "that atomic power is by far not as harmless as we have been led to believe for decades",[393] implicitly accusing those speaking in its favour of deceiving everybody else. The Jusos were convinced that "there is no absolute safety in nuclear power plants",[394] and on the state level, the SPD soon pondered the idea of not allowing the Lingen power plant to become operational.[395] This position was prominently advanced by Gerhard Schröder.[396]

With election day nearing in Lower Saxony, he was determined to adhere to the arguments of the critical experts, to turn them into the politically relevant ones, and to phase out nuclear power and prevent the new Lingen reactor from becoming operational, as he explained in an interview with the *Tagespost's* Brinkmann: "I used to believe and I said that as the power station was already under construction, one could responsibly allow it to begin operations. But the fact that the safety promises made by scientists and technicians have proved to be an illusion in the face of the Chernobyl disaster forces a rethink." German power plants, he added, did not change because of Chernobyl. "But what has changed is the safety guarantee for these power plants. I can no longer believe that as I did in the past. The consequences of an accident have become so dramatically clear to me – and I think to others as well – that an increase in nuclear power capacity cannot be justified."[397] Other nuclear power plants in Lower Saxony, Schröder demanded, should undergo

> a very strict security review. It must be carried out by a government commission consisting above all of competent scientists who have so far expressed critical views on nuclear energy. The security review must not be carried out by those who carried out the old reviews and granted the licences. This is not intended to be a vote of no confidence in principle against those who were active at the time, but it is an attempt to include in the political decisions the expertise that was previously easily pushed to one side.[398]

Experts, in other words, while still considered indispensable in questions of nuclear power and presented as the ones providing the rational foundation for political decisions, were obviously used as arguments themselves. They did not decide the fate of the "Peaceful Atom". Its application was not the result of a contest of scientific merit, but of political will, which decided which experts and arguments to listen to. In the case of Lower Saxony in

1986, the decision fell for nuclear power when the CDU remained in control of the state government, with the Green Party faring worse in Lingen than in the state's average.[399] Consequently, the KKE was able to commence operations on 14 April 1988.[400]

From the Avatars to the Core – Whey and the Role of the Rational Versus the Political

The shift in the way experts and their interpretations were publicly regarded, which had been in the making since the public debates about a new nuclear power plant for Lingen had begun, and which had become clearly visible in the wake of Chernobyl, turned out to be a quite fundamental one. This became transparent when Lingen was confronted, in late 1987, with one of the consequences of the catastrophe in the Soviet Union: contaminated whey.

In the confusing days in the aftermath of Chernobyl, when lacking as well as conflicting information about radiation and its units of measurement, limit values and their implications had reigned supreme in the FRG, Bavarian milk cows had been allowed to graze in the open. Consequently, their milk turned out to be contaminated. To avoid throwing it all away, it was processed into a more durable product: cheese. As a by-product, whey remained, which was dried, resulting in about 5,000 tons of whey powder, which was contaminated, not least, with caesium-134 and caesium-137. Plans to sell it to Egypt soon faltered, as no place could be found to mix the whey, which had been stored with uncontaminated material in more than 200 railway carriages to reduce its average contamination. As, in addition to that, no place could be found for the whey train to stay, Federal Minister for the Environment Walter Wallmann decided to buy it. Most of the whey was then, against public protest, brought to a Bundeswehr installation in Meppen. The question of what to do with it remained, however, as none of the solutions considered was unproblematic: Burning it would have resulted in more than 1,000 tons of contaminated ash, mixing it with cement to be able to store it underground largely increased its volume – so the idea of decontaminating it was investigated. A procedure to do so – to separate the radioactive caesium, at least – had been developed and tested at laboratory scale by Professor Franz Roiner, an expert in dairy issues. Now, the contaminated whey was to be used to demonstrate that the procedure could be used on a large scale. As no creameries were willing to carry it out, the ministry investigated disused nuclear installations – and found, in the old Lingen nuclear power plant, the most suitable place. Talks began with the VEW, the government of Lower Saxony, the city of Lingen, and companies able to take care of the job of reducing the whey's radioactive contamination to the level that was usually to be expected. In Lingen, however, protests ensued and a citizens' application, backed by over 11,000 Lingeners, demanded that the whey be put into a final repository instead. The confrontation came to a head when the application was debated at a meeting of Lingen's City Council on 9 December 1987, where, besides the speakers for the citizens' application, experts of the federal and state-level Ministries for the Environment were present, as were several citizens and reporters, including television crews.[401]

The way the whey was debated here showed the new equilibrium that had resulted from the transformation to the way experts and their culture of knowledge were publicly regarded, especially so in view of the development of opinions based on information derived from an empirical process. It helps to highlight how this process of forming an opinion had turned out to be a highly political one, dominated not by experts but by majorities – majorities that were, at that point in time in Lingen, commanded by the proponents of nuclear power.

At the beginning of the council meeting, the experts from the Ministries of the Environment presented the plan to decontaminate the whey in Lingen after going through a licensing procedure wherein the legal acceptability of the procedure and its environmental impact would be set forth. They explained that the plant to be installed would serve as a demonstration plant, showing the practicability of the procedure. Afterwards, they promised, it would be taken down and transported out of Lingen, as would the whey and the substances that had been separated from it. In the words of Dr Matting[402] from the Federal Ministry for the Environment:

> Although we do not harbour any doubt that the decontamination of the whey powder can be accomplished according to Professor Roiner's process, because this process is known to the milk and dairy industry (at least in its basic components), [and because, CG] Professor Roiner did experiments on the laboratory scale in his institute with the results confirming what had been expected, we would now like to demonstrate the effectiveness, harmlessness, but also technical optimisation with a pilot plant. Not only demonstrate – we want to prove it with such an installation.[403]

The argument presented here was that the attempt to decontaminate the whey was the least bad option to go forward, Lingen was the best suited place to carry it out, and the procedure would be seen as a way to test the suitability of the newly developed process, which was expected to work based upon experience, and a licensing process would see to its acceptability within the legal limits. In other words, the plan was presented as based upon a chain of rational, expert-backed arguments.

Subsequently, Heinrich Kampmann, the speaker of the Bürgergemeinschaft gegen Strahlenmolke,[404] a Ms Riegel and a Mr Krämer presented the citizens' application's arguments against the decontamination attempt. First, Kampmann argued that

> [w]e do not want any additional dangers for us, for our children and grandchildren, as no one can guarantee that the Roiner process is

> harmless and we do not have any trust in the according assurances made by politicians . . . In no plant has a larger trial . . . been done. Anyone with technical knowledge knows that a laboratory test cannot be easily multiplied to show effective results for a pilot or large-scale plant. We therefore do not want the pilot plant because the hazard potential is not clear at all.[405]

Second, further "dangerous technologies" for Lingen were rejected in principle. Third, the objectors did "not want to remain the indulgent people of the Emsland". Fourth, they did not believe that the plant would be deconstructed once it had been proven to be functional.[406] The fifth reason given was that another "problem industry" would contribute to the destruction of the Emsland's "image", keeping away customers and tourists, thereby endangering jobs. Finally, as the European Economic Community was said to be harbouring a 700,000-ton stock of whey already, the substance was seen as worthless, so the objectors proposed to simply put it into a final repository, possibly around the old nuclear power plant.[407] The number of citizens who had backed the application, he argued, could not be ignored by the council. To these arguments, Riegel added that all information so far had been provided "by the Ministers for the Environment, by Prof. Roiner and the VEW, which are all 'pro' decontamination in Lingen. Any neutral assessment and independent evaluation are lacking. The 'contra' is missing."[408] Krämer built upon this argument, concluding

> that an objective assessment of the risk potential has not yet been made. You, ladies and gentlemen of the council, have been elected by the citizens of Lingen to make considered, well-informed decisions for the benefit of the citizens. Is it not therefore also in your interest if, in addition to experts who advocate the disposal of whey, scientists who are critical of the disposal concept are also consulted?[409]

Taken together, the three presented their case mainly based upon three interlocking arguments: First, a great number of citizens with voting rights did not want the process to be carried out, because second, it brought with it unknown dangers, which had not been adequately assessed by one-sided, pro-nuclear experts; and, third, would result in an unacceptable, not least economic, burden for the Emsland.

During the following discussion, SPD and Green councillors backed up these arguments. Green councillor Astrid Mussil-Jungk contested the expertise presented, emphasising that "declarations of inoffensiveness and harmlessness of low-level radiation" were wrong, and pointed to critical experts' assessments in the form of an information brochure by the BUND which she had handed over to councillors.[410] The SPD's Elke Müller backed her up regarding the challenge brought to the existing expertise, arguing: "As far as I know, there is no scientifically proven knowledge about how additional radioactive low-level radiation affects humans, their hereditary disposition and future generations over a longer period of time. As long as this knowledge is unavailable, no one can claim with absolute certainty that health damages will not occur." Besides, on a more fundamental level, Müller criticised the fact that Roiner himself had produced the report on the process developed by him, implying that such a practice invited wrong decisions.[411] Her party colleague Koop also admonished the CDU for concentrating on "official reports" only, while ignoring "reports by critical scientists", who were excluded from the licensing procedures as they were, but which the SPD had taken into account in forming their critical opinion.[412] The party's Wiedorn finally disputed the existence of a "rational reason for decontaminating the contaminated whey", and the rationality of doing so in Lingen. Presenting a deeper mistrust in the experts heard so far, he claimed that the hearing as such had been superfluous. "To a large degree, it had the character of storytime anyway, the other part consisted of confessions of faith."[413] Openly prioritising, on top of that, the political will of the scheme's opponents over any arguments presented in its favour, he stated: "The Emsland has fulfilled its contribution to the nuclear industry to an abundant degree . . . We do not want any additional burdens."[414]

In combination, the arguments presented at the council meeting to stop the whey decontamination in Lingen demonstrated the sum of all strategies attempted by the opponents of nuclear power to cope with the phalanx of experts and expertise that had been at the argumentative and procedural centre of the technology's legitimation in the debates. They disputed the rational worthiness of the arguments presented by the experts backing the scheme, brought up additional perspectives from which it seemed disadvantageous, pointed to critical experts to back up their claims and questioned the relevance of all the expertise presented by highlighting the number of citizens opposed to the decontamination project. While the first points show that the importance of expertise as the foundation of the debate was still intact, the latter part is of special importance. On the one hand, it indicates that the atom's opponents knew very well that the debate and the decision were political ones, with the experts, while important for an accepted participation, only serving in the role of an argumentative strategy – a strategy they knew and saw through. On the other hand, this analysis was anticipating the debate's outcome, as the political majority – from the federal, to the state, down to the local level – was held by the atom's advocates.

Nonetheless, the council's CDU group very soundly presented the expert-backed rationale behind their decision to reject the citizens' application. To them, the latter carried in its own argumentation the reason it had to be turned down. The approach was quite basal, referring to a fundamental level of scientific and technological rationality, while mostly leaving aside the question of expressions of political will by the critics. As councillor Heinz Rolfes reasoned, a pilot plant was necessary to show whether the process under consideration was dangerous. The citizens' application on the other hand claimed that it was dangerous and aimed to deny the plant's supporters the opportunity to verify or falsify their claim, violating a basic principle of rational decision-making in an unacceptable manner. "Neither you nor we can work with a claim that has not been proven."[415] This, to Rolfes, was the main point: Such an outright rejection was just not acceptable, especially so as there were, on the other hand, "reasonable grounds" for the decontamination attempt in Lingen.[416] All the other arguments presented against the proposal, to him, rested upon the process turning out to be dangerous – and were, therefore, to be ignored until this had been proven true or false.[417]

He was backed up by councillor von der Haar, who reaffirmed that a political decision about something that could potentially endanger human beings had to be an informed one. To him, that necessitated having experts evaluate the question – and trusting those experts,[418] which also implied not introducing other experts into the process who claimed that the official ones erred. Similarly, councillor Rudolf Andree, who emphasised "that the question of decontamination is in an evaluation phase", insisted that this "means that one engages with the problem after the careful utilisation of experts, and all members of the council depend on these experts. Whoever disputes this approached the matter in a prejudiced way". The council's pro-nuclear majority, in other words, insisted on the importance of respecting the rational arguments presented by the experts within the officially sanctioned proceedings and resting the political decision upon those arguments. As Rolfes summed up: "Politics cannot mean . . . to see oneself as being led by a wave of moods and to forego closer, factual examination."[419] This, of course, implied that the political majority decided which experts to trust concerning the interpretation of the facts chosen for examination. In this case, the majority was held by the atom's advocates. The citizens' application, backed by a strong minority, was rejected.[420]

Clearly, the debates about nuclear energy accompanying the planning, construction, and early operations of the KKE in Lingen demonstrated how, on the one hand, the existence and backing of experts as avatars of rationality was widely accepted as a primary condition for taking part in the debate. This was true to the end, although the critics of nuclear power had identified the role as a tool in the debate played by experts at the same time, in the end merely legitimising a politically aimed-for goal. Consequentially, they had seen the failure of their efforts to reframe the questions asked during the process by introducing their own experts on topics to be included in the decision-making procedure, or to question the experts presented by the technology's proponents, or even more generally, the reliability of expertise as a tool for hedging risks in a mathematical way instead of avoiding them. As Schröder quite clearly promised in the aftermath of Chernobyl, once the political decisions were taken by the opponents of nuclear power, the relevant experts would be the ones now confined to the margins of the process. Until then, however, the atom's critics knowingly continued to fight against metaphorical windmills in an attempt to supplant nuclear power with real ones.

The Rationality Strategy – Exposed, Adopted, and Widely Spread

The debates about nuclear energy in Hinkley Point and Lingen differed widely. Most obviously, their timelines varied. The Hinkley Point debates started in the 1950s, at a peak of technocratic high modernity and planning enthusiasm at the heart of the Cold War, when the belief in technological feasibility and mankind's scientific-technical problem-solving capacity seemed unbroken. In Lingen, the commercial exploitation of nuclear energy was debated from the late 1970s onwards, when all of this was being questioned, at the peak of the German anti-nuclear movement, with critical voices rising in the ranks of the governing SPD and its partner, the Free Democratic Party.[421] These differences, the inclined temporal plane, render the analysis of the interactions between the issues of nuclear energy and expert-based rationality within the debates all the more fruitful, especially in view of their common features. They highlight the importance of the transnational plane for this part of the debates.

In both cases, the solid (albeit not necessarily purposeful) anchoring established by nuclear power's proponents between the technology on the one hand and the issue of scientific-technological rationality on the other, enabled them to employ experts and their arguments in a very utilitarian and indeed tribal way, to shortcut public dispute by passing political decisions for unalterable, rationally deduced facts. In the case of Lingen, this was visible, for example, in the reverence paid to Knizia or Pestel, and in the way the PTB's observations were employed within the debate. In the case of Hinkley Point, the most obvious examples include the delegation of whole strands of arguments to experts, not least regarding the economic considerations behind the plans or the issue of radioactivity's effects, which was deferred to the AEA. In the latter case, allegedly, even the officially adopted doubts regarding radiation were questioned after the experts' position had been presented. The strategy, in other words, worked well. It allowed the advocates of atomic energy to shape the debates about the technology and, thereby, its history.

Thus, the "Peaceful Atom" had seemingly become a textbook example of a high modern technology, dominated by an expert-based culture of knowledge that was not completely understandable to non-experts. Herein, experts' explanations were virtually unchallengeable, especially by laymen, whose arguments, on the other hand, were easily refuted. Expert backing was presented as proof of validity for arguments, functioning as a hurdle for any meaningful participation in a debate that was, nonetheless, a political one. By framing it as one dominated by rationality, however, the democratic principle of respecting minority interests was annulled; compromises, characteristic for political disputes, could be ruled out in favour of seemingly clear-cut right-or-wrong rational answers given by competent experts. Not all experts were taken to be competent, however, with their statements seemingly unchallengeable – and the choice over which experts to regard in that way was embodied in the licensing processes, designed by the political majorities backing nuclear energy. Nuclear energy's proponents had, in other words, created and rendered the group of official experts virtually immune to outside challenges to push through their political goals.[422]

This strategy's fate changed once nuclear power's critics began to employ it themselves. While they were unable to complete this transition within the time frame of the first nuclear age under consideration here because they lacked the political power to directly choose the experts they deemed relevant for the official positions within the bureaucratic processes governing the technology's deployment, they made clear that this was to be expected once they had the necessary competences. At least Schröder did so in the case of Lingen. In Hinkley Point, admittedly, no similar figure following an equally dedicated approach appeared. This may well have been a consequence of the different democratic cultures. At least, it was in line with the differences observed regarding the conditions of debate, with the British allowing for successes by minorities short of completely replacing the ruling majority (cf. "Nuclear Debates"). Nonetheless, during the Hinkley Point C public inquiry, critics of nuclear power, like SHE, were able to use the opportunity of having "their" experts heard on par with those provided by the industry. In this way, they were able to effectively turn the debate on the topic of nuclear power's economy, just as Rowe had assumed in relation to the first public inquiry three decades earlier (although, in the end, the political decision was taken rather independently of this argumentative exchange).

This chapter's focus lay on the tripartite transformation process that occurred transnationally in this regard, mostly, but not exclusively, during the 1970s. In the case of Hinkley Point, I was able to demonstrate how it developed over time. First, the ability of atomic energy's proponents to all but steer the nuclear debates via the utilisation of official experts was challenged by the technology's critics presenting critical experts. While these could easily be sidelined initially, this possibility shrank when critical perspectives spread via the (trans-) national plane of debate, especially so in the wake of the debates about reprocessing of nuclear waste at Windscale, but also visible in the growing interest in leukaemia in the wake of a television report. When the rising number of increasingly informed and interconnected critics, second, formed politically oriented groups, from citizens' initiatives to formal parties, and, third, took up issues where they were unwilling to budge, not least the storage of nuclear waste, their adoption of the discursive strategy used by atom's proponents was completed. In the case of Lingen, similar learning effects could be observed, in a temporally terser form owing to the later start of the debates here. Clearly, nuclear power's critics learned from their troubles when they appeared to be arguing without expert backing that, to carry the day regarding issues that were non-negotiable to them, they had to have their own experts *and* the political power to make those experts' perspectives count. This was the case when it came to the Jusos, their failure to establish critical experts in the debates, focusing on the topic of leukaemia, and in their attempt to challenge the official experts regarding the construction of the cooling reservoir. In the end, both cases exemplify how the anchoring of the issue of the societal role of rationality to the debates about nuclear energy shaped not only the latter but also the former.

Similar interactions and dynamics occurred at all the locales under consideration here. In the case of Stade, the atom's proponents were able to dominate the local debate with the help of official experts until the end of the 1970s, when a local loss of trust changed the situation, resulting in the power plant's critics referring more and more to critical experts – regional ones as well as ones active (trans-)nationally – ultimately generating a societal split, a tribalised friend-or-foe stand-off situation in which next to no exchange of arguments was possible anymore because both sides were only willing to consider their own perspective.[423] This was exemplified in a meeting between a citizens' initiative from Stade's neighbour Jork and Lower Saxony's Minister for the Environment, Remmers, in 1989 (cf. "Atomic Trust(s)"). In Biblis, it was mostly the same, with official experts dominating the early debates, only to be challenged once (trans-)nationally spread expertise, political determination, and a goal to fight for unbudgingly came together on the side of the atom's critics, putting them on an equal footing with their counterparts in the public debates about nuclear energy (cf. "Nuclear Debates"). In Oldbury, where the critics of nuclear power were able and even welcome to state feelings averse to the technology, the decisions taken were dominated by its proponents, who clearly relied on experts and scientific data to assert their political goals. As in the other cases under consideration here, the conflict heated up only after critical expertise had become widely available during the 1970s, with the growing number of atomic energy's opponents also finding a political voice by forming citizens' initiatives or aligning themselves with newly critical political parties, making themselves heard especially clearly regarding charged topics like nearby leukaemia cases (cf. "Emotional Atoms"). Even in Torness, the shortest of the British debates considered here, the development from the dominance of the atom's proponents utilising official experts to steer the public debates to a new equilibrium in which the technology's critics opposed them, purposefully employing their own experts and, independently collecting data in an open political conflict, was clearly visible. Although, here, a specifically British way of building trust once more helped to reduce the conflict's heat in the 1980s (cf. "Atomic Trust(s)"). In sum, the different examples strengthen the assumption that the transnational plane of debate was of outstanding importance regarding the interactions between the debates about nuclear energy and the societal role of rationality. National, or even local, differences played a markedly inconspicuous role in this regard.

In the end, the observed transformations resulted in a new kind of equilibrium in both countries: On the one hand, expertise and scientific arguments, the core of atomic rationality, seemed to have been devalued. When it came to interpretable questions, which most issues around nuclear power were, the critics of nuclear power now were resolved, similar to the original stance of the technology's proponents, to only respect "their" experts' arguments, ignoring statements that did not fit their political goal. This became clear, for example, in White's questioning of Oldfield's statements on coal-burning, Brookes' changing perspective on the NII, and in Friends of the Earth's flexible approach to the greenhouse effect, in Hinkley Point, and, in the case of Lingen, in the doubts about the possibility of constructing reliable dams in principle. Most prominently, it was clear from Schröder's plan to sideline the established, official experts in a planned review of Lower Saxony's nuclear power plants, to the benefit of the critical ones. Seemingly, the technocratic age of high modernity, with its reverence for experts and their rational culture of knowledge, had come to an end by way of the nuclear debates.

On the other hand, from the beginning, this "age" had not been one in which rationality and experts dominated the course of the nuclear debates, but where they had been used by the technology's proponents to this end. And their role in the respective discursive strategy had not changed. On the contrary, the reference to experts to undergird one's political position with scientific, rational arguments not only remained as widespread as ever, it had been taken up as a standard by all participants in the nuclear debates. Rationality's presence within them had thereby become even more common. More research was called for in the case of leukaemia around power plant sites, as had happened in the context of the Hinkley Point C public inquiry, and quite fundamental questions of epistemology were brought up in a public meeting of Lingen's City Council in the debates about contaminated whey. Seen in this way, the scientific and technical experts' rational culture of knowledge had not been pushed aside. On the contrary, it had grown and been adopted by laypeople engaging in the nuclear debates – even beyond the core topic of atomic energy.[424]

In sum, then, the rational atom had become, at the same time, devalued but also indispensable and omnipresent. Of course, with both sides of the argument commanding similarly validated positions, the dispute was seemingly reduced to a tribal one of political dominance over a split society,[425] in other words, which side could command the largest network of citizens trusting it. These networks are the focus of the next chapter, at the beginning of volume two of this book.

Notes

1 Herbert, Ulrich: Europe in High Modernity. Reflections on a Theory of the 20th Century, in: Journal of Modern European History 5 (2007), 1, pp. 5–21, especially pp. 18–19.

2 McNeill, J. R.: Something New Under the Sun. An Environmental History of the Twentieth-Century World. New York 2001, pp. 325–256; Kupper, Patrick: Die '1970er Diagnose'. Grundsätzliche Überlegungen zu einem Wendepunkt der Umweltgeschichte, in: Archiv für Sozialgeschichte 43 (2003), pp. 325–348; cf. Pfister, Christian: Das '1950er Syndrom'. Die umweltgeschichtliche Epochenschwelle zwischen Industriegesellschaft und Konsumgesellschaft, in: Pfister,

Christian (Ed.): Das 1950er Syndrom. Der Weg in die Konsumgesellschaft. Bern 1995, pp. 51–95; Engels, Naturpolitik, pp. 28, 209–214, 275–276.

3 Doering-Manteuffel et al., Boom, p. 33; cf. Hobsbawm, Eric: Das Zeitalter der Extreme. Weltgeschichte des 20. Jahrhunderts. Frankfurt/Main 1996 (Original 1994), pp. 324–337.

4 Fraunholz, Uwe, Thomas Hänseroth, Anke Woschech: Hochmoderne Visionen und Utopien. Zur Transzendenz technisierter Fortschrittserwartungen, in: Fraunholz, Uwe, Anke Woschech (Eds.): Technology Fiction. Technische Visionen und Utopien in der Hochmoderne. Bielefeld 2012, pp. 11–24; Metzler, Demokratisierung; cf. Ellul, Jacques: The Technological Society. Toronto 1964 (Original 1954).

5 Bell, Daniel: The Coming of Post-Industrial Society. A Venture in Social Forecasting. New York 1973; Inglehart, Ronald: The Silent Revolution. Changing Values and Political Styles Among Western Publics. Princeton 1977, especially pp. 3–18. Cf. Engels, Naturpolitik, pp. 15–17.

6 For a historiography of the term and concept of objectivity in contrast to subjectivity: Daston, Lorraine, Peter Galison: Objectivity. New York 2010, especially pp. 27–42, 45–46; for the replacement of religion by science through 'social scientists': Fuller, Governance, pp. 99–101, cf. Fuller, Governance, pp. 7–8.

7 Cf. Erickson et al., Reason, especially pp. 2–4, 183–188; Carson, Cathryn: Heisenberg in the Atomic Age. Science and the Public Sphere. New York 2010, pp. 435–437; Jasanoff, Sheila: Imagined and Invented Worlds, in: Jasanoff, Sheila, Sang-Hyun Kim (Eds.): Dreamscapes of Modernity. Sociotechnical Imaginaries and the Fabrication of Power. Chicago 2015, pp. 321–341, here pp. 324–325.

8 Beck, Ulrich: Risikogesellschaft. Auf dem Weg in eine andere Moderne. Frankfurt/Main 1986, especially p. 253; Giddens, Modernity, especially p. 150; Weingart, Peter: Verwissenschaftlichung der Gesellschaft – Politisierung der Wissenschaft, in: Zeitschrift für Soziologie 12 (1983), 3, pp. 225–241; Weingart, Peter: Die Stunde der Wahrheit? Zum Verhältnis der Wissenschaft zu Politik, Wirtschaft und Medien in der Wissensgesellschaft. Weilerswist 2001; cf. Nowotny, Kernenergie; Douglas et al., Risk, p. 10; van der Pligt et al., Public; Hobsbawm, Extreme, pp. 671–672; Welsh, Mobilising.

9 Introducing the concept: Fried, Johannes, Thomas Kailer: Einleitung. Wissenskultur(en) und gesellschaftlicher Wandel. Beiträge zu einem forschungsstrategischen Konzept, in: Fried, Johannes, Thomas Kailer (Eds.): Wissenskulturen. Beiträge zu einem forschungsstrategischen Konzept. Berlin 2003, pp. 7–19, especially pp. 9–13; cf. Detel, Wolfgang: Wissenskulturen und epistemische Praktiken, in: Fried, Johannes, Thomas Kailer (Eds.): Wissenskulturen. Beiträge zu einem forschungsstrategischen Konzept. Berlin 2003, pp. 119–132, especially pp. 120, 129–132. Experts, as understood here, cannot be separated from scientists; cf. Szöllösi-Janze, Margit: Wissensgesellschaft in Deutschland. Überlegungen zur Neubestimmung der deutschen Zeitgeschichte über Verwissenschaftlichungsprozesse, in: Geschichte und Gesellschaft 30 (2004), pp. 277–313, here p. 282.

10 Cf. Irwin et al., Introduction, p. 7. This does not draw into question the applicability of the more general observations mentioned above in principle.

11 For this widespread claim, cf. Kitschelt, Kernenergiepolitik, p. 214; Radkau, Fragen, p. 119; Weisker, Expertenvertrauen, pp. 405–406; Kornwachs, Klaus: Was ist neu an der Neuen Technologie?, in: Kehrt, Christian, Peter Schüssler, Marc-Denis Weitze (Eds.): Neue Technologien in der Gesellschaft. Akteure, Erwartungen, Kontroversen und Konjunkturen. Bielefeld 2011, pp. 27–47, here p. 44; Steineck, Raji C.: Nuclear Power. A Techno-Ethical Perspective, in: Bohn, Thomas et al. (Eds.): The Impact of Disaster. Social and Cultural Approaches to Fukushima and Chernobyl. Göttingen 2015, pp. 25–49, here p. 25.

12 Beck observed the "paradoxical"– situation that critics of modernity needed to believe in science, Beck, Risikogesellschaft, p. 96; cf. Beck, Risikogesellschaft, pp. 263–266; Touraine, Alan: The Crisis of 'Progress', in: Bauer, Martin (Ed.): Resistance to New Technology. Nuclear Power, Information Technology, and Biotechnology. Cambridge 1995, pp. 45–55; Welsh, Mobilising, p. 34; Engels, Naturpolitik, pp. 14–17, 28.

13 Cf. Welsh, Nuclear Nation, pp. 2–3; Jasanoff, Sheila: Science and Norms in Global Environmental Regimes, in: Hampson, Fen Osler, Judith Reppy (Eds.): Earthly Goods. Environmental Change and Social Justice. Ithaca 1996, pp. 173–197, especially p. 179; Ash, Mitchell G.: Wissenschaft und Politik als Ressourcen für einander, in: vom Bruch, Rüdiger, Brigitte Kaderas (Eds.): Wissenschaften und Wissenschaftspolitik. Bestandsaufnahmen zu Formationen, Brüchen und Kontinuitäten im Deutschland des 20. Jahrhunderts. Stuttgart 2002, pp. 32–51, especially pp. 32–36, 50–51; Ash, Mitchell G.: Wissenschaft und Politik. Eine Beziehungsgeschichte im 20. Jahrhundert, in: Archiv für Sozialgeschichte 50 (2010), pp. 11–46, especially pp. 11–17; Laucht, Elemental, pp. 151–156; Seefried, Zukünfte, pp. 161–164; 'The National, the Spectacular, and the Complex Atom: State of Research'.

14 Cf., for the argument of critical scientists 'restabilising' scientific evidence practices: Esselborn, Stefan, Karin Zachmann: Evidence against the 'Nuclear State'. Contesting Technoscience through *Gegenwissenschaft* in the 1970s and 1980s, in: Zachmann, Karin et al. (Eds.): Evidence Contestation. Dealing with Dissent in Knowledge Societies. New York 2023, pp. 193–223.

15 Cf. Topçu, Sezin: From Resistance to Co-Management? Rethinking Scientization in the Contestation of the Technosciences, in: Engaging Science, Technology, and Society 8 (2022) 1, pp. 128–149.

16 Cf. 'The National, the Spectacular, and the Complex Atom: State of Research' and 'Nuclear Debates'.

17 Cf. Irwin et al., Introduction, pp. 2–5; van Laak, Dirk: Weiße Elefanten. Anspruch und Scheitern technischer Großprojekte im 20. Jahrhundert. Stuttgart 1999, p. 186; Wynne, Public, pp. 212, 214; Cotton, Matthew: Nuclear Power and Environmental Justice. The Case for Political Equality, in: Kaijser, Arne et al. (Eds.): Engaging the Atom. The History of Nuclear Energy and Society in Europe from the 1950s to the Present. Morgantown 2021, pp. 202–229, here p. 213; for the relationship of power and knowledge as such: Foucault, Michel: Discipline and Punish. The Birth of the Prison. New York 1979 (Original 1975), especially pp. 27–28.

18 Cf. Hamblin, Fukushima, p. 293.

19 Cf. Jasanoff, Science, p. 196.

20 Walker, Mark: Die Uranmaschine. Mythos und Wirklichkeit der deutschen Atombombe. Berlin 1990, p. 272; cf. Wynne, Rationality, p. 17; Weingart, Verwissenschaftlichung, p. 226; Wörndl, Barbara: Die Kernkraftdebatte. Eine Analyse von Risikokonflikten und sozialem Wandel. Wiesbaden 1992, pp. 63–64, 76–78; Radkau, Technik in Deutschland, p. 362; Schüring, Advertising, p. 375.

21 For the belief in the calculability of all things as a feature of rationalisation: Weber, Soziologie, pp. 348–351; for its dominance after World War II: Weingart, Stunde, p. 12; Doering-Manteuffel et al., Boom, pp. 42–46; cf. Hall, A. Rupert: Die Geburt der naturwissenschaftlichen Methode 1630–1720. Von Galilei bis Newton. Gütersloh 1965 (Original 1963), especially p. 12; Welsh, Mobilising; Jasanoff, Century, p. 382. The most famous example for the presentation of nuclear energy as a primarily scientific and rational topic is: Haber, Heinz: The Walt Disney Story of Our Friend the Atom. New York 1956.

22 Cf. Nelkin et al., Besieged, p. 89; Mitchell, Rationality, p. 142; Esselborn, Stefan, Karin Zachmann: Nuclear Safety by Numbers. Probabilistic Risk Analysis

as an Evidence Practice for Technical Safety in the German Debate on Nuclear Energy, in: History and Technology 36 (2020), 1, pp. 129–164, here p. 130; Kalmbach, Afterlives, p. 36; this phenomenon had been established in disputes about environmental consequences of industries around 1900, de-politicising them: Brüggemeier, Tschernobyl, pp. 123–124.

23 For the emergence of nuclear experts after the Second World War: Johnston, Sean F.: The Neutron's Children. Nuclear Engineers and the Shaping of Identity. Oxford 2012, pp. 2, 6, 170–185, 250–251.

24 Cf. Graul, Gefährdung, especially p. 432; Manstein, Würgegriff; cf. Radkau, Angstabwehr, p. 36; furthermore: Kaijser et al., Nuclear Energy, p. 4.

25 Bundesminister Strauß über seine Aufgaben als Minister für Atomfragen, Manuskript, NWDR, 21.10.1955, Archiv für Christlich Soziale Politik, NLStrauss, ZA 1955/1, www.fjs.de/fileadmin/user_upload/FJS/Documents/politiker/NLStraussZA_1955_1.pdf (23.1.2023).

26 Cf. 'Nuclear Debates'.

27 Cf. Krohn, Wolfgang, Peter Weingart: 'Tschernobyl' – das größte anzunehmende Experiment, in: Michel, Karl Markus, Tilman Spengler (Eds.): GAU – Die Havarie der Expertenkultur. Berlin 1986, pp. 1–25, here pp. 2–3, 8–9; Wörndl, Kernkraftdebatte, pp. 181–182; Weingart, Stunde, pp. 143–148; Collins, Harry: Are We All Scientific Experts Now? Cambridge 2014, pp. 1–48; Nida-Rümelin, Julian: Zur Ethik der Kernenergienutzung: Eine deontologische Perspektive, in: Ostheimer, Jochen, Markus Vogt (Eds.): Die Moral der Energiewende. Risikowahrnehmung im Wandel am Beispiel der Atomenergie. Stuttgart 2014, pp. 38–48, here pp. 39–40; furthermore: Nowotny, Kernenergie, pp. 30–32; Douglas et al., Risk, pp. 49–66; Keck, Otto: Experten und Interessen oder Eier für den Schnellen Brüter, in: Michel, Karl Markus, Tilman Spengler (Eds.): GAU – Die Havarie der Expertenkultur. Berlin 1986, pp. 87–108; Wandschneider, Dieter: Das Gutachtendilemma – Über das Unethische partikularer Wahrheit, in: Gatzemeier, Matthias (Ed.): Verantwortung in Wissenschaft und Technik. Mannheim 1989, pp. 114–129, here p. 118; Carrera et al., Kernenergie, p. 176; Bauer, Susanne, Karena Kalmbach, Tatiana Kasperski: From Pripyat to Paris, from Grassroots Memories to Globalized Knowledge Production. The Politics of Chernobyl Fallout, in: Macdowell, Laurel Sefton (Ed.): Nuclear Portraits. Communities, the Environment, and Public Policy. Toronto 2017, pp. 149–189, here p. 174.

28 For the strategy: Meyer-Abich et al., Zukunft, pp. 20–21, 44, 72; Camilleri, State, pp. 13–15; Engelstädter, Heinz: Wissenschaft und Moral, in: Flach, Günter, Klaus Fuchs-Kittowski (Eds.): Ethik in der Wissenschaft – die Verantwortung der Wissenschaftler. Zum Gedenken an Klaus Fuchs. Berlin 2008, pp. 131–150, here pp. 135–136; Carson, Heisenberg, pp. 161–217; cf. Wandschneider, Gutachtendilemma 1989, pp. 114–129, here p. 114–115; furthermore: Bell, Daniel: Die nachindustrielle Gesellschaft. Frankfurt/Main 1975, p. 84.

29 Cf. Nowotny, Kernenergie, pp. 117–152; Paul, Anti-AKW-Bewegung, pp. 15–16; Johnston, Children, p. 14; Jasanoff, Century, p. 384; furthermore: Beker, Zumutbarkeit, pp. 76–78; for an example of only 'official' experts being perceived as experts: Slovic, Paul: Perception of Risk from Radiation (Original 1996), in: Slovic, Paul (Ed.): The Perception of Risk. London 2011, pp. 264–274, here p. 264.

30 That is not to say that they believed in the pro-nuclear experts' statements: Wynne, Brian: May the Sheep Safely Graze? A Reflexive View of the Expert-Lay Knowledge Divide, in: Lash, Scott, Bronislaw Szerszynski, Brian Wynne (Eds.): Risk, Environment and Modernity. Towards a New Ecology. London 1996, pp. 44–83, here p. 48.

31 For an insider's view on becoming a 'critical expert': Traube, umschalten, pp. 15–16; for the 'label' viewed by critics of nuclear power: Neumann, Wolfgang: Die Anti-AKW-Bewegung und die kritische Wissenschaft, in: Paul, Reimar (Ed.): . . . und auch nicht anderswo! Die Geschichte der Anti-AKW-Bewegung. Göttingen 1997, pp. 220–228; for critical institutions' importance: Oberloskamp, Ambiguities, pp. 429–430; to Frank Bösch, critical experts' acceptance grew after debates about the Three Mile Island accident: Bösch, Frank: Taming Nuclear Power. The Accident Near Harrisburg and the Change in West German and International Nuclear Policy in the 1970s and early 1980s, in: German History 35 (2017), 1, pp. 71–95, here p. 84.

32 Cf. Bell, Gesellschaft, p. 265; Nowotny, Kernenergie.

33 Beck saw a conflict between experts and laypeople transformed into one among experts: Beck, Risikogesellschaft, p. 276; for a tendency of overestimating experts when assessing new technologies; Jung, Öffentlichkeit, pp. 90–96.

34 Critics recognised the proponents' strategy earlier on, but without consequences for the public debates; for a rather inverse relation between attitudes and frequency: Kepplinger, Kernenergie in den Massenmedien.

35 Cf. Jungk, Atomstaat, pp. 54–57; Patterson, Nuclear, p. 6; Wynne, Rationality, p. 24; Irwin et al., Introduction, p. 3; Welsh, Mobilising, pp. 22–23.

36 Cf. Camilleri, State, pp. 106–108; Roose, Jochen: Der endlose Streit um die Atomenergie. Konfliktsoziologische Untersuchung einer dauerhaften Auseinandersetzung, in: Feindt, Peter H., Thomas Saretzki (Eds.): Umwelt- und Technikkonflikte. Wiesbaden 2010, pp. 79–103, here pp. 89–90.

37 Cf. Patterson, Nuclear, pp. 83–85; Radkau, Geschichte der Kerntechnik, pp. 230–231; Pringle et al., Nuclear Barons, pp. 317–318; Rüdig, Anti-Nuclear, pp. 53–58; Stölken-Fitschen, Ilona: Bombe und Kultur, in: Salewski, Michael (Ed.): Das Zeitalter der Bombe. Die Geschichte der atomaren Bedrohung von Hiroshima bis heute. Munich 1995, pp. 258–281; Walker, J. Samuel: Permissible Dose. A History of Radiation Protection in the Twentieth Century. Berkeley 2000, pp. 18–29; Nehring, Cold, especially pp. 156–161; Gassert, Popularität, pp. 51–52; Hamblin, Fukushima, p. 288; Hogg et al., Introduction, p. 492; van Lente, Introduction, pp. 11–13; Weart, Rise, pp. 96–171; Gassert, Philipp: Die Entstehung eines neuen Umweltbewusstseins, in: Greiner, Bernd, Tim B. Müller, Klaas Voß (Eds.): Erbe des Kalten Krieges. Hamburg 2013, pp. 343–363, here pp. 344–350; Hogg, British, pp. 119–120; to Lawrence S. Wittner, a loss of credibility already occurred regarding nuclear weapons' fallout: Wittner, Nuclear, pp. 439–440.

38 Cf. Radkau, kurze Geschichte, pp. 10–11, 14–15; Okrent, David: Nuclear Reactor Safety. On the History of the Regulatory Process. Madison 1981, pp. 103–127, 296–297; for early US protests: Rucht, Von Wyhl nach Gorleben, p. 79; Rüdig, Anti-Nuclear, pp. 109–110.

39 Hamblin, Jacob Darwin: Arming Mother Nature. The Birth of Catastrophic Environmentalism. Oxford 2013, pp. 151–178; Meadows, Donella H. et al.: The Limits to Growth. A Report for the Club of Rome's Project on the Predicament of Mankind. New York 1972; Seefried, Zukünfte, pp. 270–287.

40 Cf. Kuhn, Thomas S.: The Structure of Scientific Revolutions. Chicago 1970, pp. 111–112; Weingart, Stunde, p. 243; for the strategy of turning to the public to overcome obstacles within science: Brooks, Harvey: Some Notes on the Fear and Distrust of Science, in: Deutsch, Karl W., Andrei S. Markovits (Eds.): Fear of Science – Trust in Science. Conditions for Change in the Climate of Opinion. Cambridge 1980, pp. 97–110, here p. 100; Weingart, Verwissenschaftlichung, p. 236; for contemporary critique regarding Gofman and Tamplin: Burn, Nuclear, pp. 62–64.

41 Radkau, Kernkraft-Kontroverse, p. 318; cf. Müller, Kernenergie. Vol. 2, pp. 29–30; for a 'critical mass' as precondition for a 'movement': Brüggemeier, Tschernobyl, pp. 191–192; for an anti-nuclear movement developing from local groups in the United States, Breach, Windscale, pp. 73–75.

42 Weart, Rise, pp. 184–188; Rüdig, Anti-Nuclear, pp. 65–74, 105; Walker, Dose, pp. 36–44, 149–150.

43 Strohm, Holger: Friedlich in die Katastrophe. Eine Dokumentation über Kernkraftwerke. Hamburg [4]1977; exemplifying its spread: Bürgerinitiative Umweltschutz Unterelbe (Ed.): Atomkraft? Nein Danke. Informationen zur Atomenergie. Hamburg 1977, p. 46; cf. Weisker, Emotion, p. 210; for Strohm and the Friends of the Earth: Radkau, Kernkraft-Kontroverse, p. 312; Radkau, Zukunft, pp. 164–165; for his former expert status: Gassert, Kernenergie, p. 120.

44 Goerke, Untersuchung Part I, pp. 106–107.

45 Herring, Energy, pp. 90–99, 149.

46 Royal Commission, Nuclear Power and the Environment, pp. 200–205, quote p. 204; for its reception: Williams, Decisions, pp. 281–284; cf. Camilleri, State, p. 78; Eiser, J. R., Joop van der Pligt, R. Spears: Evaluation of Nuclear Power and Renewable Alternatives as Portrayed in UK Local Press Coverage, in: Environment and Planning A 18 (1986), pp. 1629–1647, here p. 1629; Rüdig, Low Profile, pp. 76–79; Welsh, Mobilising, pp. 120–123, 136–139; to Rowland Pocock, the British debate remained mostly academic: Pocock, Nuclear, pp. 222–224.

47 As had been expected by nuclear scientists: Goldschmidt, Bertrand: The Atomic Adventure. Its Political and Technical Aspects. Oxford 1964, pp. 206–207.

48 Cf. Fischer, Joschka: Vorwort, in: Fischer, Joschka (Ed.): Der Ausstieg aus der Atomenergie ist machbar. Reinbek 1986, pp. 7–11; Schulz, Rüdiger: Chernobyl. A Jolt to the Mind – Media Reporting and Population Reactions in the Federal Republic of Germany, in: Wober, J. Mallory (Ed.): Television and Nuclear Power. Making the Public Mind. Norwood 1992, pp. 145–157, here pp. 154–155; Joppke, Mobilizing, pp. 177–180; Bösch, Taming, pp. 78, 80; Brüggemeier, Tschernobyl, pp. 18–19; Weingart, Stunde, pp. 127–128; cf. Eiser et al., Evaluation, p. 1644; Giddens, Modernity, p. 134; Heßler, Kulturgeschichte, pp. 175–191; Müller et al., Nuclear, p. 1.

49 Cf. Traube, umschalten, p. 45; Rüdig, Anti-Nuclear, especially p. 31; Kepplinger, Impacts, p. 357; Engels, Naturpolitik, pp. 322–329; Rucht, Soziale Bewegungen; Uekötter, Atomare, pp. 101–102.

50 Wolfgang Rüdig pointed to the British government's willingness to hold large public inquiries and to forgo plans for final repositories during the 1980s as signs of a specific political style: Rüdig, Low Profile, pp. 70–71, 85–86, 94–96; cf. 'Nuclear Debates'.

51 Inglehart, Ronald: Kultureller Umbruch. Wertewandel in der westlichen Welt. Frankfurt/Main 1989, p. 11, sees a rising demand for and assertion of political influence by young adults since the late 1960s.

52 Jens Ivo Engels observed a tendency for unnegotiable positions being taken up in the environmental protection movement of the 1970s: Engels, Naturpolitik, pp. 273–275; cf. Guggenberger, Bernd: Bürgerinitiativen in der Parteiendemokratie. Von der Ökologiebewegung zur Umweltpartei. Stuttgart 1980, pp. 49–51, 54–59; Bösch, Zeitenwende, pp. 302–303; Meyer, Kleine Geschichte, p. 12; Jochen Roose wondered whether the intransigence had to do with nuclear energy's dangers: Roose, Streit, pp. 94–95.

53 Cf. Kalmbach, Karena: Europe and Chernobyl. Contested Localisations of the Accident's Environmental, Political, Social and Cultural Impact, in: Wenkel, Christian et al. (Eds.): The Environment and the European Public Sphere. Perceptions, Actors, Policies. Winwick 2020, pp. 54–69, here p. 64; for an

interpretation of phases of varying appreciation of experts regarding nuclear power: Arndt, Verunsicherung, pp. 254–256.

54 Cf. Weinberg, Alvin M.: A Nuclear Power Advocate Reflects on Chernobyl, in: Bulletin of the Atomic Scientists 43 (1986), 1, pp. 57–60; Giddens, Modernity, pp. 39–40; Ash, Beziehungsgeschichte, pp. 45–46; Welsh, Mobilising, pp. 22–23; for its role in nuclear power, cf. Mania, Kettenreaktion, pp. 109–110.

55 Gassert, Entstehung, pp. 355–358; Hamblin, Arming, pp. 179–216.

56 Der Bundesminister für Forschung und Technologie, Kernenergie, p. 91; cf. Tempel, Karl G.: Kernenergie in der Bundesrepublik Deutschland. Berlin 1981, pp. 30–50, 94–96; Wörndl, Kernkraftdebatte, pp. 132–139; this argument, then, was not developed after Chernobyl: Weingart, Stunde, p. 276; nor did it only become popular in 1979: Bösch, Zeitenwende, pp. 11, 17; Uekötter, Atomare, pp. 200–201, 267–270.

57 Cf., for Germany: Hasenöhrl, Zivilgesellschaft, pp. 406–407; Kretschmer, Wackersdorf, p. 208; for Britain: Brown, Invisible, pp. 211–212.

58 Cf. 'Nuclear Debates'; Engels, Naturpolitik, pp. 180–184, 330–331; Nehring, Atomzeitalter, pp. 238–239; Altenburg, Wandel, pp. 250–251, arguing for the Wyhl protests and the Freiburg Eco-Institute as starting points for the utilisation of 'critical' experts; furthermore: Esselborn et al.: Nuclear, pp. 129–130.

59 Cf., for the aftermath of Chernobyl, Uekötter, Atomare, pp. 246–247; furthermore Meiske, Martin: Die Geburt des Geoengineerings. Großbauprojekte in der Frühphase des Anthropozäns. Göttingen 2021, p. 265.

60 Cf. Beck, Risikogesellschaft, pp. 40, 95–96, 256, 263–266; for the observation of a change of priority in the anti-nuclear argumentation from critical scientists to lay experts in France: Topçu, Sezin: Confronting Nuclear Risks. Counter-Expertise as Politics within the French Nuclear Energy Debate, in: Nature and Culture 3 (2008), 2, pp. 225–245. For the spread of professional terms: Jung, Öffentlichkeit, pp. 82–88; Jung, Emotionalität, pp. 156–159; cf. Engels, Naturpolitik, p. 331.

61 Cf. Wörndl, Kernkraftdebatte, pp. 97–99; Engels, Naturpolitik, pp. 330–331; Reinecke, Christiane: Wissensgesellschaft und Informationsgesellschaft, Version 1.0, in: Docupedia-Zeitgeschichte, 11.2.2010, https://docupedia.de/zg/Wissensgesellschaft (23.1.2023). Pointing to an inverted connection: Frank, Susanne: Krisenkommunikation Kernenergie. Ein neuer intra-extra-medialer Ansatz zur Untersuchung von Wirklichkeitskonstruktion durch Massenmedien. Saarbrücken 2008, here p. 13.

62 Cf. Reinecke, Wissensgesellschaft; Elke Seefried made a similar argument regarding futures research: Seefried, Zukünfte, pp. 11–12, 14–15; criticising the concept: Vogel, Jakob: Von der Wissenschafts- zur Wissensgeschichte. Für eine Historisierung der 'Wissensgesellschaft', in: Geschichte und Gesellschaft 30 (2004), pp. 639–660.

63 Cf., for the 'honest' reading, Jasanoff, Science, p. 173.

64 Before Lingen's debate about the Emsland nuclear power plant started, the city had been home to the Kernkraftwerk Lingen, a demonstration power plant combining a boiling water reactor and an oil-fired superheater. It had been publicly welcomed, commenced operations in 1968, and was shut down in January 1977 because of various problems: Fahrt zum Main und zur Donau: In Kahl und Gundremmingen sind jetzt alle Vorurteile gegen Atomkraftwerke beseitigt. Vertreter vom Kreis, der Stadt Lingen und Darme besichtigen zwei Projekte in Bayern, in: *Emsland-Nachrichten*, 11.4.1984; confidential note on the visit by the Stadtoberrechtsrat Lingen, 14.4.1964, 10 Kr/A, Stadtarchiv Lingen (StadtA LIN), OStDir 84; Kitschelt, Kernenergiepolitik, pp. 194–197; Müller, Kernenergie. Vol. 2, pp. 373–379.

65 Cf. 'Nuclear Debates'.
66 Cf. A Vast Project, in: *Bridgwater Mercury (BM)*, 4.12.1956; Nuclear Power Station Project. To Serve County's Growing Demand for Electricity. Start at Bridgwater Bay Site in Near Future?, in: *BM*, 4.12.1956; statements by councils in Somerset Heritage Centre (SHC), A\BPR 6/1/1; Evison to Ministry of Fuel and Power, 17.1.1957, SHC, A\BPR 6/1/1.
67 A Programme of Nuclear Power, February 1955 (draft), Paragraphs 25 and 26, p. 6, National Archives (NA), CAB 129/73/c/55/31/31.
68 A Vast Project, in: *BM*, 4.12.1956.
69 There May Be Two Nuclear Stations! Press Conference Announcement, in: *BM*, 5.3.1957.
70 County Planning Officer to Lord Hylton, 13.12.1956, SHC, A\BPR 6/1/1.
71 Nuclear Power Station Project. To Serve County's Growing Demand for Electricity. Start at Bridgwater Bay Site in Near Future?, in: *BM*, 4.12.1956.
72 Nuclear Power Station Project. To Serve County's Growing Demand for Electricity. Start at Bridgwater Bay Site in Near Future?, in: *BM*, 4.12.1956.
73 Landowner Is Opposed to Power Station Plan. Mr. Colin Mackenzie Says 'No' to Surveyors, in: *BM*, 5.2.1957; for his arguments, cf. Mackenzie to Somerset County Council, 16.1.1957, SHC, A\BPR 6/1/1.
74 Cf. Atomic Power Station, in: *BM*, 22.1.1957; Page to Whittaker, 15.2.1957, SHC, A\BPR 6/1/1.
75 Stogursey and the Power Station. Housing Site at Wick Park?, in: *BM*, 19.2.1957.
76 More Details of Nuclear Power Station. Special Meeting at Stogursey, in: *BM*, 15.1.1957.
77 Nuclear Power Station Project. Stormy Protests – and Mild Approvals – at Bridgwater Public Inquiry, in: *BM*, 21.5.1957; cf. Report to the Minister of Power, 31.5.1957, p. 27, NA, HLG 79/1172; E. E. Sharp to Mr. Street, 17.10.1956, NA, HLG 79/1171; for Rickards: Personal Notes, in: *The Solicitors' Journal* 116 (1972), p. 482.
78 Notes of a Conference held in the Shire Hall, Taunton, 11.12.1956, pp. 1–7, SHC, A\BPR 3/3/49.
79 Namely, the Weston-Super-Mare Town Council, R. G. Lickfold to Rickard, 15.1.1957, SHC, A\BPR 3/3/49.
80 Minutes, Meeting, Somerset County Council, 26.3.1957, p. 244, SHC, C/CC/M 2/22.
81 Early Approval Anticipated. Hinkley Point Nuclear Power Station Plan, in: *BM*, 26.3.1957; cf. Power Station: M.P. Questioned. 'Bound to Affect Lives of Local People', in: *BM*, 9.4.1957; 'No Danger from Radio Activity'. Conservancy's Reply to Stogursey Council, in: *BM*, 30.4.1957; Letter by the Nature Conservancy, 14.3.1957, SHC, A\BPR 6/1/2; Importance of Nature Reserve. Official at Stogursey Council Meeting, in: *BM*, 21.5.1957.
82 Atomic Power Station, in: *BM*, 22.1.1957; cf. Page to Whittaker, 15.2.1957, SHC, A\BPR 6/1/1.
83 Bridgwater District Resident: Possible Source of Danger, in: *BM*, 12.3.1957.
84 Rowe, Newton A. (Parsonage Side, Otterhampton): Warmer for Bathing?, in: *BM*, 19.3.1957.
85 Report of a Public Inquiry at Town Hall Bridgwater, 14.-15.5.1957, NA, HLG120/230.
86 Cf. 'The National, the Spectacular, and the Complex Atom: State of Research'; for the then dominant figures: Gowing, Independence. Vol. 2, pp. 3–35.
87 Report to the Minister of Power, 31.5.1957, pp. 4–5, NA, HLG 79/1172.
88 Report to the Minister of Power, 31.5.1957, p. 12; for similarities during the Hunterston public inquiry: Welsh, Nuclear Nation, pp. 14–15.

89 Report to the Minister of Power, 31.5.1957, pp. 1, 14, NA, HLG 79/1172.
90 Report to the Minister of Power, 31.5.1957, pp. 33–34.
91 Farmers Have Fears. Atomic Fall-Out at Hinkley?, in: *BM*, 29.10.1957.
92 'No Interference' Say Farmers. Bridgwater Views on Staghunting, in: *BM*, 21.1.1958.
93 Bridgwater District Resident: Atomic Stations, in: *BM*, 10.12.1957.
94 Lafford, A. V. (Chilton Polden): Potential Danger, in: *BM*, 28.10.1958; exemplary 'utterances' were the Central Electricity Generating Board's brochures, presenting nuclear energy and the power station, SHC, A\BPR 4/6/1.
95 Rowe, Newton A. (Otterhampton): To the Editor, in: *BM*, 31.12.1957; cf., for a similar problem at Bradwell: Welsh, Nuclear Nation, p. 12; for a more detailed discussion, cf. this sub-chapter's final section and 'Nuclear Futures'.
96 Rowe, Newton A. (Otterhampton): Plutonium from Hinkley, in: *BM*, 24.6.1958; cf. Baker, Nuclear, p. 31.
97 A 'Windscale' at Hinkley? No, Says Resident Engineer, in: *BM*, 5.11.1957.
98 No Potential Danger at Hinkley, in: *BM*, 10.7.1962.
99 Onward: Hinkley Project, in: *BM*, 24.12.1957.
100 No Danger – No Objection. Radioactive Waste to Go into Channel, in: *BM*, 19.3.1963; cf. Radio-Active Waste Will Not Be Harmful, in: *BM*, 23.4.1963; Forms of Pollution, in: *BM*, 23.4.1963.
101 Radioactive Waste Will Be Dumped in Channel. 'No Risks' Say the Ministry, in: *BM* 19.10.1971.
102 Nuclear Station Misgiving: 'There Should Have Been an Inquiry', in: *BM*, 6.11.1973.
103 Happier about Hinkley Security, in: *BM*, 25.11.1975.
104 B&P Club Members Visit Power Station, in: *BM*, 27.9.1977.
105 Concerned: Fantastic, in: *BM*, 29.11.1960.
106 John, A. B. et al.: Radio-Active Fall-Out, in: *BM*, 30.6.1959; for Pauling: Jolly, J. Christopher: Linus Pauling and the Scientific Debate over Fallout Hazards, in: Endeavour 26 (2002), 4, pp. 149–153.
107 Lederman, H. (Bridgwater): Radio-Active Waste, in: *BM*, 30.4.1963; cf. Lederman, H. J. (Cannington): Hinkley Article 'Well Timed', in: *BM*, 29.8.1978.
108 Lederman, H. (Bridgwater): Radio-Active Fall Out, in: *BM*, 19.7.1966.
109 Patterson, Going Critical, pp. 25–46.
110 Patterson, Walter Cram: The Windscale Report. A Nuclear Apologia, in: Bulletin of the Atomic Scientists 34 (1978), 6, pp. 44–46, here p. 44–45; Patterson, Going Critical, pp. 107–109; Hall, Nuclear, pp. 148–150.
111 Royal Commission, Nuclear Power and the Environment, p. 1.
112 Environmental Hazards from Radioactivity?, in: *Thornbury Gazette*, 30.3.1974.
113 Patterson, Going Critical, pp. 111–127; Williams, Decisions, pp. 289–290; cf. Rüdig, Low Profile, p. 71; Hall, Nuclear, pp. 150–154; Anthony Wedgwood Benn ascribed the article to Friends of the Earth member Bryn Jones: Benn, Diaries, p. 333.
114 Cf. Smith, J. H.: Flowers, Brian Hilton, Baron Flowers (1924–2010), Theoretical Physicist, Public Servant, and University Administrator, in: Oxford Dictionary of National Biography, www.oxforddnb.com/view/10.1093/ref:odnb/9780198614128.001.0001/odnb-9780198614128-e-102721 (26.7.2024).
115 Royal Commission, Nuclear Power and the Environment, pp. 200–205, quote p. 204.
116 Breach, Windscale; Wynne, Rationality; Patterson, Going Critical, pp. 110–127; Hall, Nuclear, pp. 154–164.
117 Patterson, Windscale, p. 44.

118 Cf. Breach, Windscale, pp. 73, 95–121.
119 Stephen Jones Reports . . . It's Power, Not Peril, at Hinkley Point, in: *BM*, 25.2.1975; earlier on, the United Kingdom Atomic Energy Authority had been responsible, cf. 'Nuclear Debates'.
120 Cf. Hinkley Gets the 'All Clear', in: *BM*, 10.5.1975; for Rees: McKeown, Rory: Margaret Rees Dies Aged 93, in: *BM*, 14.1.2010.
121 Lederman, H. J. (Cannington): Hinkley Article 'Well Timed', in: *BM*, 29.8.1978.
122 Selley, Norman (for Bridgwater Ecology Party, Bridgwater): 'The China Syndrome' – Could It Happen Here?, in: *BM*, 13.11.1979; There may be 2,000 jobs at Hinkley, in: *BM*, 26.9.1978.
123 Stevenson, Ian (Secretary, Bridgwater Ecology Party, Combwich): Take Closer Look at Nuclear Programme, in: *BM*, 11.12.1979.
124 New Anti-Nuclear Alliance, in: *BM*, 16.10.1979.
125 Grunberg, Danielle R. (SANA, Somerset Anti Nuclear Alliance, Bridgwater): Nuclear Waste: Rethink Needed, in: *BM*, 5.2.1980.
126 New Anti-Nuclear Alliance, in: *BM*, 16.10.1979.
127 Tom Burke to Address Public Meeting, in: *BM*, 12.2.1980.
128 New Anti-Nuclear Alliance, in: *BM*, 16.10.1979.
129 Atomic Waste Move – Group Will Protest, in: *BM*, 17.6.1980.
130 Nuclear Waste Protest Planned, in: *BM*, 8.7.1980; for Grunberg: Sutcliffe, Jill: Stop Hinkley Are Very Sorry to Learn of the Death of One of Our Founders, Danielle Grunberg, in: Stop Hinkley. Campaigning against Nuclear Power in the South West, http://stophinkley.org/archive/DaniGrunberg.htm (23.1.2023).
131 Atomic Waste Move – Group Will Protest, in: *BM*, 17.6.1980.
132 Drill Sites May Soon Be Known, in: *BM*, 16.9.1980.
133 Donkin, Brod: Townspeople in the Dark on Nuclear Meeting, in: *BM*, 9.9.1980.
134 Jones, Neville: Top Post for MP, in: *BM*, 11.1.1983; for King: King of Bridgwater, Lord, in: Dod's Parliamentary Companion 177 (2009), p. 719.
135 MP Opposed to Puriton as Site for Nuclear Waste Depository, in: *BM*, 9.12.1980.
136 For his position: Stevenson, Ian (Secretary, Bridgwater Ecology Party, Combwich): Take Closer Look at Nuclear Programme, in: *BM*, 11.12.1979.
137 Oldfield, M. A. (Bridgwater): Greater Danger to Obstruct Research, in: *BM*, 23.9.1980; the paper referred to was, in all probability: Manning, P. T.: A Comparison of the Radiological Impacts of the Coal and Nuclear Power Fuel Cycles for Power Generation. London 1980.
138 Oldfield, Mike (Bridgwater): Sanity Call in Waste Argument, in: *BM*, 16.12.1980.
139 Halder, R. W. (Bridgwater): Plea to Stick to Facts, in: *BM*, 11.11.1980; Halder was in all probability 'Bill' Halder, who began an apprenticeship in Hinkley Point in 1976: End of an Era for Hinkley Point B – Switch-Off Time after 45 Years, in: *West Somerset Free Press*, 26.2.2022.
140 Halder, R. W. (Bridgwater): Nuclear Waste, in: *BM*, 16.12.1980; Halder, R. W. (Bridgwater): Putting the Nuclear Issues in Perspective, in: *BM*, 10.2.1981; Halder, R. W. (Bridgwater): Radioactive Waste: Methods of Disposal, in: *BM*, 13.2.1981.
141 Donkin, Brod: Test Bores Move Alerts Critics of Radio-Active Waste Dumping, in: *BM*, 7.10.1980.
142 Waste Sites Despoilation Discounted, in: *BM*, 14.10.1980.
143 Nuclear Waste Storage Decision Not for Another 100 Years?, in: *BM*, 11.11.1980.
144 Clarke, Frederick J. (Ecology Party, Chilton Polden): More Money Is the Answer for Energy Research, in: *BM*, 16.12.1980.
145 Jervis, Robert H. (North Petherton): Answers Are Wanted to Vital Questions, in: *BM*, 27.1.1981.
146 Donkin, Brod: Test Bores Move Alerts Critics of Radio-Active Waste Dumping, in: *BM*, 7.10.1980.

147 Boldero, Mrs Joy (Puriton): Nuclear Waste: Keep It above the Ground, in: *BM*, 18.11.1980.
148 Jervis, Robert H. (North Petherton): 'The Main Worry' of Burying This Waste, in: *BM*, 30.12.1980.
149 Jervis, Robert H. (North Petherton): Nuclear Waste – Fears of 'Bury and Be Damned' Attitude, in: *BM*, 9.12.1980.
150 Powell, Geoffrey (Highbridge): Answer to Waste Issue, in: *BM*, 21.10.1980; Patterson, Nuclear, pp. 66–67, 73–74.
151 The specific locations were near Woolavington and Brent Knoll: Donkin, Brod: Drilling Plans Submitted as Nuclear Waste Fears Grow . . ., in: *BM*, 28.10.1980.
152 Ecology Party Has Made 'Positive Contribution', in: *BM*, 28.10.1980; for Porritt, who was active in the Ecology Party from the 1970s and, from 1984, heading Britain's Friends of the Earth: Hilton, Matthew et al.: A Historical Guide to NGOs in Britain. Charities, Civil Society and the Voluntary Sector since 1945. Houndmills 2012, pp. 271–272.
153 White, P. A. (Highbridge): Nuclear Dumps, Further Points, in: *BM*, 30.9.1980.
154 Donkin, Brod: Test Bores Move Alerts Critics of Radio-Active Waste Dumping, in: *BM*, 7.10.1980; Briscombe was twice charter trustee mayor of Bridgwater: Smedley, Brian: Mayor of Bridgwater, www.westoverward.co.uk/about-westover-ward/mayor-of-bridgwater/ (23.1.2023).
155 Survey Reveals Big 'No', in: *BM*, 9.12.1980.
156 Petitions Plan by 'Anti' Group, in: *BM*, 7.10.1980.
157 Pandis Joins Protest, in: *BM*, 4.11.1980.
158 Brassil, Mrs. Jill (Puriton): Protest Was Not Aligned, in: *BM*, 16.12.1980.
159 Donkin, Brod: No Drilling in Sedgemoor Area Planners Tell County Council, in: *BM*, 9.12.1980; cf. No Drilling, Say the Planners, in: *BM*, 9.12.1980.
160 'Great Victory' Claimed as Borehole Bid Is Sent Back, in: *BM*, 3.2.1981.
161 Councillors Explain Their Voting Decision, in: *BM*, 3.2.1981.
162 'Great Victory' Claimed as Borehole Bid Is Sent Back, in: *BM*, 3.2.1981.
163 Cf. Slovic, Paul: Perceived Risk, Trust and Democracy (Original 1993), in: Slovic, Paul (Ed.): The Perception of Risk. London 2011, pp. 316–326, here p. 318.
164 Bore-Holes Inquiry Refused. Anti-Nuclear Group Angry and Worried, in: *BM*, 31.3.1981.
165 'Bore-Holes' Anger, in: *BM*; 31.3.1981.
166 'Bore-Holes' Anger, in: *BM*; 31.3.1981; County Move Should Force Public Inquiry on Test Boreholes, in: *BM*, 7.4.1981.
167 Now Surface Storage of Nuclear Waste Is Favoured Option, in: *BM*, 23.6.1981.
168 Bore-Holes Row Is Finished, in: *BM*, 22.12.1981.
169 Jervis, Robert H. (Bridgwater Anti-Nuclear Group, North Petherton): Bore-Holes Decision Welcomed, in: *BM*, 29.12.1981; for a national perspective on the conflict about waste storage: Rüdig, Low Profile, p. 84.
170 Ayres, Brian (Bridgwater Anti Nuclear Group, Bridgwater): Out of Sight, in: *BM*, 12.4.1983; Foyle, Richard (Stockland): We Should All Be Concerned, in: *BM*, 3.1.1984.
171 Nuke Protesters' Rail Yard Demo, in: *BM*, 2.7.1985.
172 Beckett, Bill: New Fear Over N-Waste Site, in: *BM*, 13.12.1983; Ling, Cyril: Nirex Name Polden Hills as Nuclear Graveyard. Nuke Waste Dump on Our Doorsteps, in: *BM*, 18.11.1987; Minutes, Meeting, Somerset County Council, 25.11.1987, p. 125, SHC, C/CC/M 2/29.
173 Ling, Cyril: Nirex Name Polden Hills as Nuclear Graveyard. Nuke Waste Dump on Our Doorsteps, in: *BM*, 18.11.1987; Ian Welsh and Wolfgang Rüdig argued that the government abandoned the plans because of the general elections in June – as the designated areas were held by Conservatives: Welsh, NIMBY, p. 16; Rüdig, Low Profile, p. 89; for Nirex' alleged application of a deficit model:

Lehtonen, Markku, Matthew Cotton, Tatiana Kasperski: Trust and Mistrust in Radioactive Waste Management. Historical Experience from High- and Low-Trust Contexts, in: Kaijser, Arne et al. (Eds.): Engaging the Atom. The History of Nuclear Energy and Society in Europe from the 1950s to the Present. Morgantown 2021, pp. 170–201, here p. 182.

174 Phelps, Kathleen (Labour County Councillor, Bridgwater North, Bridgwater): Hinkley: Treat Us Like Adults, in: *BM*, 30.11.1982; cf., for the monitoring practice: materials annexed to the invitation to the 16th meeting, Hinkley Point Local Liaison Committee, 22.11.1973, SHC, C/FB/50/1.

175 Phelps, Kathleen (County Councillor, Bridgwater): Why We Didn't Meet Press, in: *BM*, 14.12.1982.

176 Hinkley Report: Make It Easier to Understand, Says Councillor, in: *BM*, 10.1.1984.

177 Redford, Peter: 'Confession' Plays on Nuclear Fears, in: *BM*, 4.3.1987; he was backed by others: Wood, David (Combwich): All Parties Should Support This Inquiry, in: *Burnham & Highbridge Gazette*, 3.11.1987; Inquiry Must Be Independent, in: *BM*, 4.3.1987. Seconded by: A Concerned Reader: This Case Calls for a Thorough Investigation, in: *BM*, 4.3.1987; Hanna, Dave (Bridgwater): This Situation Must Not Be Played Down, in: *BM*, 4.3.1987; SHE Plans Big Hinkley Rally on Good Friday, in: *BM*, 11.3.1987; Challenger, Allan (Secretary, Hinkley Inquiry Campaign, Combwich): Public Safety Demands a Complete Investigation, in: *BM*, 8.4.1987.

178 Chicken, Nuclear, pp. 47, 64–67, 240; cf. Burn, Nuclear, p. 233; Camilleri, State, p. 70; Flood, Nuclear Dream, pp. 13, 34–35.

179 Report Rejects Bad Welds Claim at Hinkley Point, in: *BM*, 22.7.1987; cf. for the investigation: Hinkley – Inquiry Far from Over, in: *BM*, 11.3.1987; 'Hypocrisy' Jibe in Council Row over Nuke Plant, in: *BM*, 25.3.1987.

180 Assuming a high level of trust in experts in the British nuclear debates, especially after Chernobyl: Bauer et al., Pripyat, p. 162; cf. 'Atomic Trust(s)'; for the concept of veteran experts 65 years and older serving as volunteer nuclear emergency responders: Agar, Jon: Sacrificial Experts? Science, Senescence and Saving the British Nuclear Project, in: History of Science 51 (2013), 1, pp. 63–84.

181 Redford, Peter: Brookes Finds an Ally . . . Now Will They Take Me Seriously?, in: *BM*, 10.6.1987.

182 Ling, Cyril: 'Whitewash' Charge in Hinkley Row, in: *BM*, 14.10.1987; for the calls: SHE Hits 'Flimsy' Report, in: *BM*, 22.7.1987; Brookes, Bob (Bridgwater): Do They Really Want to Find These Bad Welds?, in: *BM*, 29.7.1987.

183 For the faltering privatisation of nuclear power: Welsh, Nuclear Nation, p. 1; Rüdig, Privatisierung; Roberts, Jane, David Elliott, Trevor Houghton: Privatising Electricity. The Politics of Power. London 1991; Rüdig, Low Profile, pp. 91–94; cf. Wood, Nuclear, pp. 117–119.

184 Patterson, Nuclear, p. 107.

185 Reactor 'Unsafe' Fears, in: *BM*, 15.1.1974; 'No Plans for Third Power Station', in: *BM*, 15.1.1975.

186 Fear of the Unknown, in: *BM*, 29.1.1974.

187 Worried Council to Meet Hinkley Superintendent, in: *BM*, 25.12.1973.

188 Fear of the Unknown, in: *BM*, 29.1.1974.

189 Cf. Stephen Jones Reports . . . It's Power, Not Peril, at Hinkley Point, in: *BM*, 25.2.1975.

190 'No Plans for Third Power Station', in: *BM*, 15.1.1975.

191 Jones, Neville: Third Nuclear Power Unit a Long Way Off, in: *BM*, 8.12.1981.

192 Jones, Neville: Hinkley Land OK for a Third Station, in: *BM*, 24.11.1981.

193 3,000 New Jobs – But Worries Too. Hinkley: Joy and Protests, in: *BM*, 31.8.1982.

194 HC Deb, 26.1.1981, Vol. 997, Col. 623; cf. statement by Secretary of State for Energy Nigel Lawson: HC Deb, 1.2.1982, Vol. 17, Col. 22; Hinkley C – the Case for a PWR, in: *BM*, 17.5.1988; for Lamont: Lamont of Lerwick, Lord, in: Dod's Parliamentary Companion 177 (2009), pp. 726–727; Laybourn, Keith: Lamont, Norman (Lord Lamont of Lerwick), in: Laybourn, Keith (Ed.): British Political Leaders. A Biographical Dictionary. Santa Barbara 2001, pp. 193–194.

195 Wober, J. Mallory: Toward Knowledge of the Nuclear Age, in: Wober, J. Mallory (Ed.): Television and Nuclear Power. Making the Public Mind. Norwood 1992, pp. 257–266, here p. 259.

196 Who could be considered part of a pro-nuclear transnational movement: Conze et al., Introduction, pp. 8–9.

197 3,000 New Jobs – But Worries Too. Hinkley: Joy and Protests, in: *BM*, 31.8.1982; Third Station for Hinkley Point, in: *BM*, 31.8.1982; Labour Prefers Severn Barrage, in: *BM*, 19.10.1982; Hunt, Colin (Secretary, Somerset Ecology Party): 'Green Movement' Alternatives, in: *BM*, 16.11.1982; Nuclear Opponents, in: *BM*, 23.11.1982; NALGO Vote against Hinkley Plan, in: *BM*, 14.12.1982; for an assessment of local opinion after the announcement: Straw Poll 'Tie' on Power Station, in: *BM*, 14.9.1982; cf. Jacobson, Harold E. (Bridgwater): By-Pass Bridgwater, in: *BM*, 16.11.1982; Duffy, James (Bridgwater): Think about Safety, in: *BM*, 22.11.1982; for the ongoing and escalating character of the dispute: Burrows, Glen (Bridgwater): Perils of Industry That Benefits Nobody – Apart from Itself, That Is, in: *BM*, 19.3.1986; Somers, Janice (Cannington): Be Grateful for Hinkley Point – or Go Back to Oil Lamps and Candles!, in: *BM*, 26.3.1986; for SHE's Formation Out of the Alliance: Constitution of Stop Hinkley Expansion [undated], SHC, A\ELB 2/1/2; Hinkley Coffin Protest, in: *BM*, 3.4.1984; Beckett, Bill: Hinkley Power Struggle Looms, in: *BM*, 26.2.1985; 'Face the Press' Row at Hinkley Point, in: *BM*, 12.11.1985; entry on Stop Hinkley Expansion (1987–2000), Laka-Archive, www.laka.org/docu/magazines/stop_hinkley_expansion/index.html (23.1.2023).

198 Hinkley Meetings Planned, in: *BM*, 7.9.1982.

199 Graham, Fiona: 'Community Must Revolt against Nuclear Plans', in: *BM*, 30.11.1982; cf. Eiser et al., Nuclear Neighbourhoods.

200 Cf. observations by SHE's Coordinator Crispin Aubrey: 'Arguments Will Not Stand Up to Scrutiny', in: *BM*, 24.5.1988.

201 Hinkley Probe in October, in: *BM*, 23.3.1988; County Leads Fight against Hinkley C, in: *BM*, 30.3.1988; Hinkley C Nuclear Power Station Public Inquiry Opens Today at Cannington College. Battle of Sedgemoor '88, in: *BM*, 4.10.1988; for the timing: Department of Energy, Notice, 21.3.1988, SHC, A\BAF 1/21/1; Barnes, Hinkley Point. Vol. 1, p. 3.

202 Long Time No C. Third Hinkley Plant Unlikely to Be on Stream Much before Year 2000, in: *BM*, 4.2.1987; Beckett, Bill: Battle to Stop PWR Plan, in: *BM*, 18.3.1987; Fear Over 'Risks' at Hinkley C, in: *BM*, 11.11.1987; Minutes, Meeting, Somerset County Council, 25.11.1987, pp. 143–6, SHC, C/CC/M 2/29; regarding the Stogursey and Sedgemoor Councils: Sedgemoor Says YES to Hinkley Point C, in: *BM*, 2.12.1987; Meeting, Joint Local Authorities Members' Liaison Group, 14.12.1987, p. 2, SHC, C/SIAD/6/2; Outline Statement of Case Submitted by Stogursey Parish Council (SPC 1), 1/1989, p. 1, NA, EG 4/435; for COLA's composition: Report of the Director Hinkley Point 'C' Secretariat, COLA Members Meeting, 19.9.1988, SHC, A\BPR 3/8/15; Minutes, Nuclear Free Liaison Group, 27.10/87, Bristol Archives, M/BCC/NFL/1/2. Ian Welsh counted 27 authorities: Welsh, Nuclear Nation, p. 1.

203 COLA States Its Case, in: *BM*, 8.11.1988; Rosamond, Jon: Hinkley C Warning, in: *BM*, 23.8.1988; 'Victory' Claims as Inquiry Ends, in: *BM*, 5.12.1989; for *Hinkley Headlines*: SHC, A\BPR 7/2; after the 1989 elections, the Conservatives

now dominating the county council withdrew its opposition: Donkin, Brod: New Council to Support Hinkley C?, in: *BM*, 9.5.1989; County Leaves COLA, in: *BM*, 11.7.1989; Minutes, Meeting, Somerset County Council, 2.8.1989, p. 19, SHC, C/CC/M 2/29.

204 Stop Hinkley Expansion Statement of Case, Hinkley C Public Inquiry, September 1988, p. 2, NA, EG 4/439.

205 All issues in SHC, A\BPR 7/3.

206 Outline Statement of Case, West Somerset Campaign for Nuclear Disarmament (WSCND 1), NA, EG 4/495.

207 Hinkley C Nuclear Power Station Public Inquiry Opens Today at Cannington College. Battle of Sedgemoor '88, in: *BM*, 4.10.1988; Welsh, Nuclear Nation, p. 1.

208 Objection by Mark Tester and Christina Morris, 7.11.1987, NA, EG 4/1099.

209 Objection by D. M. Crocker, 30.10.1987, NA, EG 4/1099.

210 Cormac Ashe to Margaret Thatcher, Undated, Arrived on 11.9.1987, NA, EG 4/1099.

211 Lone Group Supports the CEGB's Case, in: *BM*, 8.11.1988.

212 C. N. Hill to Secretary of State for Energy, 24.11.1987, NA, EG 4/1099; Richard Cottrell to C. J. Russell, 30.9.1987, NA, EG 4/1099.

213 MP Has Theory Refuted, in: *BM*, 15.11.1988; cf. Kalmbach, Afterlives, pp. 50–51.

214 Move to Adjourn Hinkley Inquiry, in: *BM*, 18.4.1989.

215 Hinkley C Nuclear Power Station Public Inquiry Opens Today at Cannington College. Battle of Sedgemoor '88, in: *BM*, 4.10.1988.

216 Barnes, Hinkley Point. Vol. 1, p. 6.

217 Cf. Barnes, Hinkley Point. Vol. 1, pp. i–xxvi, also: SHC, A\BPR 4/6/202. Very shortly, even the question of the role played by experts in the debates popped up, when former Secretary of State for Energy Anthony Wedgwood Benn opined that he had been misled by scientists who, themselves, had served as a tool for other's political goals: Donkin, Brod: Scargill and Benn Attack Nuclear 'Lies', in: *BM*, 11.4.1989.

218 This question about the consequences of low-level radiation was very urgently debated around British nuclear sites, especially during the 1980s, even though it is considered to have been less prominent than in other European countries, notably France, in the wake of Chernobyl: Bauer et al., Pripyat, p. 163; cf., for the experts' conflict about the existence of thresholds up to which radiation was harmless in the wake of the Windscale Fire of 1957, leading to studies and a conflict about leukaemia cases and their causes: Arnold, Windscale, pp. 139–153.

219 The report had triggered corresponding questions around most nuclear power stations: Barnes, Hinkley Point. Vol. 5, p. 1497; cf. Hall, Nuclear, pp. 165–166.

220 Beckett, Bill: Hinkley Is Cleared in Health Probe, in: *BM*, 27.11.1984; Unfortunate (Stogursey): Two Vital Questions, in: *BM*, 11.12.1984; Bowle, Cameron (Community Physician, Somerset Health Authority): Hinkley Questions Answered, in: *BM*, 1.1.1985; Power Station 'Cancer Link', in: *BM*, 5.2.1985; cf. Cutler et al., Nightmare, pp. 98–122; Corner et al., Nuclear; McDermott, Nuclear, pp. 177–183.

221 Dixey, Anne: New Fears on Cancer Deaths, in: *BM*, 23.4.1986; for the following exchange of arguments: Hinkley Cleared in Cancer Probe, in: *BM*, 14.5.1986; Rosamond, Jon: Concern over Leukaemia Cases, in: *BM*, 19.11.1986; cf. Hogg, British, pp. 9–10; for Whitney: Langdon, Julia: Sir Ray Whitney Obituary, in: *The Guardian*, 19.8.2012.

222 Beckett, Bill: Cancer Deaths in Hinkley Area 'At Double Level', in: *BM*, 11.3.1987.

223 Hinkley 'No Risk to Local Young People', in: *BM*, 14.10.1987.

224 Weston, J. (Health Physicist, CEGF [sic, CG]): The Effect of Nuclear Power Is Insignificant, in: *BM*, 18.10.1988; for Weston, cf. Reactor Is 'Fault Tolerant' Says Expert, in: *BM*, 31.1.1989.
225 Ling, Cyril: Leukaemia – No Danger Says AHA, in: *BM*, 14.6.1988.
226 Statement of Case, Severnside Campaign against Radiation (SCAR 2), November 1988, p. 1, NA, EG 4/412.
227 Outline Statement of Case, Severnside Campaign against Radiation (SCAR 1), July 1988, NA, EG 4/412.
228 Statement of Case, Severnside Campaign against Radiation (SCAR 2), November 1988, NA, EG 4/412.
229 Statement of Case, Severnside Campaign against Radiation (SCAR 2), November 1988, NA, EG 4/412, p. 27.
230 Proof of Evidence, Severnside Campaign against Radiation (SCAR 3), December 1988, p. 14, NA, EG 4/412; CEGB Challenged, in: *BM*, 28.2.1989.
231 Link between Hinkley and Leukaemia?, in: *BM*, 20.12.1988.
232 No Major Leak Says the CEGB, in: *BM*, 27.12.1988; Leukaemia Link 'Very Unlikely', in: *BM*, 21.3.1989.
233 Link 'Unlikely', in: *BM*, 17.1.1989; cf. Taylor, Richard Hamilton: Hinkley Point 'C' Power Station Public Inquiry. Proof of Evidence on Safety Criteria. s.l. 1988.
234 Radiation Risk Low, in: *BM*, 24.1.1989; cf. Radon Poses Bigger Risk, in: *BM*, 14.2.1989.
235 No Major Leak Says the CEGB, in: *BM*, 27.12.1988.
236 Greater Risks of Cancer?, in: *BM*, 7.2.1989; Stewart was known for her research into the dangers of low dose radiation; cf. Greene, Gayle: The Woman Who Knew Too Much. Alice Stewart and the Secrets of Radiation. Ann Arbor [2]2017.
237 Safety System under Attack, in: *BM*, 21.3.1989.
238 Closing Submission, Severnside Campaign against Radiation (SCAR 4), August 1989, pp. 2, 4–5, 9–10, quote p. 2, NA, EG 4/412; cf. Severnside Campaign against Radiation, in: *Hinkley Inquirer* 21, p. 3; Leukaemia Links, in: *Hinkley Inquirer* 7, p. 3; Taylor on the Hop, in: *Hinkley Inquirer* 7, p. 3; Low Level Radiation – Expert Speaks, in: *Hinkley Inquirer* 10, pp. 2–3.
239 Ling, Cyril: Leukaemia Report Is 'Alarmist', in: *BM*, 28.11.1989.
240 Phelps, Kathleen (Former member of Hinkley Point Liaison Committee, Bridgwater): Disasters Are Far from Minor!, in: *BM*, 2.5.1989.
241 Barnes, Hinkley Point. Vol. 5, p. 1504.
242 Barnes, Hinkley Point. Vol. 5, p. 1584.
243 Stop Hinkley Expansion Newsletter, Nov/Dec 1990, SHC, A\ELB 2/1/2.
244 C Output 'Twice as Expensive', in: *BM*, 15.11.1988.
245 CEGB Told: Do Your Homework, in: *BM*, 18.10.1988.
246 Coal Cheaper Admission Is Welcomed, in: *BM*, 13.12.1988.
247 Roberts et al., Privatising, p. 114; Cooke, Mortal, pp. 352–353; Taylor, Fall, p. 29; cf. Flood, Nuclear Dream, pp. 44–45; Hogg, British, pp. 162–163; for a critical discussion of cost statements based upon prices per Kilowatt hour: Rubio-Varas, Economic Context, pp. 52–53.
248 Mahaffey, James: Atomic Awakening. A New Look at the History and Future of Nuclear Power. New York 2009, p. xvii; for Britain, cf. Seldon et al., Britain, p. 43; Edgerton, Rise and Fall, p. 481.
249 CEGB Told: Do Your Homework, in: *BM*, 18.10.1988.
250 Barnes, Hinkley Point. Vol. 1, pp. 55, 97; HC Deb, 9.11.1989, Vol. 159, Col. 1171–1172; Hinkley C: The Press Speculates, in: *Hinkley Inquirer* 22, p. 1; 'Victory' Claims as Inquiry Ends, in: *BM*, 5.12.1989; £1 Billion: The Price of a Lost Cause, in: *The Scotsman*, 10.11.1989; Edwards, Rob: Taking Issue, in: *Evening News*, 14.11.1989; Wood, Nuclear, pp. 117–119; Hill, Atomic Empire, pp. 339–345.

251 O'Rourke, Zena: Hinkley C Plans Hit the Rocks in Government U-Turn. Switch Off!, in: BM, 14.11.1989; Minutes, Meeting, Nuclear Free Liaison Group, 20.11.1989, Bristol Archives, M/BCC/NFL/1/3.
252 'Victory' Claims as Inquiry Ends, in: *BM*, 5.12.1989.
253 Cf. Roberts et al., Privatising; Chick, Martin: Electricity and Energy Policy in Britain, France and the United States since 1945. Cheltenham 2007, pp. 1, 115; Jackson, Ian: Nukenomics. The Commercialisation of Britain's Nuclear Industry. Sidcup 2008, pp. 8–11; Taylor, Fall, pp. 27–30; cf. Rüdig, Privatisierung; for Lawson: Laybourn, Keith: Lawson, Nigel (Lord Lawson), in: Laybourn, Keith (Ed.): British Political Leaders. A Biographical Dictionary. Santa Barbara 2001, pp. 196–199.
254 Which was not the case at the end of the 1980s: Roberts et al., Privatising, pp. 3–4, 73–76, 161–174.
255 Barnes, Hinkley Point. Vol. 3, pp. 1018–1024, quote p. 1020.
256 Bickerstaff, Karen et al.: Reframing Nuclear Power in the UK Energy Debate. Nuclear Power, Climate Change Mitigation and Radioactive Waste, in: Public Understanding of Science 17 (2008), pp. 145–169, here p. 146; Hill, Atomic Empire, pp. 339–345; Kalmbach, Afterlives, pp. 140, 183.
257 Bickerstaff et al., Reframing, pp. 148–149, 162; Department for Business, Enterprise & Regulatory Reform: Meeting the Energy Challenge. A White Paper on Nuclear Power. January 2008. London 2008, https://assets.publishing.service.gov.uk/government/uploads/system/uploads/attachment_data/file/228944/7296.pdf (23.1.2023).
258 The idea had already been mentioned in 1973: Bürgerinitiative gegen Atomenergie Osnabrück, Fortschritt, p. 66.
259 rr: Nach Fertigstellung des Speichersees in Geeste: Für Meppen und Lingen insgesamt vier 1300-Megawatt-Kernkraftblöcke. '20 Meter hoher Damm absolut kein Sicherheitsrisiko' – Experten vor Ausschuß, in: *Lingener Tagespost (LT)*, 23.12.1977.
260 Elektromark: Kernkraftwerk in Lingen beantragt, in: *LT*, 9.6.1979; Wiborg, Klaus: Ein Land, das es nicht nur in sich hat, in: Emsländischer Heimatbund (Ed.): Emsland im Werden. Vom Handwerk bis zur Kernkraft. Sögel 1986, pp. 112–139, here pp. 136–138.
261 he: Ein Dutzend Gutachten muß erstellt werden, in: *LT*, 24.5.1979.
262 Dickopp, Alois: Betriebssitz des neuen 1200-Megawatt-Kernkraftblocks wird die Stadt Lingen. Drei-Milliarden-Projekt soll 1986 in Betrieb gehen. VEW-Hauptverwaltung: Abbruch des Kernkraftwerks Lingen reine Spekulation. Lingener Stadtrat zu Informationsgesprächen in Düsseldorf, in: *LT*, 1.7.1978; for Knizia: Altenburg, Kernenergie, pp. 116–117.
263 rs: Langfristig weiterhin 4 × 1200 MW im Emsland, aber . . . VEW/RWE wollen Dimension des Speicherbeckens reduzieren. Planungen jetzt nur noch für zwei Kraftwerksblöcke ausgelegt, in: *LT*, 21.2.1979.
264 The history of the demonstration power plant served as proof for the technology's fallibility to its critics: Bürgerinitiative gegen Atomenergie Osnabrück, Fortschritt, pp. 50–53.
265 Gegen 'Katastrophenhysterie'. ÖTV-Kreisvorstand zu Fragen der Energiepolitik, in: *LT*, 23.3.1978; for Deublein, cf. Müller, Kernenergie. Vol. 2, p. 374.
266 mü: Diskussion mit Betriebsräten der Kraftwerke. Wie stehen emsländische Politiker zu den Kernkraftprojekten? Bei der CDU Zustimmung. Bei der SPD teils Vorbehalte, teils Befürwortung, in: *LT*, 31.1.1979.
267 Regarding the organisation, cf. Seefried, Zukünfte, pp. 235–254; for its foundation: Watts, William: Foreword, in: Meadows, Donella H. et al.: The Limits to Growth. A Report for the Club of Rome's Project on the Predicament of Mankind. New York 1972, pp. 9–12; for Pestel, who was a critic of Meadows'

analysis: Füßl, Wilhelm: Pestel, Eduard, in: Neue Deutsche Biographie 20 (2001), pp. 216–217, www.deutsche-biographie.de/pnd123151376.html#ndbcontent (23.1.2023); Seefried, Zukünfte, p. 453.

268 mü: 'Wer nichts hinzulernt, befindet sich im Zustand der Vergreisung'. Minister Prof. Dr. Eduard Pestel zu Fragen der Energiepolitik, in: *Meppener Tagespost (MT)*, 6.5.1978.

269 jhd: Zwölf neue Standorte in Niedersachsen. Inbetriebnahme des KKW Lingen 1987, in: *LT*, 23.2.1980; for Eppler and his plans: Dämmplatten sind billiger als Kernkraft, in: *Der Spiegel*, 18.6.1979, p. 20; Albrecht-Baba, Alexandra, Karsten Rudolph: Eppler, Erhard, in: Vierhaus, Handbuch. Vol. 1, pp. 185–186.

270 rr: 'Gemeinsam für den Landkreis einsetzen'. Brümmer: Das Emsland kann auf Kernkraftwerke nicht verzichten. Grundlage für neue Arbeitsplätze schaffen – Mehr Humanität in Betrieben, in: *LT*, 25.1.1978; for Brümmer: Landkreis Emsland (Ed.): Von historischer Entwicklung bis zu künftigen Aufgaben. Landkreis Emsland feiert 40-jähriges Bestehen mit vielen Veranstaltungen, www.emsland.de/buerger-behoerde/aktuell/pressemitteilungen/von-historischer-entwicklung-bis-zu-kuenftigen-aufgaben.html (23.1.2023).

271 Dr. Gruhl: Katastrophale Folgen. Brümmer: Verzicht nicht möglich. Referate von Vertretern der Bürgerinitiativen – Sachliche Diskussion, in: *LT*, 30.6.1978.

272 mü: Kritik an den politischen Mandatsträgern. 'Mit Kneifen sind die Probleme nicht zu lösen'. Auf Einladung der Gewerkschaften: weitere Kernenergiedebatte, in: *MT*, 22.12.1978.

273 Jusos lehnen Atomkraft ab. 'Friedliche Nutzung der Kernenergie eine Illusion', in: *LT*, 23.1.1979.

274 Dickopp, Alois: Das Nein der Jusos, in: *LT*, 23.1.1979; for Dickopp: Koop, Robert: Nanu, in: Koop, Robert: Roberts Blog. Gelegentliche Bemerkungen zu und in Lingen (Ems), https://robertkoop.wordpress.com/tag/alois-dickopp/ (23.1.2023).

275 Vent, Hans-Karl (i.A. für 'Bürger gegen Umweltvergiftung' e. V. Lingen) (Lingen): Kernkraftwerke und Speichersee, in: *LT*, 12.5.1978.

276 Gruhl, Herbert: Ein Planet wird geplündert. Die Schreckensbilanz unserer Politik. Frankfurt/Main 1975.

277 Meadows, Limits; cf. Graf, Rüdiger: Die Grenzen des Wachstums und die Grenzen des Staates. Konservative und die ökologischen Bedrohungsszenarien der frühen 1970er Jahre, in: Geppert, Dominik, Jens Hacke (Eds.): Streit um den Staat. Intellektuelle Debatten in der Bundesrepublik 1960–1980. Göttingen 2008, pp. 207–228, here pp. 209–210.

278 Franke, BUND, pp. 7–15, 46; Markham, William T.: Environmental Organizations in Modern Germany. Hardy Survivors in the Twentieth Century and Beyond. New York 2008, p. 233; cf. Uekötter, Greenest Nation, p. 29; for a critical perspective on Gruhl: Radkau, Zukunft, p. 406; cf. Seefried, Zukünfte, p. 287.

279 Dr. Gruhl: Katastrophale Folgen. Brümmer: Verzicht nicht möglich. Referate von Vertretern der Bürgerinitiativen – Sachliche Diskussion, in: *LT*, 30.6.1978.

280 Probably: Lovins, Amory: Sanfte Energie. Das Programm für die energie- und industriepolitische Umrüstung unserer Gesellschaft. Reinbek 1978; for his reception among contemporary critics of nuclear energy: Traube, umschalten, pp. 124–125, 147; for Stricker: Immer einen flotten Spruch auf Lager, in: *Neue Osnabrücker Zeitung*, 17.4.2006.

281 tg: Sorge beim Bau weiterer Kraftwerke im Emsland. Fragenkatalog von Bürgerinitiativen abgeschickt, in: *LT*, 12.5.1979.

282 mü: Kritik an den politischen Mandatsträgern. 'Mit Kneifen sind die Probleme nicht zu lösen'. Auf Einladung der Gewerkschaften: weitere Kernenergiedebatte, in: *MT*, 22.12.1978.

283 Probleme des Speichersees nicht isoliert sehen. Stellungnahme der SPD-Stadtratsfraktion zur CDU-Erklärung, in: *LT*, 12.4.1978.
284 mü: Kritik an den politischen Mandatsträgern. 'Mit Kneifen sind die Probleme nicht zu lösen'. Auf Einladung der Gewerkschaften: weitere Kernenergiedebatte, in: *MT*, 22.12.1978.
285 'Gesundes Mißtrauen ist angebracht . . .' SPD-Fraktion entscheidet sich gegen Atomkraftwerk Lingen 2, in: *LT*, 19.12.1979; for Braese: Koop, Robert: 100 Jahre SPD Lingen, in: Koop, Robert: Roberts Blog. Gelegentliche Bemerkungen zu und in Lingen (Ems), https://robertkoop.wordpress.com/2019/11/17/100-jahre-spd-lingen/ (23.1.2023).
286 'Schwerwiegende Bedenken', in: *LT*, 16.10.1979.
287 CDW: Anhäufung nuklearer Anlagen bereitet der SPD große Sorgen. Frage nach erhöhter Leukämiegefahr erneut gestellt, in: *Grafschafter Nachrichten (GN)*, 20.4.1979.
288 PR: Sicherheitsbericht wirft Probleme auf. Stadt Schüttorf will Experten engagieren – Verwaltung überfordert, in: *GN*, 14.6.1979.
289 *GN*: Stadt holt Rat von Sachverständigen ein, in: *GN*, 20.6.1979; BM: Landkreis in vorläufiger Stellungnahme für Sozialminister. Kernkraftwerk-Bau an geplantem Standort ist nicht vertretbar. Übungsbetrieb auf Range bedeutet 'erhöhte Gefährdung', in: *GN*, 15.3.1980.
290 FG: Glubrecht-Gutachten als Entscheidungshilfe in Zweifel gestellt. GBU: Für Region wichtige Gesichtspunkte ausgeklammert, in: *GN*, 4.2.1981.
291 For the example of Nordhorn: CDW: Nordhorn bleibt bei Einwendungen gegen Kernkraftwerk Lingen. Rat der Stadt entschied nach Glubrecht-Gutachten, in: *GN*, 27.2.1981.
292 Cf. Bösch, Taming, p. 71.
293 Minutes, Meeting, Lingen City Council, 22.5.1979, pp. 4–15, StadtA LIN RP 1979a; reports were demanded by city councillors and by the Emsland District: Landkreis: Standort Lingen-Darme geeignet. 'Erheblicher struktureller Effekt für das Emsland', in: *MT*, 10.4.1980.
294 he: Ein Dutzend Gutachten muß erstellt werden, in: *LT*, 24.5.1979.
295 That argument was shared by the district's SPD: 'Schwerwiegende Bedenken', in: *LT*, 16.10.1979.
296 Minutes, Meeting, Lingen City Council, 9.10.1979, StadtA LIN, RP 1979b.
297 jhd: TOP 14 führte in Lingener Ratssitzung zur heißdiskutierten Kernenergiefrage, in: *LT*, 11.10.1979.
298 Derr, Hinrich Johann: Eigentor?, in: *LT*, 11.10.1979.
299 Minutes, Meeting, Lingen City Council, 15.6.1981, StadtA LIN RP 1981a.
300 Minutes, Meeting, Lingen City Council, 18.2.1982, StadtA LIN RP 1982a; Minutes, Meeting, Lingen City Council, 16.3.1982, StadtA LIN RP 1982a.
301 hab: Klukkert: 'Diese Diskussion hat uns um einiges weitergebracht!' Zweite Anhörung zum Thema Kernkraftwerk – Fraktionen äußerten Zufriedenheit, in: *LT*, 18.3.1982; hab: 'Bislang keine Versagensgründe'. Übereinstimmendes Urteil der beim Kernkraftwerk beteiligten Gutachter, in: *LT*, 18.3.1982.
302 Minutes, Meeting, Lingen City Council, 24.3.1982, StadtA LIN RP 1982a, quote p. 4.
303 Minutes, Meeting, Lingen City Council, 18.2.1982, StadtA LIN RP 1982a.
304 Minutes, Official Hearing, 14.9.81, Niedersächsisches Landesarchiv, Abteilung Oldenburg, Rep 410, Akz. 226 B Nr. 1, pp. II/20–21.
305 hab: Einwender beteiligten sich nicht an der Erörterung von Sachfragen. Anhörung zu geplanten Änderungen beim Kernkraftwerk dauerte nur eineinhalb Stunden, in: *LT*, 17.7.1984.
306 Brinkmann, Hans: Unergiebig, in: *LT*, 17.7.1984.
307 Kein Meisterstück, in: *Der Spiegel*, 16.3.1987.

308 For Bremen as a centre of the German anti-nuclear movement: Amedeberhan, Semhar et al.: Atomkraft? Nein Danke! Der Bremer Widerstand, in: Horn, Sabine et al. (Eds.): Protest vor Ort. Die 80er Jahre in Bremen und Göttingen. Bremen 2012, pp. 157–190.

309 Right-wing author and publisher Roland Bohlinger later came to be known for attempts to draw in doubt the source documents covering the Wannsee-Conference, cf. Mentel, Christian: Das Protokoll der Wannsee-Konferenz. Überlieferung, Veröffentlichung und revisionistische Infragestellung, in: Kampe, Norbert, Peter Klein (Eds.): Die Wannsee-Konferenz am 20. Januar 1942. Dokumente. Forschungsstand. Kontroversen. Cologne 2013, pp. 116–138, here p. 128; for Soyka and Bohlinger: Hlade, Josef: Die Beteiligung von ÄrtzInnen an der Anti-AKW Bewegung. Vom *Ärzte-Memorandum betreffend die Errichtung von Kernspaltungskraftwerken* (1969) zur Aktion 'Vorarlberger Ärzte gegen AKW' (1977), in: Nemec, Birgit et al. (Eds.): Medizin in Wien nach 1945. Strukturen, Aushandlungsprozesse, Reflexionen. Göttingen 2022, pp. 345–360, here, pp. 350–351.

310 In der Luft, in: *Der Spiegel*, 30.10.1978; cf. Bürgel, Stefan: Das Kreuz mit dem Atom. Die Debatte um die Kernenergie und die christlichen Grundwerte der CDU. Berlin 2018, pp. 194–196; for Kater: Molitor, Jochen: Katastrophe als Beruf. Die bundesdeutsche Ärzteschaft und der nukleare Ernstfall (1950–1990). Cologne 2021, pp. 151–154.

311 dpa: Mehr Leukämiefälle im Bereich des Kernkraftwerkes? Ministerium bezweifelt Untersuchungen, in: *LT*, 25.10.1978; cf. Kritik an 'Geschäft mit der Angst'. Minister: Für die Bevölkerung im Raum Lingen gibt es keine Gefahren. 'Strahlenbelastung sehr gering' – Blutkrebs bald anzeigepflichtig?, in: *LT*, 9.11.1978.

312 'Unverantwortliche Panikmache mit wissenschaftlichem Anstrich'. Gewerkschaft ÖTV zum Thema Kernkraftwerk und Leukämie, in: *LT*, 31.10.1978.

313 'Es gibt keine Zunahme von Leukämieerkrankungen' Erklärung von Oberkreisdirektor K. H. Brümmer vor dem Kreistag, in: *LT*, 7.11.1978.

314 Hiemann, Georg (Twist): Zufall oder Methode?, in: *MT*, 3.11.1978.

315 dpa: Mehr Leukämiefälle im Bereich des Kernkraftwerkes? Ministerium bezweifelt Untersuchungen, in: *LT*, 25.10.1978.

316 'Gefährdungsgrenze nicht überschritten'. Lingen: Versuchsweise auch Plutoniumelement benutzt. Sozialministerium erteilte 1970 Genehmigung, in: *LT*, 31.10.1978.

317 Hiemann, Georg (Twist): Zufall oder Methode?, in: *MT*, 3.11.1978.

318 'Es gibt keine Zunahme von Leukämieerkrankungen' Erklärung von Oberkreisdirektor K. H. Brümmer vor dem Kreistag, in: *LT*, 7.11.1978; this probably referred to Günter Landbeck from Hamburg University Clinic: Winkler, K.: Zum Tod unseres Kollegen Professor Dr. Günter Landbeck, in: Klinische Pädiatrie 204 (1992), 4, p. 189.

319 Bonn und Hannover: Keine Häufung von Leukämie. Zahl der Erkrankungen in Lingen weder über noch unter dem Durchschnitt – 'Unverantwortliche Panikmache', in: *LT*, 9.11.1978.

320 Sommer, Dieter: Lernprozeß notwendig, in: *LT*, 9.11.1978.

321 'Emotionen müssen abgebaut werden'. Wolf: Soyka vor Umweltausschuß hören, in: *LT*, 2.11.1978; later on, a similar line of argument was taken by Kater: hh: 'Unverantwortlicher Panikmacher' hielt erstes 'Leukämie'-Referat: Lebensrisiken stecken nicht nur im 'Atom'. Volkshochschule hatte zu Seminarbeginn ein volles Haus, in: *GN*, 8.2.1979; for Wolf: Wolf, Willi, in: Vierhaus, Handbuch. Vol. 2, p. 972.

322 'Emotionen müssen abgebaut werden'. Wolf: Soyka vor Umweltausschuß hören, in: *LT*, 2.11.1978.

323 Minutes, Meeting, Lingen City Council, Administrative Board, 13.11.1978, pp. 3–4, StadtA LIN, AP, Verw 1978.

324 Dickopp, Alois: Politischer Hefesatz, in: *LT*, 1.12.1978.
325 A.D.: Leukämievorwurf in der Diskussion: Strahlenemission von Kernkraftwerken – Gefahr für die Gesundheit? Sachliche und breitangelegte Diskussion im Ludwig-Windthorst-Haus – 'Höchste Zeit für Krebsregister' – Soyka: Wo wir auch hinkamen, stießen wir auf Leukämiefälle – Sozialministerium: Leukämieerkrankungen liegen im Rahmen der bundesdeutschen Durchschnittswerte, in: *LT*, 1.12.1978.
326 Mü.: 'Diskussion kommt zehn Jahre zu spät'. 'Mehr Leukämie bei Kernkraftwerken' – Experten tagen, in: *LT*, 30.11.1978; A.D.: Leukämievorwurf in der Diskussion: Strahlenemission von Kernkraftwerken – Gefahr für die Gesundheit? Sachliche und breitangelegte Diskussion im Ludwig-Windthorst-Haus – 'Höchste Zeit für Krebsregister' – Soyka: Wo wir auch hinkamen, stießen wir auf Leukämiefälle – Sozialministerium: Leukämieerkrankungen liegen im Rahmen der bundesdeutschen Durchschnittswerte, in: *LT*, 1.12.1978.
327 Dickopp, Alois: Politischer Hefesatz, in: *LT*, 1.12.1978; a few days later, Lingen City Council's environmental committee also demanded more data from the Ministry: Minutes, Meeting, Lingen City Council, environmental committee, 4.12.1978, pp. 6–7, StadtA LIN, AP, Umw 1977/81.
328 Derr, Johann Hinrich: Leukämie-Sterblichkeitsrate im Emsland deutlich unter dem Landesdurchschnitt. Podiumsdiskussion zur Frage 'Vermehrte Leukämieerkrankungen im Emsland?' – Experten nahmen Stellung zum 'Soyka-Alarm', in: *LT*, 22.2.1979; cf. Schweizerische Vereinigung für Atomenergie (Ed.): Allgemeine Strahlenbelastung des modernen Menschen. Zurich 1973, p. 16; Niedersächsischer Landtag (Ed.): Übersicht über die Verwaltung des Niedersächsischen Landtages, die Niedersächsische Landesregierung und die Dienststellen der Bezirksregierungen. Hanover 1982, p. 4; Gesellschaft für Reaktorsicherheit (Ed.): Sicherheitstechnik bei der Entsorgung radioaktiver Abfälle. Cologne 1985, p. 65; Whiteside, Ruth M. et al. (Eds.): Major Companies of Europe 1991/92. Vol. 1. Major Companies of the Continental European Economic Community. London 1991, p. 619; Streffer, Christian et al.: Radioactive Waste. Technical and Normative Aspects of Its Disposal. Berlin 2011, p. 464; for Schellong: Müller, Christian: Ein Meilenstein in der Medizin. Die Geschichte des Behandlungserfolges bei Leukämie und Lymphom im Kindesalter. Berlin 2014, p. 66; for Portheine, who in all likelihood was Friedrich Portheine: EB: Geburtstage, in: Deutsches Ärzteblatt 97 (2000), 31–32, p. A 2120.
329 Diskussionspunkte, in: *LT*, 22.2.1979.
330 Derr, Johann Hinrich: Leukämie-Sterblichkeitsrate im Emsland deutlich unter dem Landesdurchschnitt. Podiumsdiskussion zur Frage 'Vermehrte Leukämieerkrankungen im Emsland?' – Experten nahmen Stellung zum 'Soyka-Alarm', in: *LT*, 22.2.1979; cf. Leukämie-Sterblichkeit im Emsland deutlich unter Landesdurchschnitt. Experten nahmen während Podiumsdiskussion in Lingen Stellung zum 'Soyka-Alarm' – Kein Leukämie-Register, in: *GN*, 24.2.1978.
331 Doubts were even recognisable among critics of nuclear power: Cohrs, Ernst-Otto, Wolfgang Knigge: Atomenergie. Das Grundgesetz und die Wirklichkeit. Tatsachen – Hintergründe und die gezielte Irreführung der Öffentlichkeit. Ed. by Weltbund zum Schutze des Lebens. Rotenburg ^{2}1979, p. 51.
332 K.G.-Eigenbericht: 'Kindesterblichkeit im Emsland unter Durchschnitt', in: *LT*, 4.3.1980.
333 Weniger Leukämiefälle im Emsland als anderswo. Winfried Hartmann: Nicht der letzte Fall von Panikmache, in: *LT*, 7.3.1980; for Hartmann: Hartmann, Winfried. CDU, in: Simon, Barbara (Ed.): Abgeordnete in Niedersachsen 1946–1994. Biographisches Handbuch. Hanover 1996, p. 141.
334 Hager, Protest, p. 134.
335 FG: Glubrecht-Gutachten als Entscheidungshilfe in Zweifel gestellt. GBU: Für Region wichtige Gesichtspunkte ausgeklammert, in: *GN*, 4.2.1981; cf. Bürgerinitiative gegen Atomenergie Osnabrück, Fortschritt, pp. 55–63.

336 'Für Kleinkinder lebensgefährlich'. Grünen-Kandidat sprach in Lingen, in: *LT*, 5.5.1986; cf. Minutes, Meeting, Lingen City Council, 25.2.1987, StadtA LIN RP 1987a, pp. 6–7; Minutes, Meeting, Lingen City Council, 29.4.1987, StadtA LIN RP 1987a.
337 Götter, Accepted, pp. 137–139.
338 rr: Nach Fertigstellung des Speichersees in Geeste: Für Meppen und Lingen insgesamt vier 1300-Megawatt-Kernkraftblöcke. '20 Meter hoher Damm absolut kein Sicherheitsrisiko' – Experten vor Ausschuß, in: *LT*, 23.12.77.
339 rr: Nach Fertigstellung des Speichersees in Geeste: Für Meppen und Lingen insgesamt vier 1300-Megawatt-Kernkraftblöcke. '20 Meter hoher Damm absolut kein Sicherheitsrisiko' – Experten vor Ausschuß, in: *LT*, 23.12.77; at the company, Täubert was responsible for engineering, cf. Whiteside, Ruth M. et al. (Eds.): Major Energy Companies of Europe 1994. London 1994, pp. 111–112.
340 mü: Vorteile für Lingen und Meppen? Nachteile für Gemeinde Geeste? Diskussion über den geplanten See mit Vertretern von VEW und RWE, in: *MT*, 13.2.1978.
341 'Steriles Becken' – 'Wertloses Ding' – 'Fremdkörper'. Geester Rat: Nein zum geplanten Wasserspeicher. Bedenken nicht ausgeräumt – Zwei Gegenstimmen, in: *MT*, 25.3.1978; cf. mü: Kritik an den politischen Mandatsträgern. 'Mit Kneifen sind die Probleme nicht zu lösen'. Auf Einladung der Gewerkschaften: weitere Kernenergiedebatte, in: *MT*, 22.12.1978.
342 Hackmann, P. H. (Biene): Industrie- oder Wählervertreter, in: *LT*, 26.3.1979.
343 Atomkraftwerke bleiben für die SPD fragwürdig. Höhne: 'RWE/VEW-Pläne sind bedrohlich', in: *LT*, 25.4.1979; cf. Radkau et al., Aufstieg und Fall, pp. 233–235; Wehner, Versicherung; for Wiedorn: Jungeblut, Ludger: Auch Einsatz für die Ukraine. Verdienstkreuz für Hans-Joachim Wiedorn aus Lingen, in: *Neue Osnabrücker Zeitung*, 28.3.2014.
344 Probleme des Speichersees nicht isoliert sehen. Stellungnahme der SPD-Stadtratsfraktion zur CDU-Erklärung, in: *LT*, 12.4.1978; for the citizens' initiative and the quote: Vent, Hans-Karl (i.A. für 'Bürger gegen Umweltvergiftung' e. V. Lingen) (Lingen): Kernkraftwerke und Speichersee, in: *LT*, 12.5.1978.
345 Minutes, Meeting, Lingen City Council, 22.5.1979, pp. 4–15, StadtA LIN RP 1979a.
346 he: Ein Dutzend Gutachten muß erstellt werden, in: *LT*, 24.5.1979.
347 'Sachbearbeiter ablösen'. Bürgerinitiative Geeste protestiert, in: *LT*, 30.6.1979.
348 Minutes, Meeting, Lingen City Council, 7.2.1980, Ecological Report on the Reservoir, StadtA LIN RP 1980a.
349 mü: Speichersee: CDU entschied sich für südlichen Standort. Gemeinsame Informationstagung mit SPD-Fraktion und Geester Gemeinderat, in: *MT*, 5.5.1980; Minutes, Meeting, Lingen City Council, 8.5.1980, StadtA LIN RP 1980a; for further decisions regarding the reservoir's realisation: Minutes, Meeting, Lingen City Council, 2.7.1981, StadtA LIN RP 1981a; Minutes, Meeting, Lingen City Council, 8.3.1982, StadtA LIN RP 1982a.
350 Minutes, Meeting, Lingen City Council, 24.6.1982, pp. 2–6, quote p. 3, StadtA LIN RP 1982a.
351 Minutes, Meeting, Lingen City Council, 24.6.1982, pp. 2–6.
352 For an attempt to use criminal charges: hab: Grüne fordern Stillegung der Baustelle des Speicherbeckens. 'Verstöße gegen rechtliche Bestimmungen' – Strafanzeigen gegen Behörden und Firmen, in: *LT*, 11.5.1984; Brinkmann, Hans: Auf dem Prüfstand, in: *LT*, 5.11.1984.
353 Clearly, there was a place for experts among the Greens' discursive strategy; cf. Raschke, Die Grünen, p. 721.
354 hab: Grüne: Stubben im Damm. VEW: Völlig unrealistisch. Schwere Vorwürfe – 'Firma hat gewaltig geschludert' – Hinweis auf Kontrollen, in: *LT*, 5.12.1984; for Bußmann: Koop, Robert: Werner Bußmann +, in: Koop, Robert: Roberts

Blog. Gelegentliche Bemerkungen zu und in Lingen (Ems), https://robertkoop.wordpress.com/2012/11/12/werner-busmann/ (23.1.2023).
355 Brinkmann, Hans: Nerv berührt, in: *LT*, 5.12.1984.
356 Brinkmann, Hans: Nerv berührt, in: *LT*, 5.12.1984.
357 Minutes, Meeting, Lingen City Council, 6.12.1984, StadtA LIN RP 1984b, pp. 31–32.
358 Minutes, Meeting, Lingen City Council, 6.12.1984, StadtA LIN RP 1984b, pp. 31–32.
359 Minutes, Meeting, Lingen City Council, 6.12.1984, StadtA LIN RP 1984b, p. 34.
360 Minutes, Meeting, Lingen City Council, 6.12.1984, StadtA LIN RP 1984b, p. 35.
361 hab: VEW: Bezirksregierung soll Vorwürfe genau untersuchen. Grüne zur Beweisführung aufgefordert – CDU verlangt Klärung in Sachen Speichersee, in: *LT*, 7.12.1984.
362 hab: Informant entlassener Vorarbeiter, der Chef zu erpressen versuchte? Speichersee: Firmen reagieren auf Vorwürfe der Grünen mit Strafanzeige – Diskussion im Rat Lingen, in: *LT*, 8.12.1984.
363 Bunte, Johann u.a.: 'Diffamierung und Beleidigung', in: *LT*, 8.12.1984.
364 hab: 'Grüne haben für Horrorgemälde noch nie Beweis antreten können'. Scharfe Kritik von CDU und SPD im Kreistag – 'Doppelstrategie' – Grüne: Nun sieben Zeugen, in: *LT*, 12.12.1984; for Proske: Proske, Hermann. SPD, in: Simon, Barbara (Ed.): Abgeordnete in Niedersachsen 1946–1994. Biographisches Handbuch. Hanover 1996, p. 296.
365 CDU: Verängstigung der Bürger ist gründlich danebengegangen. Scharfe Kritik an den Grünen – 'Hatten Horrormärchen sofort vermutet', in: *LT*, 21.12.1984; Hövelmann: Politische Katastrophe für Grüne. CDU-Kreistagsfraktion fordert zur Distanzierung auf, in: *LT*, 21.12.1984.
366 This should not imply that Schröder was the first to put his trust in anti-nuclear experts; on the contrary, right after Chernobyl, a rather chaotic situation regarding different limit values applied in different German States was a result of the different state governments following different groups of experts. Schröder's Social Democrats, however, were the most directly relevant for the Lingen power plant's fate.
367 Cf. Sofortmaßnahmen, in: *LT*, 3.5.1986.
368 Brinkmann, Hans: Strahlung im Emsland, in: *LT*, 8.5.1986; for additional measurements taken up: Minutes, Meeting, Lingen City Council, 29.4.1987, StadtA LIN RP 1987a, pp. 4–6.
369 Rickers, Rudi: Bis zu 4150 Becquerel in Gras. Wesentlich höhere Strahlenbelastung als bislang von Behörden mitgeteilt. Luftmessung in Lingen ergab Spitzenzahlen – Hohe Kontaminationswerte bei Freilandgemüse, in: *LT*, 8.5.1986.
370 Cf. dpa: Rem, Röntgen und Curie. Strahlungsfachleute benutzen verschiedene Maßeinheiten, in: *LT*, 3.5.1986.
371 Brinkmann, Hans: Strahlung im Emsland, in: *LT*, 8.5.1986.
372 Dpa/ddp/AP/Reuter: Bonn hält deutsche Atommeiler für sicher. Grüne fordern sofortige Abschaltung, in: *LT*, 30.4.1986; Bonn beruhigt Bundesbürger. 'Keine Gefahr durch radioaktive Wolke', in: *LT*, 1.5.1986.
373 Which describes the number of radioactive decays occurring per second.
374 AP/dpa/Reuter/ddp: Bonn: Sofortmaßnahmen wegen strahlenbelasteter Frischmilch. Grüne nennen Ratschläge verantwortungslos – Auch in Lingen wieder sinkende Werte, in: *LT*, 3.5.1986; Dpa/Reuter/ddp: Milchverkauf ging schon um zehn Prozent zurück. Albrecht: Niedersachsen setzt auf Kernenergie. Bremen beschloß Ausstieg – Hamburg verschärft Werte, in: *LT*, 7.5.1986.
375 Keine akute Gefahr, in: *LT*, 6.5.1986.
376 Ratschläge zum Strahlenschutz, in: *LT*, 10.5.1986.
377 Rickers, Rudi: Landkreis: Milchvieh im Stall behalten. Radioaktivität in der Luft. Messungen im Emsland sind vorerst unverändert. Dringende Warnung: Hände weg vom Regenwasser und von jeglichem Blattgemüse, in: *LT*, 7.5.1986.

378 rr: Nur spärlich Informationen. Eine Milchprobe mit 50 Becquerel, in: *LT*, 6.5.1986.
379 Wurm, Carola: 'Säuglinge und Kleinkinder sind höherer Gefährdung ausgesetzt'. Schilddrüsenstoffwechsel bei Babys stärker – Mehr Hygiene empfohlen, in: *LT*, 8.5.1986.
380 Keine akute Gefahr, in: *LT*, 6.5.1986; SPD fragt nach Folgen des Reaktorunfalls für Lingen. Fraktion begehrt Informationen im Verwaltungsausschuß, in: *LT*, 6.5.1986; hab: Grüne: 'Verbrechen' Kreisverband äußert sich kritisch, in: *LT*, 6.5.1986.
381 pd: Seiters: Höchste Sicherheit für alle Kernkraftwerke der Welt. Unionspolitiker greift die Sowjetunion an – Besuch beim Meppener THW, in: *LT*, 10.5.1986.
382 Reinmöller, Klaus: Mit dem nassen Finger?, in: *LT*, 5.5.1986; Brugger, Alfred: Dokument des Desasters, in: *LT*, 6.5.1986.
383 Otten, Gerd (Ohrtermersch): 'Keine Gefahr?', in: *LT*, 6.5.1986.
384 Ripken, Karin, Bruni Müllner: Frauen gegen Atomkraft, in: *LT*, 23.6.1986.
385 sr: Manche haben Angst, andere sind sorglos – und viele vorsichtig. Reaktionen von Emsländern auf den Reaktorunfall – Breites Meinungsspektrum, in: *LT*, 7.5.1986; Wurm, Carola, Petra Diek: 'Alles Papenburger Treibhaussalat'. Skepsis bei den Kunden und Einnahmeeinbußen für Marktbeschicker, in: *LT*, 7.5.1986.
386 Brugger, Alfred: Das Ende der Sicherheit, in: *LT*, 30.4.1986.
387 rr: Grüne: Hilflosigkeit und eine gehörige Portion Trotteligkeit. Kritik am Vorgehen des Kreises – Erneut Abschalten der Kernkraftwerke gefordert, in: *LT*, 10.5.1986.
388 'Kernenergie-Technik wird nicht beherrscht'. Grüne und Bürgerinitiativen: Reaktorunglück bestätigt Sorgen, in: *LT*, 5.5.1986.
389 hab: 'Katastrophe wie in UdSSR extrem unwahrscheinlich'. Dr. Kröger: Ungleich höherer Sicherheitsaufwand bei deutschen Reaktoren, in: *LT*, 15.5.1986.
390 rr: Grüne: Hilflosigkeit und eine gehörige Portion Trotteligkeit. Kritik am Vorgehen des Kreises – Erneut Abschalten der Kernkraftwerke gefordert, in: *LT*, 10.5.1986.
391 'Grenzwerte zu locker gehandhabt'. 100 Zuhörer in Versammlung der Grünen, in: *LT*, 24.5.1986.
392 'KKW Lingen nicht in Betrieb nehmen'. Telkämper: Klare Aussage nötig, in: *LT*, 31.5.1986; for Telkämper: Wilfried Telkämper, in: Members of the European Parliament, www.europarl.europa.eu/meps/en/1229/WILFRIED_TELKAMPER/history/2 (23.1.2023).
393 'Bevölkerung spät und mangelhaft informiert'. SPD-Kritik an Reaktionen auf Reaktorunfall, in: *LT*, 12.5.1986.
394 Jusos. Bau in Lingen stoppen. Gegen Atomkraftnutzung, in: *LT*, 16.5.1986.
395 Rickers, Rudi: 'Wenn ich den Ausstieg will, ist der zusätzliche Einstieg unvernünftig'. Kernenergie, kommunale Finanzen und Landwirtschaft – AGS und SGK in Uelsen, in: *LT*, 23.5.1986.
396 Streit über Zukunft des Kernkraftwerkes Lingen. Schröder gegen Inbetriebnahme – Remmers fragt nach Alternativen, in: *LT*, 2.6.1986.
397 Brinkmann, Hans: Schröder will Inbetriebnahme des Kernkraftwerks Lingen verhindern. SPD-Spitzenkandidat: Weitere Atommeiler dürften nicht ans Netz gehen, in: *LT*, 2.6.1986.
398 Brinkmann, Hans: Schröder will Inbetriebnahme des Kernkraftwerks Lingen verhindern. SPD-Spitzenkandidat: Weitere Atommeiler dürften nicht ans Netz gehen, in: *LT*, 2.6.1986.
399 Brinkmann, Hans: Aufschlußreiche Resultate, in: *LT*, 17.6.1986.
400 Brinkmann, Hans: Tag von historischem Rang, in: *LT*, 16.4.1988.
401 Minutes, Meeting, Lingen City Council, 9.12.1987, StadtA LIN RP 1987b, pp. 2–7, 10, 23–25; cf. Lauerer, Matthias, Jochen Leffers: Die Reise des

verstrahlten Molkepulvers, in: *Der Spiegel*, 26.4.2016, www.spiegel.de/geschichte/tschernobyl-die-reise-des-verstrahlten-molkepulvers-a-1089233.html (23.1.2023); Minutes, Meeting, Emsland District Council, 28.9.1987, pp. 15–17, Kreisarchiv Emsland, Rep 1 EL, Nr. 1763; for leaflets distributed on the topic: StadtA LIN Allg. Slg. 1050; for Wallmann: Osswald, Klaus-Dieter, Barbara Peter: Wallmann, Walter, in: Kempf, Udo, Hans-Georg Merz (Eds.): Kanzler und Minister 1949–1998. Biografisches Lexikon der deutschen Bundesregierungen. Wiesbaden 2001, pp. 721–723.

402 This probably was Dr Arnulf Matting.

403 Minutes, Meeting, Lingen City Council, 9.12.1987, StadtA LIN RP 1987b, p. 7.

404 mm: Grüne stellen Ausfallbürgschaften für Klagen gegen Kernkraftwerk. Landtags- und Kreistagsabgeordnete stellen Finanzierungskonzept vor, in: *LT*, 11.11.1987.

405 Minutes, Meeting, Lingen City Council, 9.12.1987, StadtA LIN RP 1987b, p. 11.

406 Minutes, Meeting, Lingen City Council, 9.12.1987, StadtA LIN RP 1987b, p. 11.

407 Minutes, Meeting, Lingen City Council, 9.12.1987, StadtA LIN RP 1987b, p. 12.

408 Minutes, Meeting, Lingen City Council, 9.12.1987, StadtA LIN RP 1987b, p. 13.

409 Minutes, Meeting, Lingen City Council, 9.12.1987, StadtA LIN RP 1987b, p. 14.

410 Minutes, Meeting, Lingen City Council, 9.12.1987, StadtA LIN RP 1987b, p. 16.

411 Minutes, Meeting, Lingen City Council, 9.12.1987, StadtA LIN RP 1987b, p. 32; for Müller: Müller, Elke, in: Jahn, Bruno (Ed.): Biographisches Handbuch der deutschen Politik. Vol. 1. A-R. Munich 2004, p. 539.

412 Minutes, Meeting, Lingen City Council, 9.12.1987, StadtA LIN RP 1987b, p. 37.

413 Minutes, Meeting, Lingen City Council, 9.12.1987, StadtA LIN RP 1987b, p. 26.

414 Minutes, Meeting, Lingen City Council, 9.12.1987, StadtA LIN RP 1987b, p. 25.

415 Minutes, Meeting, Lingen City Council, 9.12.1987, StadtA LIN RP 1987b, p. 19; for Rolfes, cf. Borchert, Jens, Klaus Stolz: Die Bekämpfung der Unsicherheit. Politikerkarrieren und Karrierepolitik in der Bundesrepublik Deutschland, in: Politische Vierteljahresschrift 44 (2003), 2, pp. 148–173, here p. 161.

416 Minutes, Meeting, Lingen City Council, 9.12.1987, StadtA LIN RP 1987b, p. 23.

417 Minutes, Meeting, Lingen City Council, 9.12.1987, StadtA LIN RP 1987b, p. 20.

418 Minutes, Meeting, Lingen City Council, 9.12.1987, StadtA LIN RP 1987b, p. 29.

419 Minutes, Meeting, Lingen City Council, 9.12.1987, StadtA LIN RP 1987b, p. 36.

420 Minutes, Meeting, Lingen City Council, 9.12.1987, StadtA LIN RP 1987b, pp. 39–41; cf. Schulz, Chernobyl, p. 156.

421 This opinion-forming process within the parties was often interpreted by political competitors as a Janus-faced strategy. In historical perspective, it is an example for the parties' task of democratic opinion-forming within the constraints of representative parliaments. Cf. Weingart, Peter: Das 'Harrisburg-Syndrom' oder die De-Professionalisierung der Experten, in: Nowotny, Helga: Kernenergie. Gefahr oder Notwendigkeit. Anatomie eines Konflikts. Frankfurt/Main 1979, pp. 9–17, here pp. 12–13; Rucht, Campaigns, p. 213; Stange, Atomenergie, pp. 44–51; Wagner, Contesting Policies, pp. 280–286; Hillengaß, Atomkraft, pp. 32–37.

422 Cf. Irwin et al., Introduction, pp. 6–7.

423 For the contemporary diagnosis of such a "polarisation": Schelsky, Helmut: Mehr Demokratie oder mehr Freiheit? Der Grundsatzkonflikt der 'Polarisierung' in der Bundesrepublik Deutschland, in: Schelsky, Helmut: Systemüberwindung, Demokratisierung und Gewaltenteilung. Grundsatzkonflikte der Bundesrepublik. Munich 1973, pp. 47–82, especially p. 60; Slovic, Perceived Risk, p. 316; for the adaptation of the term originally applied to ethnical groups to ones based upon common interests by Michel Maffesoli: Tyldesley, Mike: The Thought of Sorbonne Professor Michel Maffesoli (1944–). Sociologist of Postmodernity. Lewiston 2010, pp. 59–62, 67, 69, 71; cf. Guggenberger, Bürgerinitiativen, pp. 57–59,

referring to observations by Peter Cornelius Mayer-Tasch; Bösch, Zeitenwende, pp. 302–303; Uekötter, Frank: Proxy Wars. The Deutsches Museum and the Peaceful Atom, in: Technikgeschichte 89 (2022), 1, pp. 63–86, here p. 82; Uekötter, Atomare, pp. 28, 167–168, 251, 299.

424 Cf. Beck, Risikogesellschaft, especially pp. 276, 328, 372–373.

425 Cf. Uekötter, Atomare, pp. 251, 254.

Index

Kolibri 360 GmbH
Ferdinand-Jühlke-Straße 7
99095 Erfurt, Deutschland
produktsicherheit@kolibri360.de